Springer Proceedings in Physics

Volume 210

The series Springer Proceedings in Physics, founded in 1984, is devoted to timely reports of state-of-the-art developments in physics and related sciences. Typically based on material presented at conferences, workshops and similar scientific meetings, volumes published in this series will constitute a comprehensive up-to-date source of reference on a field or subfield of relevance in contemporary physics. Proposals must include the following:

- name, place and date of the scientific meeting
- a link to the committees (local organization, international advisors etc.)
- scientific description of the meeting
- list of invited/plenary speakers
- an estimate of the planned proceedings book parameters (number of pages/ articles, requested number of bulk copies, submission deadline).

More information about this series at http://www.springer.com/series/361

Olena Fesenko • Leonid Yatsenko
Editors

Nanooptics, Nanophotonics, Nanostructures, and Their Applications

Selected Proceedings of the 5th International Conference Nanotechnology and Nanomaterials (NANO2017), August 23-26, 2017, Chernivtsi, Ukraine

Springer

Editors
Olena Fesenko
National Academy of Sciences
of Ukraine
Institute of Physics
Kyiv, Ukraine

Leonid Yatsenko
National Academy of Sciences
of Ukraine
Institute of Physics
Kyiv, Ukraine

ISSN 0930-8989 ISSN 1867-4941 (electronic)
Springer Proceedings in Physics
ISBN 978-3-030-08172-0 ISBN 978-3-319-91083-3 (eBook)
https://doi.org/10.1007/978-3-319-91083-3

Printed on acid-free paper

This Springer imprint is published by the registered company Springer International Publishing AG part of Springer Nature.
The registered company address is: Gewerbestrasse 11, 6330 Cham, Switzerland

Preface

This book highlights the most recent advances in nanoscience from leading researchers in Ukraine, Europe, and beyond. It features contributions from participants of the 5th International Research and Practice Conference "Nanotechnology and Nanomaterials" (NANO-2017), held in Chernivtsi, Ukraine, on August 23–26, 2017. This event was organized jointly by the Institute of Physics of the National Academy of Sciences of Ukraine, Ivan Franko National Chernivtsi National University (Ukraine), University of Tartu (Estonia), University of Turin (Italy), and Pierre and Marie Curie University (France). Internationally recognized experts from a wide range of universities and research institutes shared their knowledge and key results in the areas of nanocomposites and nanomaterials, nanostructured surfaces, microscopy of nano-objects, nanooptics and nanophotonics, nanoplasmonics, nanochemistry, nanobiotechnology, and surface-enhanced spectroscopy.

Nowadays, nanotechnology is becoming the most actively developing and promising field of science. Numerous nanotechnology investigations are already producing practical results that can be applied in various areas of human life from science and technology to medicine and pharmacology. The aim of the present book is to highlight the latest investigations from different areas of nanoscience and to stimulate new interest in this field. Volume I of this two-volume work covers such important topics as nanostructured interfaces and surfaces, nanooptics, nanoplasmonics, and enhanced vibrational spectroscopy.

This book is divided into three parts: Part I—*Nanoscale Physics*, Part II—*Nanooptics and Photonics*, and Part III—*Nanostructured Interfaces and Surfaces.* Parts covering Nanochemistry and Biotechnology, and Nanocomposites and Nanomaterials can be found in Volume II.

The papers published in these five parts fall under the broad categories of nanomaterial preparation and characterization, nanobiotechnology, nanodevices and quantum structures, and spectroscopy and nanooptics. We hope that both volumes will be equally useful and interesting for young scientists or PhD students and mature scientists alike.

Kyiv, Ukraine

Olena Fesenko
Leonid Yatsenko

Contents

Part I Nanoscale Physics

1 **Anion Exchange Resin Modified with Nanoparticles of Hydrated Zirconium Dioxide for Sorption of Soluble U(VI) Compounds** 3
Olga Perlova, Yuliya Dzyazko, Iryna Halutska, Nataliia Perlova, and Alexey Palchik

2 **High-Entropy Film Alloys: Electrophysical and Magnetoresistive Properties** 17
Yu. Bereznyak, L. Odnodvorets, D. Poduremne, I. Protsenko, and Yu. Shabelnyk

3 **Strain- and Adsorption-Dependent Electronic States and Transport or Localization in Graphene** 25
Taras Radchenko, Ihor Sahalianov, Valentyn Tatarenko, Yuriy Prylutskyy, Paweł Szroeder, Mateusz Kempiński, and Wojciech Kempiński

4 **The Theory of Absorption and Emission of Light by Free Electrons in Ferromagnetic Semiconductors** 43
Oleksandr Semchuk

5 **Mechanism of Active Electron Transfer in a Protein-Like Nanowire Under Real Conditions** 59
L. V. Shmeleva and A. D. Suprun

6 **Electron Correlation Effects in Theoretical Model of Doped Fullerides** 73
Yu. Skorenkyy, O. Kramar, L. Didukh, and Yu. Dovhopyaty

7 The Possible Mechanisms of Conductivity in Polyene-Like Polymers and Types of Conductivity in Maximally Feeble External Fields . . . 89
A. D. Suprun, S. V. Vasylyuk and V. N. Yashchuk

8 Nanosized Oxides of Different Compositions as Adsorbents for Hazardous Substances Removal from Aqueous Solutions and Wastewaters . . . 103
Małgorzata Wiśniewska, Monika Wawrzkiewicz, Anna Wołowicz, and Olena Goncharuk

9 Electrodes Based on Carbon Nanomaterials: Structure, Properties, and Application to Capacitive Deionization in Static Cells . . . 127
Yurii Volfkovich, Daniil Bograchev, Alexey Mikhalin, Alexey Rychagov, Valentin Sosenkin, Vitaly Milyutin, and Daewook Park

10 Gradient-Type Theory for Electro-Thermoelastic Non-ferromagnetic Dielectrics: Accounting for Quadrupole Polarization and Irreversibility of Local Mass Displacement . . . 147
Olha Hrytsyna and Vasyl Kondrat

11 Graphene Quantum Dots in Various Many-Electron π-Models . . . 161
Anatoliy Luzanov

12 Electronic Properties of Defects in Carbon Nanotubes . . . 175
H. Yu. Mykhailova, M. M. Nischenko, G. P. Prikhodko, V. V. Anikeev, and V. Yu. Koda

13 Features of Spin Transport in Magnetic Nanostructures with Nonmagnetic Metal Layers . . . 183
A. M. Korostil and M. M. Krupa

Part II Nanooptics and Photonics

14 Digital Interferometry Methods for the Surface Relief Study . . . 207
Yu. Kotsiuba, H. Petrovska, V. Fitio, and Ya. Bobitski

15 Spectrum of Electron in Quantum Well with Continuous Position-Dependent Effective Mass . . . 219
Julia Seti, Mykola Tкach, and Oxana Voitsekhivska

16 Design and Synthesis of Silver Nanoparticles with Different Shapes Under the Influence of Photon Flows . . . 231
T. Bulavinets, V. Varyshchuk, I. Yaremchuk, and Ya Bobitski

17 Internal Reflection of the Surface of a Plasmonic Substrate Covered by Active Nanoparticles 243
Eugene Bortchagovsky and Yurii Demydenko

18 Temperature Studies of Luminescence in Nanosize SnO_2 Films 265
V. Grinevych, L. Filevska, V. Smyntyna and B. Ulug

Part III Nanostructured Interfaces and Surfaces

19 Fabrication and Characterization of High-Performance Anti-reflecting Nanotextured Si Surfaces for Solar Cells 275
Stepan Nichkalo, Anatoly Druzhinin, Valeriy Yerokhov, and Oleksandr Ostapiv

20 Effects of Amino Acids Protonation on Double-Layer Parameters of the Electrode/Chlorates(VII) Interface, as well as Kinetics and Mechanism of Bi(III) Ion Electroreduction in the Aspect of the "Cap–Pair" Effect 285
Agnieszka Nosal-Wiercińska, Mariusz Grochowski, and Małgorzata Wiśniewska

21 Phase Composition and Nanoporous Structure of Core and Surface in the Modified Granules of NH_4NO_3 301
A. Artyukhov and J. Gabrusenoks

22 Interaction-Induced Polarizability Anisotropy Correlations in Ultra-Thin Layer of Neon Atoms Confined Between Graphene Walls: Computer Simulation 311
Z. Gburski

23 Sintering Methods of Inkjet-Printed Silver Nanoparticle Layers . . . 317
O. Kravchuk, R. Lesyuk, Ya. Bobitski, and M. Reichenberger

24 Influence of Thermal Treatment on Phase Formation Processes in Amorphous Alloys 341
Volodymir Lysov, Tatiana Tsaregradskaya, Oleg Turkov, and Galina Saenko

Index 353

Contributors

V. V. Anikeev G. V. Kurdyumov Institute for Metal Physics, National Academy of Sciences of Ukraine, Kyiv, Ukraine

A. Artyukhov Processes and Equipment of Chemical and Petroleum Refinery Department, Sumy State University, Sumy, Ukraine

Yu. Bereznyak Department of Electronics, General and Applied Physics, Sumy State University, Sumy, Ukraine

Ya. Bobitski Department of Photonics, Lviv Polytechnic National University, Lviv, Ukraine

Faculty of Mathematics and Natural Sciences, Institute of Technology, University of Rzeszow, Rzeszow, Poland

Daniil Bograchev A. N. Frumkin Institute of Physical Chemistry and Electrochemistry of the Russian Academy of Sciences, Moscow, Russia

Eugene Bortchagovsky V. Lashkaryov Institute of Semiconductor Physics of National Academy of Sciences of Ukraine, Kyiv, Ukraine

T. Bulavinets Lviv Polytechnic National University, Lviv, Ukraine

Yurii Demydenko V. Lashkaryov Institute of Semiconductor Physics of National Academy of Sciences of Ukraine, Kyiv, Ukraine

L. Didukh Ternopil Ivan Puluj National Technical University, Ternopil, Ukraine

Yu. Dovhopyaty Ternopil Ivan Puluj National Technical University, Ternopil, Ukraine

Anatoly Druzhinin Lviv Polytechnic National University, Lviv, Ukraine

Yuliya Dzyazko V.I. Vernadskii Institute of General and Inorganic Chemistry of the National Academy of Sciences of Ukraine, Kyiv, Ukraine

L. Filevska Department of Physics, Odessa I.I. Mechnikov National University, Odessa, Ukraine

V. Fitio Department of Photonics, Lviv Polytechnic National University, Lviv, Ukraine

J. Gabrusenoks Institute of Solid State Physics, University of Latvia, Riga, Latvia

Z. Gburski Institute of Physics, University of Silesia, Katowice, Poland

Silesian Centre of Education & Interdisciplinary Research, Chorzów, Poland

Olena Goncharuk O. O. Chuiko Institute of Surface Chemistry, National Academy of Sciences of Ukraine, Kyiv, Ukraine

V. Grinevych Department of Physics, Odessa I.I. Mechnikov National University, Odessa, Ukraine

Mariusz Grochowski Department of Analytical Chemistry and Instrumental Analysis, Faculty of Chemistry, Maria Curie-Sklodowska University, Lublin, Poland

Iryna Halutska Odessa I.I. Mechnikov National University, Odesa, Ukraine

Olha Hrytsyna Center of Mathematical Modeling of Pidstryhach Institute for Applied Problems of Mechanics and Mathematics, National Academy of Sciences of Ukraine, Lviv, Ukraine

Mateusz Kempiński Faculty of Physics, Adam Mickiewicz University, Poznań, Poland

NanoBioMedical Centre, Adam Mickiewicz University, Poznań, Poland

Wojciech Kempiński Institute of Molecular Physics, Polish Academy of Sciences, Poznań, Poland

V. Yu. Koda G. V. Kurdyumov Institute for Metal Physics, National Academy of Sciences of Ukraine, Kyiv, Ukraine

Vasyl Kondrat Hetman Petro Sahaydachnyi Academy of Army Ground Forces, Lviv, Ukraine

A. M. Korostil Institute of Magnetism National Academy of Sciences of Ukraine, Kyiv, Ukraine

Yu. Kotsiuba Department of Photonics, Lviv Polytechnic National University, Lviv, Ukraine

O. Kramar Ternopil Ivan Puluj National Technical University, Ternopil, Ukraine

O. Kravchuk Department of Photonics, Lviv Polytechnic National University, Lviv, Ukraine

M. M. Krupa Institute of Magnetism, National Academy of Sciences of Ukraine, Kyiv, Ukraine

R. Lesyuk Pidstryhach Institute for Applied Problems of Mechanics and Mathematics, National Academy of Sciences of Ukraine, Lviv, Ukraine

Institute for Physical Chemistry, University of Hamburg, Hamburg, Germany

Anatoliy Luzanov State Scientific Institution “Institute of Single Crystals”, National Academy of Sciences of Ukraine, Kharkiv, Ukraine

Volodymir Lysov Department of Physics, Taras Shevchenko National University of Kyiv, Kyiv, Ukraine

Alexey Mikhalin A. N. Frumkin Institute of Physical Chemistry and Electrochemistry of the Russian Academy of Sciences, Moscow, Russia

Vitaly Milyutin A. N. Frumkin Institute of Physical Chemistry and Electrochemistry of the Russian Academy of Sciences, Moscow, Russia

H. Yu. Mykhailova G. V. Kurdyumov Institute for Metal Physics, National Academy of Sciences of Ukraine, Kyiv, Ukraine

Stepan Nichkalo Lviv Polytechnic National University, Lviv, Ukraine

M. M. Nischenko G. V. Kurdyumov Institute for Metal Physics, National Academy of Sciences of Ukraine, Kyiv, Ukraine

Agnieszka Nosal-Wiercińska Department of Analytical Chemistry and Instrumental Analysis, Faculty of Chemistry, Maria Curie-Sklodowska University, Lublin, Poland

L. Odnodvorets Department of Electronics, General and Applied Physics, Sumy State University, Sumy, Ukraine

Oleksandr Ostapiv Lviv Polytechnic National University, Lviv, Ukraine

Alexey Palchik V. I. Vernadskii Institute of General and Inorganic Chemistry of the National Academy of Sciences of Ukraine, Kyiv, Ukraine

Daewook Park Samsung Electronics Co., Ltd, Gyeonggi-do, South Korea

Nataliia Perlova Odessa I.I. Mechnikov National University, Odesa, Ukraine

Olga Perlova Odessa I.I. Mechnikov National University, Odesa, Ukraine

H. Petrovska Department of Photonics, Lviv Polytechnic National University, Lviv, Ukraine

D. Poduremne Department of Electronics, General and Applied Physics, Sumy State University, Sumy, Ukraine

G. P. Prikhodko G. V. Kurdyumov Institute for Metal Physics, National Academy of Sciences of Ukraine, Kyiv, Ukraine

I. Protsenko Department of Electronics, General and Applied Physics, Sumy State University, Sumy, Ukraine

Yuriy Prylutskyy Taras Shevchenko National University of Kyiv, Kyiv, Ukraine

Taras Radchenko G. V. Kurdyumov Institute for Metal Physics, National Academy of Sciences of Ukraine, Kyiv, Ukraine

M. Reichenberger Technische Hochschule Nuernberg Georg Simon Ohm, Nuremberg, Germany

Alexey Rychagov A. N. Frumkin Institute of Physical Chemistry and Electrochemistry of the Russian Academy of Sciences, Moscow, Russia

Galina Saenko Department of Physics, Taras Shevchenko National University of Kyiv, Kyiv, Ukraine

Ihor Sahalianov Taras Shevchenko National University of Kyiv, Kyiv, Ukraine

Oleksandr Semchuk Chyiko Institute of Surface Chemistry, National Academy of Sciences of Ukraine, Kyiv, Ukraine

Julia Seti Yuriy Fedkovych Chernivtsi National University, Chernivtsi, Ukraine

Yu. Shabelnyk Department of Electronics, General and Applied Physics, Sumy State University, Sumy, Ukraine

L. V. Shmeleva Taras Shevchenko National University of Kyiv, Kyiv, Ukraine

Yu. Skorenkyy Ternopil Ivan Puluj National Technical University, Ternopil, Ukraine

V. Smyntyna Department of Physics, Odessa I.I. Mechnikov National University, Odessa, Ukraine

Valentin Sosenkin A. N. Frumkin Institute of Physical Chemistry and Electrochemistry of the RAS, Moscow, Russia

A. D. Suprun Taras Shevchenko National University of Kyiv, Kyiv, Ukraine

Paweł Szroeder Institute of Physics, Kazimierz Wielki University, Bydgoszcz, Poland

Valentyn Tatarenko G. V. Kurdyumov Institute for Metal Physics, National Academy of Sciences of Ukraine, Kyiv, Ukraine

Mykola Tкach Yuriy Fedkovych Chernivtsi National University, Chernivtsi, Ukraine

Tatiana Tsaregradskaya Department of Physics, Taras Shevchenko National University of Kyiv, Kyiv, Ukraine

Oleg Turkov Department of Physics, Taras Shevchenko National University of Kyiv, Kyiv, Ukraine

B. Ulug Department of Physics, Faculty of Science, Akdeniz University, Antalya, Turkey

V. Varyshchuk Lviv Polytechnic National University, Lviv, Ukraine

S. V. Vasylyuk Physics Faculty, Kyiv National Taras Shevchenko University, Kyiv, Ukraine

Oxana Voitsekhivska Yuriy Fedkovych Chernivtsi National University, Chernivtsi, Ukraine

Yurii Volfkovich A. N. Frumkin Institute of Physical Chemistry and Electrochemistry of the Russian Academy of Sciences, Moscow, Russia

Monika Wawrzkiewicz Department of Inorganic Chemistry, Faculty of Chemistry, Maria Curie-Sklodowska University, Lublin, Poland

Małgorzata Wiśniewska Department of Radiochemistry and Colloid Chemistry, Faculty of Chemistry, Maria Curie-Sklodowska University, Lublin, Poland

Anna Wołowicz Department of Inorganic Chemistry, Faculty of Chemistry, Maria Curie-Sklodowska University, Lublin, Poland

I. Yaremchuk Lviv Polytechnic National University, Lviv, Ukraine

V. N. Yashchuk Physics Faculty, Kyiv National Taras Shevchenko University, Kyiv, Ukraine

Valeriy Yerokhov Lviv Polytechnic National University, Lviv, Ukraine

Part I
Nanoscale Physics

Chapter 1
Anion Exchange Resin Modified with Nanoparticles of Hydrated Zirconium Dioxide for Sorption of Soluble U(VI) Compounds

Olga Perlova, Yuliya Dzyazko, Iryna Halutska, Nataliia Perlova, and Alexey Palchik

1.1 Sorption Materials for Removal of Soluble U(VI) Compounds from Water

Besides military industry, uranium is also applied to civilian needs. Uranium compounds are employed in geology (to determine age of rocks), as a pigment for paints, in analytical chemistry, and for other purposes [1]. However, produced uranium is mainly consumed by nuclear power plants. Therefore, mining and processing of uranium ores are important practical tasks. Efficient, accessible, and cheap methods for the removal of U(VI) compounds from liquid wastes could provide ecological hygiene in uranium extraction from mineral raw materials. Moreover, the technique has to cover uranium recuperation in order to prevent appearance of toxic ions in sources of water supply. Adsorption and ion exchange present a possibility to decrease the content of U(VI) compounds down to maximum allowable concentration [2]. The following characteristics for materials are required: significant capacity and selectivity toward uranium-containing ions, high sorption rate, and facile regeneration. Currently, attention is focused on change of chemical composition of organic (particularly biopolymers) or inorganic sorbents. This is achieved by treatment of the materials with solutions of metal chlorides (Li^+, Na^+, K^+, Mg^{2+}, Ca^{2+}, Fe^{2+}, and Zn^{2+}) [3], acids [3–5] or alkali [4–6] (sometimes simultaneously with irradiation [6]). Natural or synthetic materials are also modified with functional groups, which are able to form complexes with sorbed ions [7]. For example, silica is functionalized with phosphate groups [8]; graphene oxide is modified with phenanthroline diamide [9]. The other way

O. Perlova · I. Halutska · N. Perlova (✉)
Odessa I.I. Mechnikov National University, Odesa, Ukraine

Y. Dzyazko · A. Palchik
V.I. Vernadskii Institute of General and Inorganic Chemistry of the National Academy of Sciences of Ukraine, Kyiv, Ukraine

O. Fesenko, L. Yatsenko (eds.), *Nanooptics, Nanophotonics, Nanostructures, and Their Applications*, Springer Proceedings in Physics 210,
https://doi.org/10.1007/978-3-319-91083-3_1

is the development of composites, particularly organic–inorganic sorbents. Combination of constituents of different nature allows us to obtain materials that possess improved functional properties compared to individual components. Different substances are applied to synthesize the composites. These materials contain, for instance, iron nanoparticles [10, 11], Fe_3O_4 [12–18] (particularly functionalized with organic functional groups [18]), MgO [19], ZrO_2 [20, 21], layered double oxides of multivalent metals [22, 23], and functionalized polymers, such as polyacrylonitrile [24] (amidoxime has been proposed for its functionalization [13, 25–28]). Other composites are considered in [28–33]. The polymer constituents are poly(vinyl imidazole) [28], polyaniline [29, 30], tetraphenylmethylenediphospine dioxide [31], and poly(vinyl alcohol) [32, 33]. The polymers are modified with molybdenum disulfide [28], mesoporous carbon [29], oxidized graphene [30], carbon nanotubes [32], or silica [33].

Among polymer matrices, ion exchange resins are the most attractive materials due to their availability, low cost, significant sorption capacity, and high rate of sorption. In order to improve selectivity toward uranium, the resins are modified with inorganic ion-exchangers [34–38], organic compounds [39–43], or microorganisms [44]. However, the modified resins show lower sorption rate than pristine materials. The composites containing non-aggregated nanoparticles show higher rate of sorption of transition metal ions [38, 45] than the resin modified with large particles (aggregates and agglomerates) [37, 46]. Approaches based on the Ostwald–Freundlich equation have been proposed in order to control the size of particles in inert [47] and cation exchange polymers [38, 48]. In the last case, the equation has been adopted taking properties of the matrix into consideration.

As known, sodium carbonate and sulfuric acid are applied to uranium ore processing [1]. Anionic forms of U(VI) dominate in carbonate and sulfate solutions [49]. It is the same also for groundwater, which normally contains carbonate anionic complexes. The aim of the investigation was to develop an approach for purposeful control of the size of the particles incorporated into anion exchange polymer. The tasks of the work were to confirm the approach experimentally and to study sorption of uranium-containing anions.

1.2 Synthesis of Composite Sorbents. A Study of Their Morphology

EDE-10P anion exchange resin (Schekinoazot, RF) was used as the polymer matrix for modification. The pristine resin was marked AR-0. This material contains such functional groups as $-NR_3^+$, $=NH$, and $-NR_2$. Particles of hydrated zirconium dioxide (HZD) were precipitated in the matrix. This amphoteric inorganic ion-exchanger is characterized by anion exchange ability in neutral and acidic media [50, 51]. Under these conditions, HZD (particularly in combination with oxide of other multivalent metals [52]) shows considerable capacity and rather fast rate of sorption of anionic forms of transition metals ($HCrO_4^-$ [51, 52]) and even cations (Cu^{2+}, Cd^{2+}, Pb^{2+} [52]). Due to these remarkable properties, HZD is used for modification of ceramic [53–56] or polymer membranes [47, 57, 58]. Modified membranes based on inert polymers show higher rejection of colloidal particles than the pristine materials [47, 57]. HZD slows down transport of co-ions through polymer anion exchange membranes [58]. Moreover, the incorporated particles prevent fouling of the membranes with organics. HZD was also applied to modification of both cation [59] and anion [60] exchange resins. Modification improves selectivity of the ion-exchangers toward transition metal cations [59] and arsenate anions [60].

The ion-exchanger was impregnated with a 0.1 M $ZrOCl_2$ aqueous solution during 24 h at 25 °C. The ratio of volumes of solid and liquid was 1:20. Then the ion-exchanger and solution were separated and the grains were washed with a 0.01 M HCl to remove additionally sorbed electrolyte from macropores. The inorganic constituent was precipitated with a 1 M NH_4OH solution. The AR-1 sample was obtained in this manner.

In order to obtain the AR-2 ion-exchanger, similar modification procedure was applied. However, both impregnation and precipitation were carried out at 100 °C.

Concentration of $ZrOCl_2$ and NH_4OH solutions was 1 M, when the AR-3 and AR-4 samples were synthesized. The temperature of impregnating and precipitating solutions was kept at the level of 25 °C (AR-3) and 100 °C (AR-4).

For synthesis of the AR-5 sample, a mixture of water (20 vol %) and glycerol (80 vol %) was used for preparation of 0.1 M $ZrOCl_2$ and 0.1 M NH_4OH solutions. The water–glycerol mixture was also used for washing of the resin loaded with $ZrOCl_2$. The synthesis temperature was 25 °C.

After precipitation, the samples were washed with deionized water down to constant pH of the effluent, washed at room temperature down to constant mass, and treated with ultrasound at 30 kHz using a Bandelin ultrasonic bath (Bandelin, Hungary). This treatment was necessary to remove inorganic particles from outer surface of the grains.

TEM images were obtained with a JEOL JEM 1230 transmission electron microscope (JEOL, Japan). Preliminarily, the ion-exchangers were milled and treated with ultrasound.

1.3 Investigation of U(VI) Sorption and Desorption

Salts of uranyl acetate and sodium hydrocarbonate (Chemapol, Czech Republic) were used for preparation of modeling solutions containing 2×10^{-4} M U(VI). Anionic complexes of uranyl ($[UO_2(CO_3)]_2^{2-}$) dominated in the solution that contained also 0.02 M $NaHCO_3$ (pH 7).

Sorption was investigated under batch conditions at 20 °C. A Water Bath Shaker type 357 apparatus (Elpan, Poland) was used for shaking of the flasks during 15–180 min. The volume of the solution was 50 cm^3. The sorbent dosage is pointed for each particular case (mass of air-dry sorbent was taken into consideration).

Preliminarily, the samples were treated with deionized water for swelling. During the contact of solid and liquid, the pH was controlled with an I–160 MI pH-meter (Izmeritelnaya technika LTD, RF).

After predetermined time, the liquid was separated from the ion-exchanger, uranium(VI) in the solution was determined in the form of complex with Arsenazo III [61] using KFK-2MP photocolorimeter (Zagorsk OMZ, RF). Removal degree of uranium (*RD*) was determined as $\frac{C_0-C}{C_0} \times 100\%$, where C_0 and C are concentrations of the solution before and after sorption. Sorption capacity (*A*) was calculated via $\frac{V\,(C_0-C)}{m}$, where V is the solution volume, m is the mass of weighted sample.

The *RD* and *A* values that correspond to equilibrium conditions were also determined. These samples, which were loaded with U(VI) as fully as possible (the sorbent dosage was 1 g dm^{-3}, the solution volume was 50 cm^3), were regenerated. Preliminarily, the resins were dried at room temperature down to constant mass. Desorption was carried out under batch conditions, the content of the flasks was stirred intensively as mentioned earlier. Deionized water or 1 M one-component solutions of NaOH or $NaHCO_3$ were used for regeneration, their volume was 50 cm^3. Regeneration with an $NaHCO_3$ solution was performed two times. Desorption was carried out during 2 h. Then the solid and liquid were separated, the content of U(VI) in the effluent was determined. Desorption degree (*DD*) was calculated as $\frac{C_d}{C_0-C} \times 100\%$, where C_d is the content of uranium in the effluent.

1.4 Precipitation of HZD in Anion Exchange Polymer: Theory

Let us consider precipitation of hydrated oxide (we will write down its formula as $Cat(OH)_z$ for simplicity) inside anion-exchange resin. During precipitation, dissolution of small particles and their reprecipitation on larger particles occur. Decrease of the particle surface causes reduction of Gibbs energy. In this case, the Ostwald–Freundlich equation [62] is valid:

$$\ln \frac{\bar{C}_{\mathrm{Cat(OH)}_z}}{C_{\mathrm{Cat(OH)}_z,\infty}} = \frac{\beta V_{\mathrm{m}} \sigma \operatorname{Cos} \phi}{\mathrm{RTr}}. \tag{1.1}$$

Here $\bar{C}_{\mathrm{Cat(OH)}_z}$ and $C_{\mathrm{Cat(OH)}_z,\infty}$ are the concentrations of dissolved compound in ion-exchanger and saturated solution, respectively (regarding insoluble compounds, these values are extremely low), β is the shape factor of particles, V_m is the molar volume of the compound, σ is the surface tension of the solvent, ϕ is the wetting angle (≈1 for hydrophilic compound), and r is the particle radius.

Thus, $\bar{C}_{\mathrm{Cat(OH)}_z} = \overline{[\mathrm{Cat}^{z+}]} = \frac{K_{sp}}{[\mathrm{OH}]^z}$, here the square brackets correspond to equilibrium molar concentration, K_{sp} is the solubility product, z is the charge number. Dissociation degree of NH_4OH (α) can be determined according to Ostwald dilution law [63]:

$$\alpha = K^{0.5}_{\mathrm{NH_4OH}} C^{0.5}_{\mathrm{NH_4OH}} \tag{1.2}$$

where $K_{\mathrm{NH_4OH}}$ is the dissociation constant of NH_4OH, $C_{\mathrm{NH_4OH}}$ means concentration. The precipitating solution provides certain equilibrium concentration of OH^- ions.

$$[\mathrm{OH}^-] = [\mathrm{NH_4}] = \alpha C_{\mathrm{NH_4OH}} \tag{1.3}$$

Taking formula (1.2) into consideration, Eq. (1.3) can be written as

$$[\mathrm{OH}^-] = K^{0.5}_{\mathrm{NH_4OH}} C^{1.5}_{\mathrm{NH_4OH}} \tag{1.4}$$

OH^- ions are partially consumed for $Cat(OH)_z$ deposition:

$$\mathrm{Cat}^{z+} + z\mathrm{OH}^- \leftrightarrow \mathrm{Cat(OH)}_z \downarrow \tag{1.5}$$

This consumption is equal to zC_{Cat}. In the first approximation, it is possible to suppose that the volume of additionally sorbed electrolyte (zirconium hydroxocomplexes in our case) corresponds to the volume of the ion-exchanger (V_i). Resulting concentration of OH^- ions is

$$[\mathrm{OH}^-] = K^{0.5}_{\mathrm{NH_4OH}} C^{1.5}_{\mathrm{NH_4OH}} - \frac{zC_{\mathrm{Cat}} V_i}{V_{\mathrm{NH_4OH}}}, \tag{1.6}$$

where $V_{\mathrm{NH_4OH}}$ is the volume of the precipitating solution. Thus,

$$\bar{C}_{\mathrm{Cat(OH)}_z} = \frac{K_{sp}}{\left(K^{0.5}_{\mathrm{NH_4OH}} C^{1.5}_{\mathrm{NH_4OH}} - \frac{zC_{\mathrm{Cat}} V_i}{V_{\mathrm{NH_4OH}}}\right)^z} \tag{1.7}$$

Substituting this expression into Eq. (1.1), it is possible to obtain

$$r = \frac{\beta V_m \sigma}{RT \ln \frac{K_{sp}}{C_{\mathrm{Cat(OH)}_z,\infty} \left(K^{0.5}_{\mathrm{NH_4OH}} C^{1.5}_{\mathrm{NH_4OH}} - \frac{zC_{\mathrm{Cat}} V_i}{V_{\mathrm{NH_4OH}}}\right)^z}} \tag{1.8}$$

The particles, the size of which is lower than the r value, are dissolved and reprecipitated as larger particles. In accordance with expression (1.8), increase in temperature provides formation of smaller particles. Decrease in concentration of precipitant and additionally sorbed electrolyte as well as reducing of surface tension of solvent give the same results.

1.5 Visualization of Incorporated Particles

As seen from the TEM images (Fig. 1.1), the AR-1 sample contains single nanoparticles, the size of which is about 5 nm and even smaller. Some nanoparticles are placed close to each other, but they are isolated. HZD nanoparticles are evidently located in nanosized pores (clusters and channels), similarly to nanoparticles of zirconium hydrophosphate embedded to cation exchange resin [38, 64]. The nanoparticles are stabilized by pore walls.

Large particles (≈200–300 nm) have been found for the AR-2 sample. Their shape is close to globular. It is seen that the particles consist of smaller nanoparticles (≈30 nm). These aggregates can be located in hydrophobic pores of the ion exchange polymer (voids between gel regions). As opposed to Eq. (1.8), increase in synthesis temperature results in enlargement of the particles. Aggregation is probably caused by thermal motion of polymer chains. As a result, pore walls cannot perform stabilization function under these conditions. From the point of view of thermodynamics, enlargement of HZD particles can be caused by increase of the $C_{\mathrm{Cat(OH)}_z,\infty}$ and $K_{\mathrm{NH_4OH}}$ values.

Increase in concentration of the impregnating solution leads to enlargement of HZD particles as opposed to Eq. (1.8). However, hydrolysis of zirconium-containing ions becomes stronger when the impregnating solution is diluted. When $ZrOCl_2$ salt is dissolved in aqueous media, zirconium ions form soluble polymerized hydroxocomplexes. Among the ligands (OH^- and H_2O), water molecules

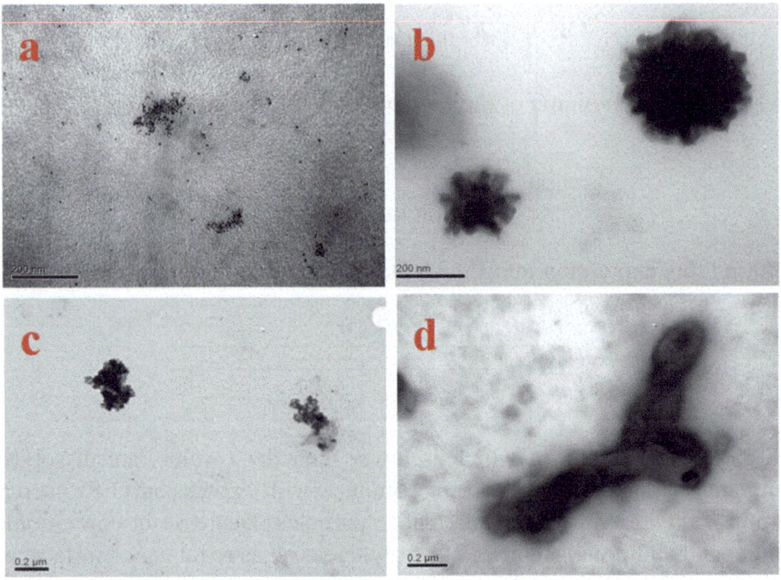

Fig. 1.1 TEM images of AR-1 (**a**), AR-2 (**b**), AR-3 (**c**), and AR-5 (**d**) samples

dominate [65]. Dilution of the solution enhances hydrolysis. As a result, the amount of OH^- ligands around zirconium atoms increases; they replace water molecules. It means reducing of molar volume of HZD and decrease of particle size in accordance with Eq. (1.8). The data obtained for the AR-4 samples are similar to those for the AR-2 resin.

According to Eq. (1.8), decrease of surface tension reduces particle size ($\sigma = 59.4$ and 72.8 mN m^{-1} at 20 °C for glycerol and water, respectively). However, large aggregated particles of irregular shape are formed in the glycerol–water mixture (AR-5). These particles are of micron size and can be located in structure defects. Molar volume of the precipitant becomes larger when it is deposited from glycerol-containing solution due to formation of insoluble Zr complexes containing both glycerol and OH ligands. Indeed, both dark and light spots are visible in TEM image. Dark spots correspond to regions where water dominates. Lights spots are attributed to regions where glycerol dominates in HZD.

1.6 Uranium Sorption Under Batch Conditions

Removal degree of U(VI) from modeling solutions is plotted in Fig. 1.2a, b for all tested samples (sorbent dosage was 1 g dm^{-3}). It has been found for all composites (except AR-5) that incorporated nanoparticles accelerate sorption. Both the pristine resin and composites containing particle, the size of which is from several nanometers up to several hundred nanometers, show practically complete removal of U(VI) from the solutions after 150–250 min. In the case of the AR-5 sample, RD = 85% after 250 min. The most complete removal is reached only after 1400 min. It is seen from Fig. 1.2c that the sorption rate strongly depends on dosage of the sorbent. Increase of the sorbent amount in the solution accelerates sorption.

The models of film and particle diffusion [66], chemical reactions of the pseudo-first [67] and pseudo-second order [68] were applied to experimental data. As found, the AR-0, AR-2, and AR-4 composites obey the model of the pseudo-first order:

$$\ln\left(A_\infty - A_t\right) = \ln A_\infty - K_1 t. \tag{1.9}$$

At the same time, the model of the pseudo-second order:

$$\frac{t}{A} = \frac{1}{K_2 A_\infty^2} + \frac{1}{A_\infty} \cdot t \tag{1.10}$$

is applied to other samples. Here A_t and A_∞ are the capacity after certain time and under equilibrium conditions, respectively, K_1 and K_2 are the constants. The calculated data are given in Fig. 1.3 and summarized in Tables 1.1 and 1.2. The experimental and calculated A_∞ values are rather close to each other. This and also high correlation coefficients show adequacy of the models.

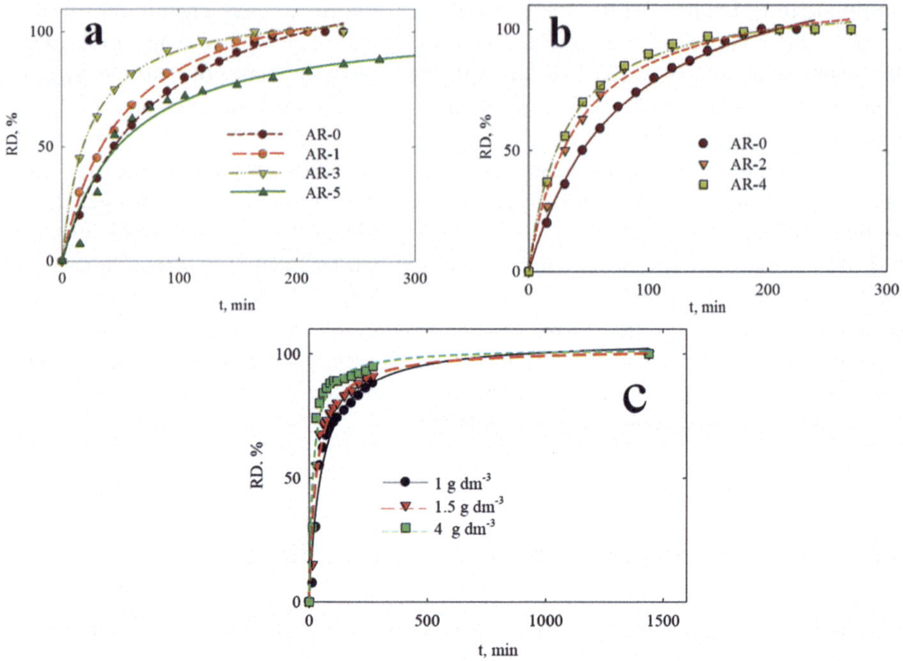

Fig. 1.2 Removal degree of uranium(VI) from modeling solution over time. Sorbent dosage was 1 g dm^{-3} (**a**, **b**, all samples) or varied (**c**. AR-5)

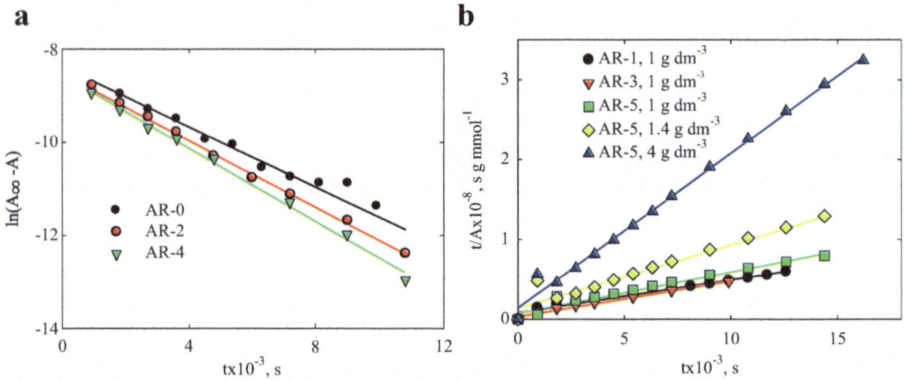

Fig. 1.3 Application of the model of chemical reaction of pseudo-first (**a**) and pseudo-second order (**b**). The sorbent dosage was 1 g dm^{-3} (**a**) or varied as the legend shows (**b**)

The highest constant for the chemical reaction of the pseudo-first order has been found for the AR-4 sample. It is 1.3 times higher than that for the pristine resin. Regarding the K_2 magnitudes, the AR-3 sample shows the highest constant (among the data for sorbent dosage of 1 g dm^{-3}). It should be stressed that the sorbents,

Table 1.1 Uranium(VI) sorption: model of chemical reaction of pseudo-first order

Sample	$A_\infty \times 10^4$, mol g^{-1}		$K_1 \times 10^4$, s^{-1}	R^2
	Experimental	Calculated		
AR-0	2.11	2.02	2.92	0.98
AR-2	1.05	1.01	3.64	0.99
AR-4	1.05	1.01	3.73	0.99

Table 1.2 Uranium(VI) sorption: model of chemical reaction of pseudo-second order

Sample	Dosage, g dm^{-3}	$A_\infty \times 10^4$, mol g^{-1}		K_2, g mol^{-1} s^{-1}	R^2
		Experimental	Calculated		
AR-1	1.0	2.10	2.66	1.19	0.99
AR-3	1.0	2.10	2.41	2.92	0.99
AR-5	1.0	2.10	2.21	1.19	0.99
	1.5	1.24	1.28	3.61	0.99
	4.0	0.52	0.53	18.5	0.99

which were obtained using more concentrated $ZrOCl_2$ solution, demonstrate the highest sorption rate. These samples contain mainly aggregated nanoparticles, the size of which is several hundred nanometers. At the same time, the lowest K_2 value has been found for the sample containing particles of micron size (AR-5). Sorption can be accelerated only by means of increasing the sorbent dosage.

1.7 Regeneration of Sorbents

Certain samples loaded with U(VI) were regenerated with different reagents (Table 1.3). No sufficient uranium removal from the resins was observed during their treatment with deionized water. However, treatment with NaOH solution allows us to reach rather high desorption degrees (41–62%). The lowest desorption degree has been found for the AR-1 sample, which contains non-aggregated nanoparticles. The pristine resin demonstrates the highest *DD* value. The most complete regeneration is achieved when $NaHCO_3$ solution is used. After double washing with this solution, the *DD* magnitude is 100% (AR-1) or close to it (other composites). In this case, ion exchange is enhanced by complex formation. The *DD* values decreases in the following order: AR-1 > AR-3 > AR-5. A size of incorporated particles reduces as follows: AR-5 > AR-3 > AR-1. It means the most facile regeneration is characteristic for the ion-exchangers containing non-aggregated nanoparticles. The lowest desorption degree is reached for the pristine resin.

In order to use the sorbents further, their hydrocarbonate forms have to be transformed into OH-forms by means of washing with alkali solution. This is necessary to provide high sorption rate.

Table 1.3 Regeneration of sorbents

Reagent	DD, %			
	AR-0	AR-1	AR-3	AR-5
H_2O	1	2	4	3
NaOH	62	41	61	53
$NaHCO_3$ (1 time)	66	95	83	84
$NaHCO_3$ (2 times)	89	100	98	95

1.8 Conclusions

The Ostwald–Freundlich equation was adapted to HZD precipitation in anion exchange polymer matrix. Experimental verification shows that the chemical composition of incorporated particles has to be taken into consideration. Non-aggregated nanoparticles can be obtained at room temperature and concentration of impregnating solution of 0.1 M. Increase of concentration of $ZrOCl_2$ solution and elevation of temperature provide formation of particles, the size of which is several hundred nanometers. Depending on type of incorporated particles, the rate of sorption of uranium-containing anions obeys the model of chemical reaction of pseudo-first or pseudo-second order. Among other tested sorbents, the samples that contain particles in voids between gel regions show the highest sorption rate.

Insertion of glycerol to the solvent provides precipitation of large particles of micron size. This sample demonstrates the slowest sorption rate; it is lower in comparison even with the pristine resin.

As found, the most suitable reagent for regeneration is $NaHCO_3$ solution. Decrease in size of incorporated particles has been established to facilitate desorption of uranium-containing anions.

References

1. Morrell JS, Jackson MJ (eds) (2013) Uranium processing and properties. Springer Science + Business Media, New York
2. WHO (2008) Guidelines for drinking-water quality, 3rd ed. Incorporating the first and second addenda, vol 1. Recommendations. Geneva
3. Shakur HR, Saraee KRE, Abdi MR et al (2016) Selective removal of uranium ions from contaminated waters using modified-X nanozeolite. Appl Radiat Isot 118:43–55
4. Yakout SM (2016) Evaluation of mineral and organic acids on the selective separation of radioactive elements (U and Th) using modified carbon. Desalin Water Treat 57(7):3292–3297
5. Yi ZJ, Yao J, Kuang YF et al (2016) Uptake of hexavalent uranium from aqueous solutions by using coconut husk activated carbon. Desalin Water Treat 57(4):1749–1755
6. Menacer S, Lounis A, Guedioura B et al (2016) Uranium removal from aqueous solutions by adsorption on Aleppo pine sawdust, modified by NaOH and neutron irradiation. Desalin Water Treat 57(34):16184–16195
7. Loureiro JM, Kartel MT (eds) (2009) Combined and hybrid adsorbents: fundamentals and applications. Springer, Berlin

8. Das A, Sundararajan M, Paul B et al (2017) Assesment of Phosphate Functionalised Silica Gel (PFSG) for separation and recovery of uranium from Simulated Silicide Fuel Scraps Dissolver Solution (SSFSDS). Colloids Surf A Physicochem Eng Asp 530:124–133
9. Li F, Yang Z, Weng H et al (2018) High efficient separation of U(VI) and Th(IV) from rare earth elements in strong acidic solution by selective sorption on phenanthroline diamide functionalized graphene oxide. Chem Eng J 332:340–350
10. Li ZJ, Wang L, Yuan LY et al (2015) Efficient removal of uranium from aqueous solution by zero-valent iron nanoparticle and its graphene composite. J Hazard Mater 290:26–33
11. Sun YB, Ding CC, Cheng WC et al (2014) Simultaneous adsorption and reduction of U(VI) on reduced graphene oxide-supported nanoscale zerovalent iron. J Hazard Mater 280:399–408
12. Fan FL, Qin Z, Bai J et al (2012) Rapid removal of uranium from aqueous solutions using magnetic $Fe_3O_4@SiO_2$ composite particles. J Environ Radioact 106:40–46
13. Zhao Y, Li J, Zhao L et al (2014) Synthesis of amidoxime-functionalized $Fe_3O_4@SiO_2$ core-shell magnetic microspheres for highly efficient sorption of U(VI). Chem Eng J 235:275–283
14. Tan L, Zhang X, Liu Q et al (2015) Synthesis of Fe_3O_4-TiO_2 core-shell magnetic composites for highly efficient sorption of uranium (VI). Colloids Surf A Physicochem Eng Asp 469:279–286
15. Zhang J, Guo Z, Li Y et al (2016) Effect of environmental conditions on the sorption of uranium on $Fe_3O_4@MnO_2$ hollow spheres. J Mol Liq 223:534–540
16. Loukanov A, Udono H, Takakura R et al (2017) Monitoring and extraction of uranium in polluted acid mine drainage by super-paramagnetic nanoparticles coated with carbon nanodots. J Radioanal Nucl Chem 314(2):1149–1159
17. El-Sherif RM, Lasheen TA, Jebril EA (2017) Fabrication and characterization of CeO_2-TiO_2-Fe_2O_3 magnetic nanoparticles for rapid removal of uranium ions from industrial waste solutions. J Mol Liq 241:260–269
18. Xu M, Han X, Hua D (2017) Polyoxime-functionalized magnetic nanoparticles for uranium adsorption with high selectivity over vanadium. J Mater Chem A 5:12278–12284
19. Han R, Zou W, Wang Y et al (2007) Removal of uranium(VI) from aqueous solutions by manganese oxide coated zeolite: discussion of adsorption isotherms and pH effect. J Environ Radioact 93(3):127–143
20. Yaroshenko NA, Perlova OV, Sazonova VF et al (2012) Sorption of uranium compounds by zirconium-silica nanosorbents. Russ J Appl Chem 85(6):849–855
21. Perlova OV, Sazonova VF, Yaroshenko NA et al (2014) Kinetics of sorption of uranium (VI) compounds with zirconium-silica nanosorbents. Russ J Phys Chem A 88(6):1012–1016
22. Yang D, Song S, Zou Y et al (2017) Rational design and synthesis of monodispersed hierarchical SiO_2@layered double hydroxide nanocomposites for efficient removal of pollutants from aqueous solution. Chem Eng J 323:143–152
23. Zou Y, Wang P, Yao W et al (2017) Synergistic immobilization of UO_2^{2+} by novel graphitic carbon nitride@layered double hydroxide nanocomposites from wastewater. Chem Eng J 330:573–584
24. Cakir P, Inan S, Altas Y (2014) Investigation of strontium and uranium sorption onto zirconium-antimony oxide/polyacrylonitrile (Zr-Sb oxide/PAN) composite using experimental design. J Hazard Mater 271(3):108–119
25. Yuan D, Chen L, Xiong X et al (2016) Removal of uranium (VI) from aqueous solution by amidoxime functionalized superparamagnetic polymer microspheres prepared by a controlled radical polymerization in the presence of DPE. Chem Eng J 285:358–362
26. Yu HW, Yang SS, Ruan HM et al (2015) Recovery of uranium ions from simulated seawater with palygorskite/amidoxime polyacrylonitrile composite. Appl Clay Sci 111:67–75
27. Bai J, Yin X, Zhu Y et al (2016) Selective uranium sorption from salt lake brines by amidoximated Saccharomyces cerevisiae. Chem Eng J 283:889–895
28. Shen L, Han X, Qian J et al (2017) Amidoximated poly(vinyl imidazole)-functionalized molybdenum disulfide sheets for efficient sorption of a uranyl tricarbonate complex from aqueous solutions. RSC Adv 7:10791–10797

29. Liu Y, Li Q, Cao X et al (2013) Removal of uranium(VI) from aqueous solutions by CMK-3 and its polymer composite. Appl Surf Sci 285:258–266
30. Shao D, Hou G, Li J et al (2014) PANI/GO as a super adsorbent for the selective adsorption of uranium (VI). Chem Eng J 255:604–612
31. Turanov AN, Karandashev VK, Masalov VM et al (2013) Adsorption of lanthanides(III), uranium(VI) and thorium(IV) from nitric acid solutions by carbon inverse opals modified with tetraphenylmethylenediphospine dioxide. J Colloid Interface Sci 405:183–188
32. Abdeen Z, Akl ZF (2015) Uranium (VI) adsorption from aqueous solutions using poly(vinyl alcohol)/carbon nanotube composites. RSC Adv 5:74220–74229
33. Al Keshtkar AR, Irani M, Moosavian A (2013) Removal of uranium (VI) from aqueous solutions by adsorption using a novel electrospun PVA/TEOS/APTES hybrid nanofiber membrane: comparison with casting PVA/TEOS/APTES hybrid membrane. J Radioanal Nucl Chem 295(1):563–571
34. Rahmani-Sani A, Hosseini-Bandegharaei A, Hosseinib S-H et al (2015) Kinetic, equilibrium and thermodynamic studies on sorption of uranium and thorium from aqueous solutions by a selective impregnated resin containing carminic acid. J Hazard Mater 286:152–163
35. Elabd AA, Zidan WI, Abo-Aly MM et al (2014) Uranyl ions adsorption by novel metal hydroxides loaded Amberlite IR 120. J Environ Radioact 134:99–108
36. Zidan WI, Abo-Aly MM, Elhefnawy OA et al (2014) Batch and column studies on uranium adsorption by Amberlite XAD-4 modified with nano-manganese dioxide. J Radioanal Nucl Chem 304(2):645–653
37. Dzyazko YS, Perlova OV, Perlova NA et al (2017) Composite cation-exchange resins containing zirconium hydrophosphate for purification of water from U(VI) cations. Desalin Water Treat 69:142–152
38. Perlova N, Dzyazko Y, Perlova O et al (2017) Formation of zirconium hydrophosphate nanoparticles and their effect on sorption of uranyl cations. Nanoscale Res Lett 12:209. https://doi.org/10.1186/s11671-017-1987-y
39. Gao Q, Hu J, Li R et al (2016) Radiation synthesis of a new amidoximated UHMWPE fibrous adsorbent with high adsorption selectivity for uranium over vanadium in simulated seawater. Radiat Phys Chem 122:1–8
40. Zhang S, Zhao X, Li B et al (2016) "Stereoscopic" 2D super-microporous phosphazene-based covalent organic framework: design, synthesis and selective sorption towards uranium at high acidic condition. J Hazard Mater 314:95–104
41. Heshmati H, Torab-Mostaedi M, Gilani HG et al (2015) Kinetic, isotherm, and thermodynamic investigations of uranium(VI) adsorption on synthesized ion-exchange chelating resin and prediction with an artificial neural network. Desalin Water Treat 55(4):1076–1087
42. Zhou L, Zou H, Huang Z et al (2016) Adsorption of uranium (VI) from aqueous solution using magnetic carboxymethyl chitosan nano-particles functionalized with ethylenediamine. J Radioanal Nucl Chem 308(3):935–946
43. Şimşek S, Yilmaz E, Boztuğ A (2013) Amine-modified maleic anhydride containing terpolymers for the adsorption of uranyl ion in aqueous solutions. J Radioanal Nucl Chem 298 (2):923–930
44. Su S, Liu Q, Liu J et al (2017) Enhancing adsorption of U(VI) onto EDTA modified L. cylindrica using epichlorohydrin and ethylenediamine as a bridge. Sci Rep 7:44156. https://doi.org/10.1038/srep44156
45. Dzyazko YS, Ponomareva LN, Volfkovich YM et al (2012) Effect of the porous structure of polymer on the kinetics of Ni^{2+} exchange on hybrid inorganic-organic ionites. Russ J Phys Chem 86(6):913–919
46. Dzyazko YS, Ponomaryova LN, Volfkovich YM et al (2014) Ion-exchange resin modified with aggregated nanoparticles of zirconium hydrophosphate. Morphology and functional properties. Microporous Mesoporous Mater 198:55–62
47. Myronchuk VG, Dzyazko YS, Zmievskii YG et al (2016) Organic-inorganic membranes for filtration of corn distillery. Acta Periodica Technologica 47:153–165

48. Dzyazko YS, Volfkovich YM, Ponomaryova LN et al (2016) Composite ion-exchangers based on flexible resin containing zirconium hydrophosphate for electromembrane separation. J Nanosci Technol 2(1):43–49
49. Cormelis R, Caruso JA, Crews H et al (2005) Handbook of elemental speciation II. Species in the environment, food, medicine and occupational health. Wiley, Chichester
50. Amphlett CB (1964) Inorganic ion exchangers. Elsevier, Amsterdam
51. Dzyazko YS, Rozhdestvenskaya LM, Vasilyuk SL et al (2009) Electrodeionization of Cr(VI)-containing solution. Part I: chromium transport through granulated inorganic ion-exchanger. Chem Eng Commun 196(1–2):3–21
52. Maltseva TV, Kudelko EO, Belyakov VN (2009) Adsorption of Cu(II), Cd(II), Pb(II), Cr(VI) by double hydroxides on the basis of Al oxide and Zr, Sn, and Ti oxides. Russ J Phys Chem A 83 (13):2336–2339
53. Dzyaz'ko YS, Belyakov VN, Vasilyuk SL et al (2006) Anion-exchange properties of composite ceramic membranes containing hydrated zirconium dioxide. Russ J Appl Chem 79(5):769–773
54. Dzyazko YS, Volfkovich YM, Sosenkin VE et al (2014) Composite inorganic membranes containing nanoparticles of hydrated zirconium dioxide for electrodialytic separation. Nanoscale Res Lett 9:271. https://doi.org/10.1186/1556-276X-9-271
55. Marti-Calatayud MC, Garcia-Gabaldon M, Perez-Herranz V et al (2015) Ceramic anion-exchange membranes based on microporous supports infiltrated with hydrated zirconium dioxide. RSC Adv 5:46348–46358
56. Dzyazko YS, Rudenko AS, Yukhin YM et al (2014) Modification of ceramic membranes with inorganic sorbents. Application to electrodialytic recovery of Cr(VI) anions from multicomponent solution. Desalination 342:52–60
57. Pang R, Li X, Li J et al (2014) Preparation and characterization of ZrO_2/PES hybrid ultrafiltration membrane with uniform ZrO_2 nanoparticles. Desalination 332:60–66
58. Dzyazko YS, Rozhdestvenska LM, Vasilyuk SL et al (2017) Composite membranes containing nanoparticles of inorganic ion exchangers for electrodialytic desalination of glycerol. Nanoscale Res Lett 12:438. https://doi.org/10.1186/s11671-017-2208-4
59. Mal'tseva TV, Kolomiets EA, Vasilyuk SL (2017) Hybrid adsorbents based on hydrated oxides of Zr(IV), Ti(IV), Sn(IV), and Fe(III) for arsenic removal. J Water Chem Technol 39(4):214–219
60. Kolomiets EA, Belyakov VN, Palchik AV et al (2017) Adsorption of arsenic by hybrid anion–exchanger based on titanium oxyhydrate. J Water Chem Technol 39(2):80–84
61. Savvin SB (1961) Analytical use of arsenazo III: determination of thorium, zirconium, uranium and rare earth elements. Talanta 8:673–685
62. Myerson AS (ed) (2002) Handbook of industrial crystallization. Butterworth-Heinemann, Woburn
63. Ostwald WF (1888) Zur Theorie der Lösungen. Z Phys Chem 2(1):36–37
64. Dzyazko YS, Ponomareva LN, Volfkovich YM et al (2013) Conducting properties of a gel ionite modified with zirconium hydrophosphate nanoparticles. Russ J Electrochem 49(3):209–215
65. Kostrikin AV, Spiridonov FM, Komissarova LN (2010) On the structure and dehydration of hydrous zirconia and hafnia xerogels. Russ J Inorg Chem 55(6):866–875
66. Helfferich F (1995) Ion exchange. Dover, New York
67. Lagergren S (1898) About the theory of so called adsorption of soluble substances. Kungliga Svenska Vetenskapsakademiens Handlingar 24(4):1–39
68. Ho YS, McKay G (1999) Pseudo-second order model for sorption processes. Process Biochem 34(5):451–465

Chapter 2
High-Entropy Film Alloys: Electrophysical and Magnetoresistive Properties

Yu. Bereznyak, L. Odnodvorets, D. Poduremne, I. Protsenko, and Yu. Shabelnyk

2.1 Introduction

In the last decade, research was carried out on the crystal structure and mechanical properties of a new class of materials—high-entropy alloys (HEA)—whose improved properties were first noticed by the authors [12]. Since these alloys are formed from 5 to 13 elements with face centric cubic (fcc), base centric cubic (bcc), or hexagonal (hcp) lattice, they have high entropy of mixing (ΔS_{mix}), and have more stable phase as a bcc or fcc solid solution (s.s.) compared to intermetallic compounds and other complex structures (see, e.g., [4]).

In a bulk AlCrFeCoNiCu high-entropy alloy can simultaneously stabilize the fcc and bcc s.s [2, 4, 5] or bcc s.s. α-Fe(Cr) and intermetallic AlNi [11] depending on the concentration of some component (e.g., atoms Ni).

The authors [4] concluded that stabilizing the fcc or bcc completely determined by the average concentration of valence electrons per atom located in the valence band of the alloy. At the concentration less than 7.2 el/at bcc phase formed, at 7.2–8.2 el/at two-phase composition fcc + bcc stabilized and at concentrations greater than 8.2 el/at—fcc phase s.s. HEA.

The peculiarity of our experiments consists in the fact that formation of s.s. made by layered or simultaneous deposition of some individual components of thickness up to 8 nm (total thickness up to 60 nm), which, thanks to condensation-stimulated diffusion and low thickness, causes mixing of atoms and forming of s.s.

Yu. Bereznyak · L. Odnodvorets · D. Poduremne · I. Protsenko (✉) · Yu. Shabelnyk
Department of Electronics, General and Applied Physics, Sumy State University, Sumy, Ukraine
e-mail: i.protsenko@aph.sumdu.edu.ua

O. Fesenko, L. Yatsenko (eds.), *Nanooptics, Nanophotonics, Nanostructures, and Their Applications*, Springer Proceedings in Physics 210,
https://doi.org/10.1007/978-3-319-91083-3_2

The aim of our work was to study phase state, electrophysical and magnetoresistivity properties of high-entropy film alloys, which was established in our previous work [10].

2.2 Methodology and Techniques Experiment

The film samples are condensed by controlling the thickness of the layers by quartz crystal method. For the diffraction and electron-microscopic studies, C-substrate (S) was used, and for the resistance and magnetoresistance measured, sitall-substrates were used.

Strain coefficient was measured by the method [9] using polystyrene substrates. The concentration of component was calculated and refined by energy dispersive X-ray (EDX) analysis. The calculation was performed on the ratio

$$C_i = \frac{D_i d_i \mu_i^{-1}}{\sum_{i=1}^{n} D_i d_i \mu_i^{-1}},$$

where D is density, μ is molar mass, d is thickness of individual layers at the layered deposition or effective thickness at the simultaneous deposition.

It is known (see, e.g., [6, 13]) that the formation of HEA occurs under the following conditions: the value of ΔS_{mix} should be greater than $\Delta S_{\text{mix}} = 1.61$ $R = 13.38$ J/(mole·K) (for five-component alloys) or $\Delta S_{\text{mix}} = 1.75\ R = 14.54$ J/(mol·K) (for six-component alloys) and parameters of atomic size differences

$$\delta = \sqrt{\sum_{i=1}^{n} c_i \left(1 - \frac{r_i}{\vec{r}}\right)^2},$$

where r_i is the atomic radius of i-component; $\vec{r} = \sqrt{\sum_{i=1}^{n} c_i r_i}$, the average atomic radius, should be less than 6.6% (in our case, these conditions are fulfilled).

Calculation of mixing entropy (ΔS_{mix}) as a criterion of HEA was performed on ratio [13] (see also [4])

$$\Delta S_{\text{mix}} = -R \sum_{i=1}^{n} c_i \ln c_i,$$

where i and c_i are the number and atomic concentrations of i-components.

Table 2.1 shows general characteristics of the samples. Note that the No 1′ and No 2′ samples with seven components are different from the six-component No 1 and No 2 samples only by the additional layer of Ti. Also note that the elemental

Table 2.1 General characteristics of HEA films

№	Sample (thickness, nm)	c_i, at. %	ΔS_{mix}, J/(mol·K)	TCR·10^3, K^{-1} at $T = 300$ K
1	Cr(7.5)/Al(4.5)/Co(7.3)/Cu(4.8)/Ni (7.0)/Fe(7.0)/S	19/8/21/12/19/19/S	14.58	1.80
1′	Ti(2.2)/Cr(7.5)/Al(4.5)/Co(7.3)/Cu (4.8)/Ni(7.0)/Fe(7.0)/S	4/18/9/20/12/19/18/S	15.33	1.75
2	Al(4.0)/Cu(3.2)/Co(4.5)/Cr(5.2)/Fe (4.7)/Ni(4.0)/S	11/13/19/20/19/17/S	14.76	2.00
2′	Ti(2.5)/Al(4.0)/Cu(3.2)/Co(4.5)/Cr (5.2)/Fe(4.7)/Ni(4.0)/S	6/11/12/18/18/19/16/S	15.54	1.75
3	Cr(10)/Fe(9)/Ni(10)/S	33/30/37/S	9.12	1.20
4	Cu(32)/Co(9)/Cr(10)/$Ni_{0,8}Fe_{0,2}$(19)/Al (12)/S	32/15/16/19/5/14/S	14.02	1.80
5	Cu(10)/Co(9)/Cr(10)/$Ni_{0,8}Fe_{0,2}$(19)/Al (12)/S	18/17/17/24/8/15/S	14.38	1.22

composition of No. 1 and No. 2 samples corresponds to samples studied by the authors of [4, 11]. We will indicate that the permalloy (Py) $Ni_{0,8}Fe_{0,2}$ as a separate layer has been selected in order to determine whether Py is stored as a separate layer in the HEA, or dissociation of the NiFe complexes in the film occurs. As we have established, there is a dissociation of these complexes (Sect. 2.3.1). According to [1], the magnitude of the enthalpy of mixing ΔH_{mix} can make a prognosis of the phase state of the HEA: solid solution, ordered phase, or chemical compound.

The thermal coefficient of resistance (TCR) was calculated based on the temperature dependence on the resistivity (ρ) (the second temperature cycle at the cooling) and the ratio

$$\beta = \frac{\rho(T) - \rho(300)}{\rho(300)(T - 300)}.$$

Integral value of strain coefficient determined based on depending $\Delta R/R(0)$ versus on the value of longitudinal strain ε_l on the ratio

$$\gamma_l = \frac{R(\varepsilon_l) - R(0)}{R(0)\varepsilon_l}.$$

The value of magnetoresistance (MR) calculated on the basis of field dependence $R(B)$ on the ratio

$$MR = \frac{R(B) - R(0)}{R(0)}.$$

To measure $\rho(T)$, $R(\varepsilon_l)$, and R(B) and calculate β, γ_l and MR used appropriate computerized complex.

2.3 Results and Discussion

2.3.1 Crystalline Structure

Electron microscopic study indicates that in the initial state (after condensation) the crystal structure are high-disperse (the average crystallite size L ≤ 10 nm). Beside with the high-disperse fraction is formed a structure with L ≅ 50 nm after annealing to 850 K (Fig. 2.1). The character of electron diffraction indicates that at the stage of condensation in HEA film, two fcc and one bcc phases (so-called B2 phase) were formed. A similar phase composition was observed by [4]. The fcc1 and fcc2 phases formation associated with the unfinished process of s.s. HEA formation. Although with annealing of the samples to 850 K homogenization alloy appears and s.s. HEA forms with fcc lattice based on the fcc1 and fcc2 (the lattice parameter $a = 0.3604$ nm) (Fig. 2.1, Table 2.2). Most likely, one of the fcc phases initially formed as s.s. HEA, the other was a metastable phase based on Al and Ni [11], which decomposited during annealing. According to conclusions [11] bcc phase can be magnetic s.s. α-Fe (Cr). The gray background between the diffraction lines (111) and (200) indicates its specific nonuniformity and vacancy defect.

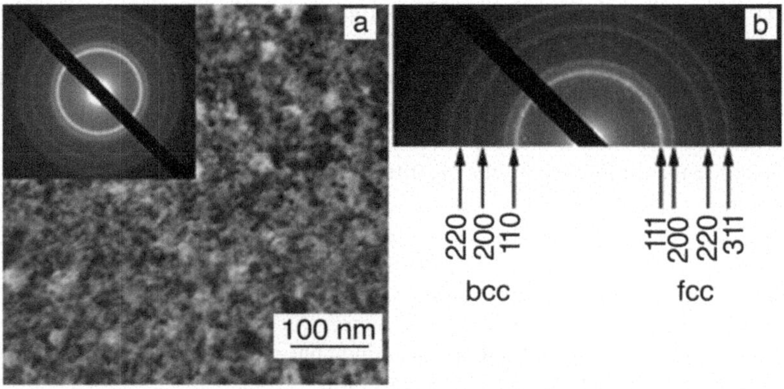

Fig. 2.1 Microstructure and diffraction pattern from system Cr(7.5)/Al(4.5)/Co(7.3)/Cu(4.8)/Ni (7.0)/Fe(7.0)/S. Annealing temperature, K: 600 (**a**) and 850 (**b**)

Table 2.2 Decryption of diffraction pattern from HEA

№	d_{hkl}, nm	hkl	a, nm	Phase
1	0.2175	111	0.3616	s.s.
2	0.2088	200	0.3600	s.s.
3	0.1420	200	0.2840	bcc
4	0.1272	220	0.3597	s.s.
5	0.1010	220	0.2850	bcc
6	0.10870	311	0.3605	s.s.
			a(s.s.) = 0.3604 nm	

Annealing of samples contributes to the healing of defects and some ordering s.s. and leads to the formation of a small number of Al_2O_3 oxides, ($d_{\text{hkl}} = 0.237$ nm) and Cr_2O_3 ($d_{\text{hkl}} = 0.261$ and 0.247 nm), whose reflexes are of a point character and of low intensity (not shown in Table 2.2) due to the low effective thickness layers of Cr and Al. In addition, we note that two very weak lines (200) and (220) are recorded from the bcc phase (phase B2), which was observed by the authors [4], and the most intense line (110) is not fixed. Note that the average value of the parameter fcc lattice s.s. is in very good agreement with the corresponding parameter for s.s. Cu(Ni) $a = 0.352$–0.366 nm [3]. Also, we note that the addition of a thin layer of Ti (specimens 1'and 2′) or Py (see 4 and 5) does not affect their phase composition.

It is known (see, e.g., [6, 13]) that the formation of HEA is due to the performance of the following conditions: the value of ΔS_{mix} six-component alloy should be more $\Delta S_{\text{mix}} = 1.75\,R = 1454$ J/(mol·K) for equiatomic alloy and parameters of difference in atomic size. The fact that in our case annealed HEA actually has only one fcc phase is explained by the corresponding value of ΔS_{mix} and δ. It should also be noted that the described phase composition does not depend on the method of film material formation—simultaneous or layered condensation—which is explained by the active mixing of the components in the process of deposition.

2.3.2 Electrophysical Properties

Figure 2.2 shows the typical dependence of resistivity and TCR (insert) for HEA. The typical features of the samples are: the relatively large value of the TCR (order of magnitude 10^{-3} K^{-1}) and the defect concentration of the crystalline structure of the «vacancy-intersite atom» type, which is indicated by the reduction of the specific resistance during ignition in the I cycle of «heating ↔ cooling».

During the study of strain coefficient (SC) of HEA, we, probably, first observed the strain effect feature. At the strain in the range $\Delta\varepsilon_l = (0$–$1)$ % at the $\varepsilon'_{ltr} \cong 0.5\%$ transition from elastic to plastic strain occurs (we call it plastic strain type I). At the I strain cycle on the range $\Delta\varepsilon_l = (0 - 1)\%$, the value strain coefficient $\gamma_{l\text{I}} \cong 12.0$, and range $\Delta\varepsilon_{l1} \cong (0 - 0.5)\%$, $\gamma_{l\text{II,III}} \cong 25$. In II and III cycles of strain were observed linear dependence of $\Delta R/R(0)$ versus ε_l, value of strain coefficient $\gamma_l \cong 12.5$. At this stage, sold, most likely after the relaxation effect quasiplastic strain of I type. With further strain to $\varepsilon_l = 2\%$ is the transition to plastic strain of II type at $\varepsilon''_{ltr} \cong 1\%$. This transition is accompanied by an increase of value strain coefficient to $\gamma_{l\text{II,III}} \cong 12.5$ to $\gamma_{l\text{IV}} \cong 90$, which is a significant value for metal films. At the V and VI strain cycles, there is also a kind of strain to quasiplastic because there was a relaxation of plastic strain of the II type. The described two-stage feature of the strain effect was not observed in the case of single-layer or multilayer films (see, e.g., reviews [7, 8]) with their plastic deformation.

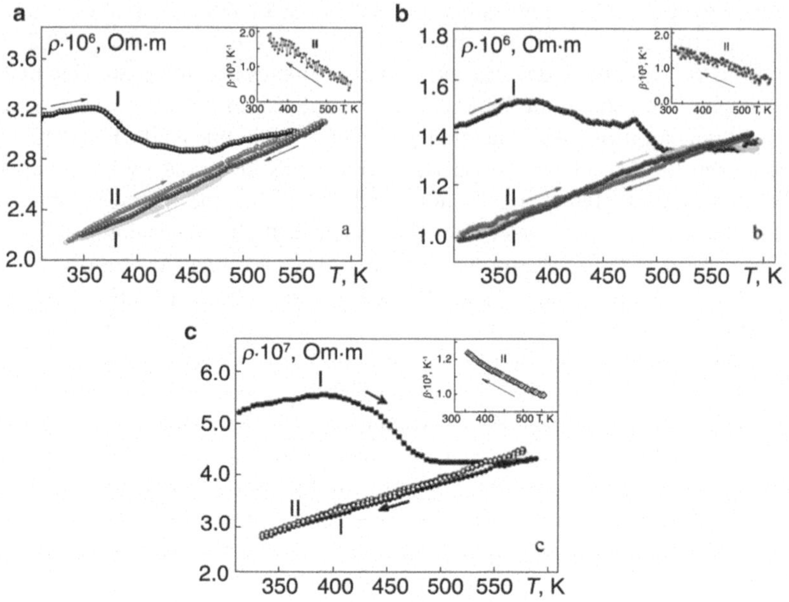

Fig. 2.2 Dependence of resistivity and TCR (insert) versus temperature for samples No. 2 (**a**), No. 2′(**b**), and No. 5 (**c**). I, II—numbers of annealing cycles

2.3.3 Magnetoresistivity Properties

Research of MR was performed in geometry CIP (current j in the film plane) at the three mutual orientation of the magnetic field: longitudinal ($\vec{B} \parallel \vec{j}$), transverse ($\vec{B} + \vec{j}$), and perpendicular ($\vec{B} \perp j$) at an operating current from 0.5 to 1 mA. Typical dependences of MR on the induction B is shown in Figs. 2.3 and 2.4.

Since bcc HEA is typical ferromagnetic [11], without elements of granular state, it is anisotropic magnetoresistant (AMR), which is clearly shown in Fig. 2.3a and Fig. 2.4 (curve 1) on one side and Fig. 2.3b, c and Fig. 2.4 (curves 2 and 3) on the other side. The effect of annealing to 800 K leads to some increase in the amplitude of the MR with longitudinal and transverse geometries of measurement, while in the case of perpendicular geometry, the amplitude MR practically does not change (Fig. 2.3), although the coercivity decreases significantly. The relatively small amplitude is due to the small thickness (up to 60 nm) and the volume of samples. A very noticeable difference in the shape of the magnetoresistive dependence and the value of the coercive force on Fig. 2.3c and Fig. 2.4 (curve 3) can be explained by the stabilization of planar anisotropy in film alloys.

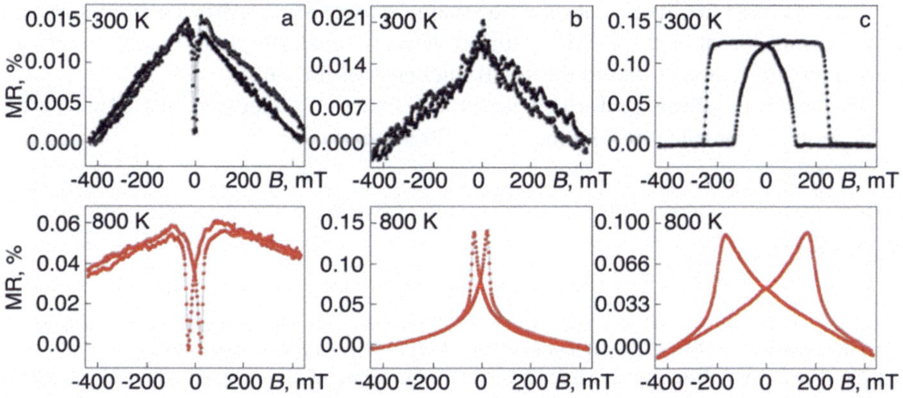

Fig. 2.3 Dependence of MR versus induction B for the sample Al(4.0)/Cu(3.2)/Co(4.5)/Cr(5.2)/ Fe (4.7)/Ni(4.0)/S with three mutual orientations of induction and electric current: longitudinal (**a**), transverse (**b**), and perpendicular (**c**)

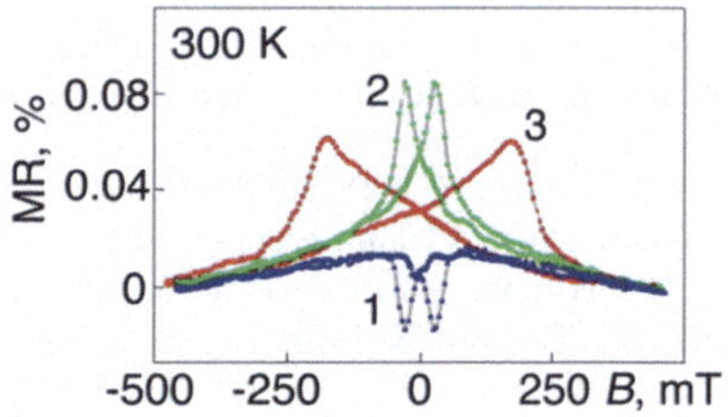

Fig. 2.4 Dependence of MR versus induction B for the sample Cu(10)/Co(9)/Cr(10)/ $Ni_{0.8}Fe_{0.2}$(19)/Al(12)/S with three mutual orientations of induction and electric current: longitudinal (*1*), transverse (*2*), and perpendicular (*3*)

2.4 Conclusions

The crystalline structure, electrophysical (resistivity, TCR, and SC) and magnetic resistive (AMR) properties of HEA films based on Cr, Fe, Al, Cu, Ni, and Co are investigated. It has been established that the films consisted of three phases: fcc1 (s.s. HEA), fcc2 (possibly metastable AlNi) and bcc (possibly s.s. α-Fe(Cr)) after condensation. In the process of annealing fcc2 the phase breaks down, and s.s. α-Fe(Cr) actually disappears, since even weak lines are not fixed on it diffraction pattern.

In the study of electrophysical properties (the strain effect), for the first time, there was a two-stage plastic deformation, each of which corresponds to a very large value of the strain coefficient (up to 90 units); the obscure cause of high sensitivity resistance of s.s. HEA to strain.

The results of the investigation of the magnetoresistive properties unambiguously indicate that the anisotropic MR with a relatively small amplitude up to 0.13% is implemented, which is due to the small thickness of the films.

The work has been performed under the financial support of the Ministry of Education and Science of Ukraine (2018–2020 years).

References

1. Cheng K-H, Weng C-H, Lai C-H, Lin S-J (2009) Study on adhesion and wear resistance of multi-element (AlCrTaTiZr)N coatings. Thin Solid Films 517(17):4989–4993
2. Del Grosso MF, Bozzolo G, Mosca HO (2012) Modeling of high entropy alloys of refractory elements. Physica B 407(16):3285–3287
3. Kalinichenko SM, Tkach OP, Hrychanovska TM, Odnodvorets LV (2015) Thermoresistive properties of the thin film solid solutions based on Cu and Ni. J Nano- Electron Phys 7(4), 04048-1-04048-5
4. Karpets MV, Makarenko OS, Myslyvchenko OM, Krapivka MO, Gorban VF, Samelyuk AV (2014) Properties of multicomponent high-entropy alloy alcrfeconi alloyed copper. Probl Frict Wear 2(63):103–111
5. Li BS, Wang YP, Ren MX, Yang C, Fu HZ (2008) Effects of Mn, Ti and V on the microstructure and properties of AlCrFeCoNiCu high entropy alloy. Mat Sci Eng A 498 (1–2):482–486
6. Pazukha IM, Makukha ZM, Shabelnyk YM, Protsenko IY (2012) Tensoresistive properties of thin film systems based on Ag and Co. J Nano- Electron Phys 4(3), 03020-1-03020-4
7. Protsenko I, Odnodvorets L, Chornous A (1998) Electroconductivity and tensosensibility of multilayer films. Metallofiz Nov Tekhnol 20(1):36
8. Protsenko SI, Odnodvorets LV, Protsenko IY (2014) Nanocomposites, nanophotonics, nanobiotechnology and application. Springer Proc Phys 156:345–374
9. Tyschenko KV, Protsenko IY (2012) Nonlinear effects in piezoresistive features of Fe and Ni based film alloys. Metalofiz Nov Teknol 34(7):907–917
10. Vorobiov SI, Kondrakhova DM, Nepijko SA, Poduremne DV, Shumakova NI, Protsenko IY (2016) Crystalline structure, electrophysical and magnetoresistive properties of high entropy film alloys. J Nano- Electron Phys 8(3), 03026-1-03026-5
11. Wang YP, Li BS, Fu HZ (2009) Solid solution or intermetallics in a high-entropy alloy. Adv Eng Mater 11(8):641–644
12. Yeh JW, Chen SK, Lin SJ, Gan JY, Chin TS, Shun TT, Tsau CH, Chang SY (2004) Nanostructured high-entropy alloys with multiple principal elements: novel alloy design concepts and outcomes. Adv End Mater 6(5):299–303
13. Zhang Y, Zhou Y (2007) Solid solution formation criteria for high entropy alloys. Mater Sci Forum 561–565:1337–1339

Chapter 3
Strain- and Adsorption-Dependent Electronic States and Transport or Localization in Graphene

Taras Radchenko, Ihor Sahalianov, Valentyn Tatarenko, Yuriy Prylutskyy, Paweł Szroeder, Mateusz Kempiński, and Wojciech Kempiński

3.1 Introduction

Among various types of structural (point or extended) defects in the physics of graphene, adsorbed atoms or molecules are probably the most important examples [1]. They act as the lattice imperfections and strongly affect electronic, optical, thermal and mechanical properties of graphene. Many characteristics, such as electron states, electrical conductivity and degree of localization of electrons (and their spins), are governed by such defects [2]. Adsorption or introduction of specific defects [3, 4], their configurations (ordering) [5–9], and application of different strains (particularly, uniaxial stretching) [10–13] can serve as ways to solve the problem of gapless graphene for a wide practical application in nanoelectronic devices.

T. Radchenko (✉) · V. Tatarenko
G.V. Kurdyumov Institute for Metal Physics, National Academy of Sciences of Ukraine, Kyiv, Ukraine
e-mail: tarad@imp.kiev.ua

I. Sahalianov · Y. Prylutskyy
Taras Shevchenko National University of Kyiv, Kyiv, Ukraine

P. Szroeder
Institute of Physics, Kazimierz Wielki University, Bydgoszcz, Poland

M. Kempiński
Faculty of Physics, Adam Mickiewicz University, Poznań, Poland

NanoBioMedical Centre, Adam Mickiewicz University, Poznań, Poland

W. Kempiński
Institute of Molecular Physics, Polish Academy of Sciences, Poznań, Poland

O. Fesenko, L. Yatsenko (eds.), *Nanooptics, Nanophotonics, Nanostructures, and Their Applications*, Springer Proceedings in Physics 210,
https://doi.org/10.1007/978-3-319-91083-3_3

We follow the methodology of the Kubo–Greenwood formalism (see, e.g., reviews [14, 15] and references therein), where transport properties are governed by the movement of electrons. If there are no defects on graphene surface, the electrons can propagate without any backscattering, resembling classical ballistic particles. Therefore, such a transport regime is called ballistic. The presence of adsorbed atoms or molecules acting as scattering centres results in diffusive transport regime, when electron diffusion coefficient becomes time-independent, and Ohm's law is valid. Finally, with the course of time, charge carriers start to localize, diffusion coefficient decreases, and localization regime occurs.

Being significantly influenced by adsorption of various atoms and molecules [16, 17], localization process is a crucial issue in the physics of graphene when considering its application in multiple areas such as energy storage, molecule sensing, photovoltaics and nanoelectronics. This phenomenon can be well observed using the electron paramagnetic resonance (EPR). EPR detects the unpaired spins localized within the structure of material and allows for the observation of their interaction with other spins and the crystal lattice, and was shown as very useful in investigations of graphene-based materials [18, 19].

This chapter summarizes and generalizes the recent theoretical [20, 21] and experimental [22] results obtained for electron behaviours in the afore-mentioned diffusive transport and localization regimes taking place in both unstrained and strained imperfect graphene. Computational results on electron states and quantum transport in diffusive regime are obtained within the framework of both the tight-binding model and the Kubo–Greenwood approach capturing all (ballistic, diffusive and localization) regimes [14, 15, 23–25]. Experimental observations via EPR and conductivity measurements are interpreted using the granular metal model [26], implying appearance of a strong localization of charge carriers due to the existence of potential barriers with charge carrier hopping, which are sensitive to various factors such as temperature, adsorbates and external fields.

The theoretical part of the study is motivated by, first, disagreements in the literature regarding the stability of differently (randomly, correlatively or orderly) distributed adatoms of various kinds on graphene surface [27–30], and, second, contradictions concerning impact of the strain on electronic properties of (even perfect, i.e., defect-free) graphene [31–42], all the more so for realistic graphene samples containing different point (or/and extended) defects, particularly due to the fabrication technology. Experimental part of the work, focusing on graphene oxide and its reduced form, is motivated by, firstly, the current popularity of such materials due to the relative simplicity and repeatability of the manufacturing procedure, and, secondly, their structure imperfection (strong wrinkling, edging, etc.), which, being responsible for electronic transport characteristics and localization phenomena, results in more localization sites [22].

3.2 Defect-Configuration-Dependent Charge Carrier Transport

3.2.1 Modelling Electronic Transport, Bond Deformations and Defects

In the Kubo–Greenwood model, the energy- (E) and time- (t) dependent diffusivity $D(E,t) = \left\langle \Delta\widehat{X}^2(E,t)\right\rangle/t$, where the wave-packet mean-quadratic spreading along x-direction is $\left\langle \Delta\widehat{X}^2(E,t)\right\rangle = \mathrm{Tr}\left[\left(\widehat{X}(t)-\widehat{X}(0)\right)^2\delta(E-\widehat{H})\right]/\mathrm{Tr}\left[\delta(E-\widehat{H})\right]$ [14, 15] with $\widehat{X}(t)=\widehat{U}^{\dagger}(t)\widehat{X}\widehat{U}(t)$ being the position operator in the Heisenberg representation, $\widehat{U}(t)=\exp\left(-i\widehat{H}t/\hbar\right)$—time-evolution operator, and tight-binding Hamiltonian (with hopping integrals up to the first three coordination shells) define Bernal-stacked few-layer honeycomb lattice [43, 44]. $\widehat{H}=\Sigma_{l=1}^{N_{\mathrm{layer}}}\widehat{H}_l+\Sigma_{l=1}^{N_{\mathrm{layer}}-1}\widehat{H}'_l$, where N_{layer} is a number of layers, H_l is a Hamiltonian contribution of lth layer and H_l' describes hopping parameters between neighbour layers (vanishes in case of 1 layer) [43, 44], that is, $\widehat{H}=-\gamma_0^1\Sigma_{\langle i,j\rangle}c_i^{\dagger}c_j-\gamma_0^2\Sigma_{\langle\langle i,j\rangle\rangle}c_i^{\dagger}c_j-\gamma_0^3\Sigma_{\langle\langle\langle i,j\rangle\rangle\rangle}c_i^{\dagger}c_j+\Sigma_iV_ic_i^{\dagger}c_i$; $c_i^{\dagger}$ and c_i are standard creation or annihilation operators acting on a quasi-particle at the site i. The summation over i runs the entire honeycomb lattice, while j is restricted to the nearest-neighbours (in the first term), next nearest-neighbours (second term) and next-to-next nearest-neighbours (third term) of ith site; $\gamma_0^1=2.78$ eV is inlayer hopping for the nearest-neighbouring C atoms occupying i and j sites at a lattice-parameter distance a = 0.142 nm between them [43, 44]; $\gamma_0^2=0.085\gamma_0^1$ and $\gamma_0^3=0.034\gamma_0^1$ are intralayer hoppings for the next (second) and next-to-next (third) nearest-neighbouring sites at the second and third coordination shells, respectively [35] (Fig. 3.1a); and V_i is on-site potential defining defect strength at a

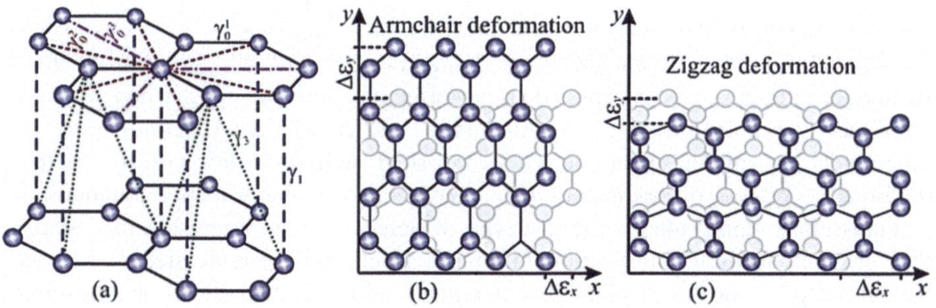

Fig. 3.1 Intra- ($\gamma^1{}_0$, $\gamma^2{}_0$, $\gamma^3{}_0$) and interlayer (γ_1, γ_3) hopping parameters for two layers (AB) of Bernal-stacked multilayer graphene (**a**) Two types of uniaxial tensile strain (by ≈30%) along armchair- (**b**) or zigzag-type (**c**) edges for single graphene layer

given graphene-lattice site i due to the presence of different sources of disorder [43, 44]. The Slonczewski–Weiss–McClure model of electron states [45–47] describes interlayer connection as follows: $\widehat{H}'_l = -\gamma_1 \Sigma_j \left(a^{\dagger}_{l,j} b_{l+1,j} + \text{H.c.} \right) - \gamma_3 \Sigma_{j,j'} \left(b^{\dagger}_{l,j} a_{l+1,j'} + \text{H.c.} \right)$ with $\gamma_1 = 0.12\gamma_0^1$, $\gamma_3 = 0.1\gamma_0^1$ [44] defining interlayer-hopping amplitudes (Fig. 3.1a).

We considered two types of uniaxial tensile strain: along so-called armchair (Fig. 3.1b) and zigzag (Fig. 3.1c) directions (edges). In both cases, the uniaxial strain induces deformation of lattice, that is, of bond lengths, and hence changes hopping amplitudes between different sites.

Following [35, 36, 48], where random strain is modelled by the Gaussian function, we can obtain dependence of the bond lengths on the deformation tensor components and then relate hopping parameters of the strained (γ) and unstrained (γ_0^1) graphene via exponential decay: $\gamma(l) = \gamma_0^1 e^{-\beta(l/a-1)}$ with a strained bond length l, and decay rate $\beta \approx 3.37$ [35, 36] being extracted from experimental data [49] along with Poisson's ratio $\nu = 0.15$ selected between that measured for graphite [50] and calculated for graphene [51].

We model several kinds of disorder with various point defects. First are resonant (uncharged) impurities [43, 44], when C atom from graphene layer is chemically (covalently) bonded with H (C or O) atom from adsorbed organic molecule. Modelling of resonant impurities was carried out with the Hamiltonian part [43, 44] $\widehat{H}_{\text{imp}} = \upsilon_d \Sigma_i^{N_{\text{imp}}} d_i^{\dagger} d_i + V \Sigma_i^{N_{\text{imp}}} \left(d_i^{\dagger} c_i + \text{H.c.} \right)$, where N_{imp} denotes number of the resonant impurities, and band parameters $V \approx 2\gamma^1{}_0$ and $\upsilon_d \approx \gamma^1{}_0/16$ were obtained from density-functional theory calculations [52]. These parameters, being previously adopted for unstrained graphene [43, 44], serve as the typical values for resonant impurities (CH_3, C_2H_5OH, CH_2OH as well as hydroxyl groups).

Vacancies are considered as the second important type of defects. A vacancy is regarded as a site with zero hopping parameters to other sites.

Screened charged impurity ions (adatoms, admolecules) on graphene or/and dielectric substrate of it constitute third-type defects. They are commonly described by the Gaussian-type on-site potential [43, 44]: $V_i = \Sigma_{j=1}^{N_{\text{imp}}^V} U_j e^{-|\mathbf{r}_i - \mathbf{r}_j|^2/(2\xi^2)}$, where $\mathbf{r}_i$ is a radius-vector of ith site, $\{\mathbf{r}_j\}$ define positions of $N^V{}_{\text{imp}}$ impurity atoms (Gaussian centres), ξ is interpreted as an effective potential length and potential amplitude U_j is uniformly random in the range $[-\Delta, \Delta]$ with Δ—maximum potential magnitude. Varying these parameters, we consider such impurities with shortly ($\xi = 0.65a$, $\Delta = 3\gamma^1{}_0$) or more distantly ($\xi = 5a$, $\Delta = \gamma^1{}_0$) acting effective potential.

Gaussian hopping [44] is the last type of defects we are interested in. Usually, they originate from the substitutional impurities causing the atomic-size misfit effect as local in-plane or out-of-plane displacements of atoms, and short- or long-range distortions in graphene lattice due to the curved ripples or wrinkles. In this case, modified distribution of the hopping integrals between different (i, j) sites reads as [44] $\gamma_{i,j} = \gamma + \Sigma_{k=1}^{N_{\text{hop}}^{\gamma}} U_k^{\gamma} e^{-|\mathbf{r}_i - \mathbf{r}_j - 2\mathbf{r}_k|^2/(8\xi_\gamma^2)}$ with $N^{\gamma}{}_{\text{hop}}$ (Gaussian) straining centres at the $\mathbf{r}_k$ positions, ξ_γ is an effective potential length and hopping amplitude $U^{\gamma}{}_k \in [-\Delta_\gamma, \Delta_\gamma]$.

The distortion centres are also considered with shortly ($\xi_\gamma = 0.65a$, $\Delta\gamma = 1.5{\gamma^1}_0$) or more distantly acting ($\xi_\gamma = 5a$, $\Delta\gamma = 0.5{\gamma^1}_0$) hoppings. The summation in expressions for Gaussian impurities and hoppings is restricted to the sites belonging to the same layer (possibility for the overlapping of Gaussian distributions in different layers is omitted).

However, sometimes Gaussian (and even Coulomb) scattering potentials are not the most appropriate to describe scattering by various point defects [20]. Therefore, in our calculation, for example, for K adatoms, we used scattering potential adapted from independent self-consistent ab initio calculations [53].

Correlation between impurity adatoms is approximately modelled introducing pair distribution function dependent on correlation length that defines minimal possible distance between any two neighbouring adatoms [8, 20, 23]. Adatomic ordering corresponds to a certain stoichiometric-type superstructure [8, 20].

The dc conductivity σ can be extracted from the electron diffusivity $D(E,t)$ undergoing saturation and reaching the maximum, $\lim_{t\to\infty} D(E,t) = D_{\max}(E)$, when diffusive transport regime occurs. Then, the semi-classical conductivity at zero temperature is [14, 15, 23] $\sigma = e^2\widetilde{\rho}(E)D_{\max}(E)$, where $\widetilde{\rho}(E) = \rho/\Omega = \mathrm{Tr}\left[\delta\left(E - \widehat{H}\right)\right]/\Omega$ is electron density of sates (DOS) per unit area Ω (and per spin), and $-e < 0$ denotes the electron charge. The DOS can be used to calculate the electron density as $n_e(E) = \int_{-\infty}^{E} \widetilde{\rho}(E)\, dE - n_{\text{ions}}$, where $n_{\text{ions}} = 3.9\cdot 10^{15}$ cm^{-2} is density of positive ions compensating the negative charge of p-electrons in graphene. Note that, for the defect-free graphene, at a neutrality (Dirac) point, $n_e(E) = 0$. Combining calculated $n_e(E)$ with $\sigma(E)$, we compute the density-dependent conductivity $\sigma = \sigma(n_e)$. Numerical details for computing DOS, D, and σ are described in appendixes to [23].

3.2.2 Calculated Density of States, Diffusivities and Conductivities

Before proceeding to graphene with defects of various types, we initially considered defect-free graphene subjected to different values of relative uniaxial tension along the above-mentioned two directions. The numerically calculated DOS curves [21] agree with independent analytical results [35]. A spectral gap appearance requires threshold deformations of $\approx$23% along zigzag direction, while there is not any gap opening for any deformations along armchair edge. The band-gap opening originates from an additional displacement of both graphene sublattices with respect to each other that occurs most pronouncedly at a deformation along zigzag direction.

High energy values (far from the Dirac point, at $E = 0$) are less experimentally realizable. Therefore, they are not depicted in Fig. 3.2, where DOS is calculated for single- and bilayer unstrained and strained graphene with 0.1% random defects.

The DOS curves for graphene monolayer and bilayer (Fig. 3.2) as well as for trilayer, quadruple-layer and quintuple-layer [21] are similar, which is an indication of the band-structure similarity, independently of the number of layers. The cause of

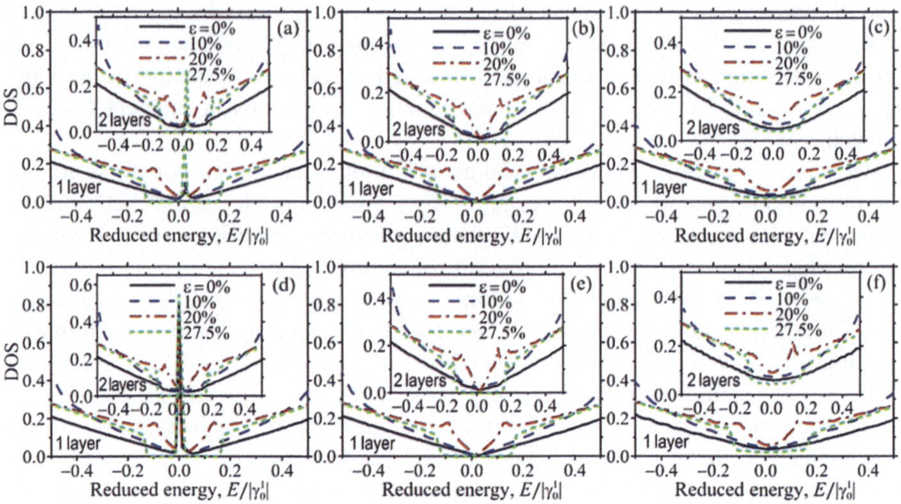

Fig. 3.2 DOS for zigzag (un)strained ($0\% \leq \varepsilon \leq 27.5\%$) single- (main panels) and bilayer (insets) graphene with 0.1% *randomly* distributed point defects: resonant impurities (**a**), short- (**b**) and long-range (**c**) Gaussian impurities, vacancies (**d**), short- (**e**) and long-range (**f**) Gaussian hoppings

such similarity lies in the energy band parameters defining intra- and interlayer hopping integrals (see Fig. 3.1): intralayer nearest-neighbour hopping integral ($\gamma^1{}_0$) is circa 10 times larger than both the interlayer parameters, that is, interlayer hoppings are much weaker than the intralayer ones.

As Fig. 3.2a, d shows, resonant impurities (O- or H-containing molecules) and vacancies similarly alter the DOS of the strained graphene: they bring an increase in spectral weight (central peak) near the Dirac point. The central peak, being attributed to impurity (or vacancy) band, increases and broadens as the resonant impurity (or vacancy) concentration rises [21]. The principle distinction between O- or H-containing molecules and vacancies concerning their effects on the spectrum consists in position of the central peak (impurity/vacancy band) in the DOS curves: it is centred at a neutrality point in case of vacancies, whereas it is shifted from it for the hydroxyl groups due to the nonzero (positive) on-site potential modelling them. In contrast to the resonant impurities and vacancies, the Gaussian potentials and hoppings do not induce low-energy impurity (vacancy) band around the neutrality point as shown in Fig. 3.2b, c, e, f.

Like for the perfect graphene [21, 35, 36], the spectrum is also strongly gapless for small and even moderate strains of impure graphene (Fig. 3.2). The gap overcoming requires the threshold (zigzag) deformations over $\varepsilon \approx 20\%$ for non-long-range-acting impurities or vacancies (Fig. 3.2a, b, d, e), whereas 'long-range' potentials (hoppings) smear gap region and transform it into quasi- or pseudo-gap—plateau-shaped deep minimum in DOS near the Dirac point (Fig. 3.2c, f).

Figure 3.3a, b show DOS around the Fermi level ($E = 0$) as a function of tensile strain parameter $\varepsilon \in [0\%, 30\%]$ for single-layer graphene with a fixed concentration

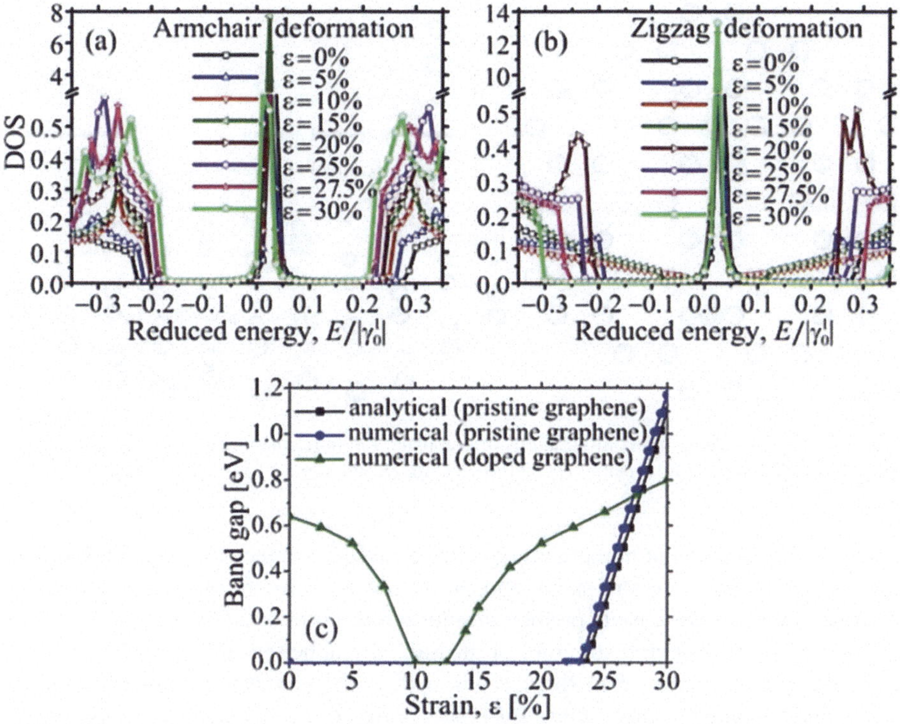

Fig. 3.3 (**a**, **b**) DOS for graphene monolayer with 3.125% of *ordered* resonant impurities (O- or H-containing molecules) for different (up to 30%) values of the tension strain along armchair (**a**) and zigzag (**b**) directions. (**c**) Comparison of analytically [35] and numerically calculated band-gap energies vs. the uniaxial deformation along zigzag direction for monolayer graphene without defects (squares and circles) and with 3.125% of ordered hydroxyl groups (triangles)

of the *ordered* H or O adatoms. Band gap decreases slowly (however, permanently), if armchair deformation increases. However, in case of zigzag strain, the band gap initially (for $0\% \leq \varepsilon \leq 10\%$) becomes narrower and narrower up to the total disappearance, but then, at a certain threshold strain value ($\varepsilon_{\min} \approx 12.5\%$), the gap reappears, grows up, and can be even wider than it was before the stretching (see also next figure). Importantly, this threshold value $\varepsilon_{\min}$, when the band gap opens, is lower in comparison with those that have been estimated earlier for perfect defect-free graphene layers subjected to the uniaxial zigzag strain ($\varepsilon_{\min} \approx 23\%$ [35]), shear strain ($\varepsilon_{\min} \approx 16\%$ [37]), and almost coincides with the value expected combining shear with armchair uniaxial deformations ($\varepsilon_{\min} \approx 12\%$ [37]).

Comparing band-gap energies calculated analytically in [35] and numerically computed for pristine as well as for doped graphene subjected to uniaxial tensile deformation along zigzag-edge direction (Fig. 3.3c), one can see a pronounced non-monotony of the curve for strained graphene with ordered pattern of defects. Such abnormal non-monotonic behaviour of the strain-dependent band gap mainly originates from the simultaneous contribution of two factors: impurity ordering and

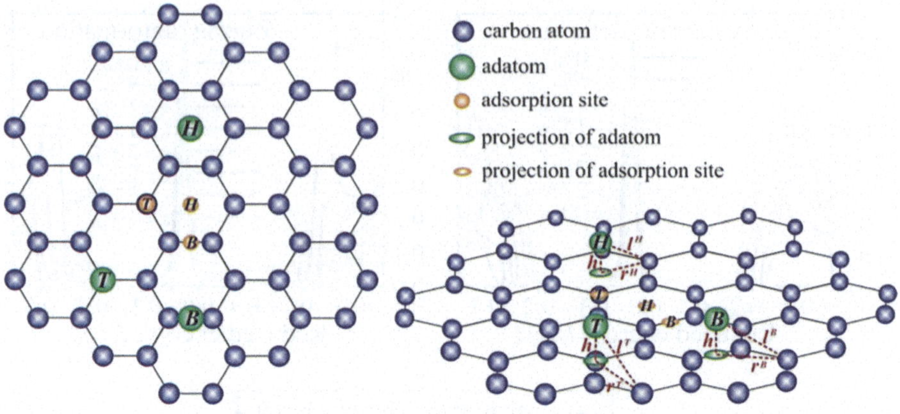

Fig. 3.4 Typical adsorption configurations in graphene: top (left) and perspective (right) views of graphene lattice with hollow centre (H), bridge centre (B) and top (T) adsorption sites

applied strain. Note that numerically obtained curve for defect graphene in Fig. 3.3c also becomes linear for strains beyond the ≈20% and crosses other two curves for pristine graphene close to its predicted failure limit point (≈27.5% [54]).

Due to the honeycomb structure of unstrained graphene lattice, possible adsorption sites can be reduced to three types with high-symmetry favourable (stable) positions; so-called hollow centre (H-type), bridge centre (B-type) and atop or top (T-type) adsorption sites are illustrated in Fig. 3.4. Taking into account discrepancies in the literature [27–30] on the energy stability (favourableness) of adsorption sites, we study how the positioning of dopants on each H-, B- and T-site type affects the electrotransport properties of unstrained graphene in comparison with the cases of their location on two other types of the sites.

In case of a random adatomic configuration, the steady diffusive regime, when electron diffusivity reaches maximum and saturates (Fig. 3.5a), occurs for a shorter time as compared with correlation (Fig. 3.5b) and ordering (Fig. 3.5c) cases. Maximal value in a temporal evolution of diffusivity for ordered impurities substantially exceeds its value for correlated defects and much more for randomly distributed ones. This is a 'hint' that corresponding conductivity should also be higher as compared to other ones. Really, a considerable increase in conductivity due to the correlation and, much more, to the ordering of adatoms as compared with their random distribution is seen from graphs in Fig. 3.6a–c, where the electron-density-dependent conductivities are calculated. The graphs in Fig. 3.6d–f allow seeing how different types of adsorption sites affect the conductivity for each type of distribution. If adatoms are randomly distributed, conductivity depends on types of adsites: H-, B- or T-type ones (Fig. 3.6d). For correlated distribution, conductivity depends on how adatoms manifest: as substitutional (being on T-sites) or interstitial (being on H- or B-sites) atoms (Fig. 3.6e). If adatoms form ordered superstructures, with equal periods, the conductivity is practically independent of adsorption type, especially at low electron densities (Fig. 3.6f).

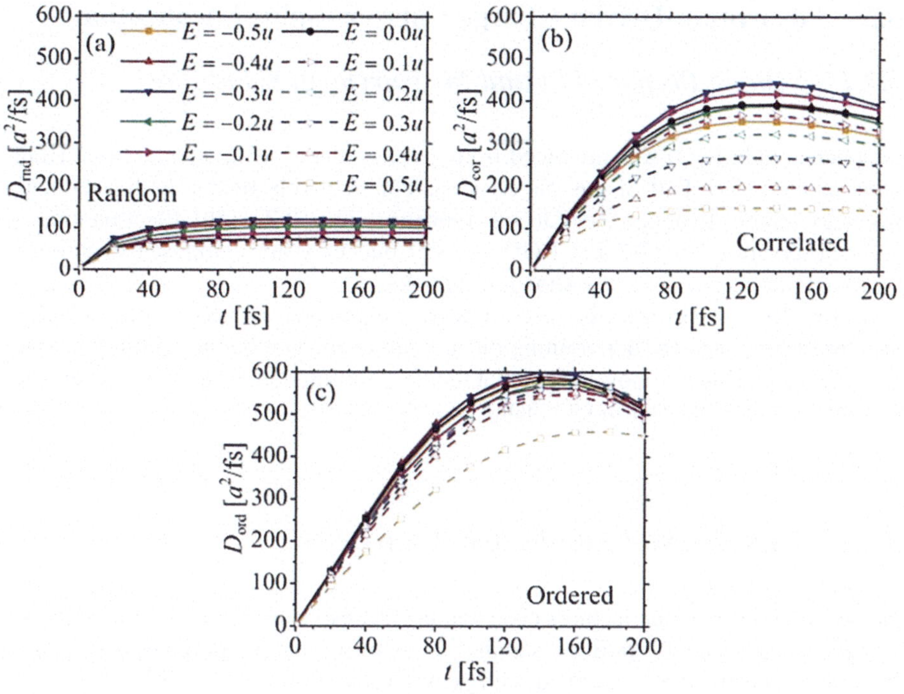

Fig. 3.5 Time-dependent diffusivity within the energy range $E \in [-0.5u, 0.5u]$ for random (**a**), correlated (**b**) and ordered (**c**) K adatoms located on hollow (H) sites (see also previous figure)

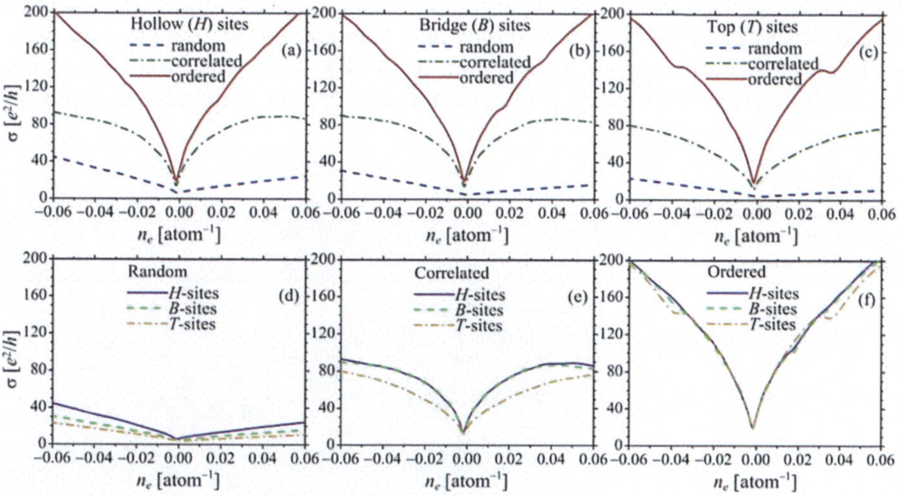

Fig. 3.6 Conductivity vs. the electron density for 3.125% of random, correlated and ordered potassium adatoms occupying hollow (H), bridge (B) or top (T) adsorption sites

3.3 Adsorption-Driven Charge Carrier (Spin) Localization

3.3.1 Sample Preparation and Measurement Conditions

Graphene oxide (GO) was produced from graphite flakes using the modified Hummers method [55]. Part of this material was consecutively treated with a reducing agent, hydrazine, to obtain the reduced graphene oxide (RGO) [56]. Electron microscopy observation for GO and RGO showed that they are composed of strongly wrinkled microscale sheets as shown in Fig. 3.7.

Before EPR experiment, both samples were examined with X-ray photoelectron spectroscopy (XPS) for determining the amount of oxygen bound within the structure. XPS experiments showed [22] that the level of functionalization of carbon with oxygen was much higher in GO than RGO.

3.3.2 Experimental Results and Analysis

Below, we present some features observed in EPR experiments during the sequence of the stages: stage 1—purified sample; 2—open to air; 3—purified; 4—open to helium; 5—purified; 6—saturated with heavy water (D_2O).

In EPR of RGO experiment, purified RGO showed no EPR signal from carbon, even in the lowest temperatures. There was also no signal from other paramagnetic centres (e.g., Mn ions). Lack of EPR signal of pure RGO in the whole temperature range suggests that electrons are highly delocalized even at the lowest temperatures. The EPR spectra could be observed only after saturation of the sample with guest molecules and decreasing the temperature below 100 K. Comparison of the EPR spectra of RGO after saturation is presented in Fig. 3.8. Opening the sample tube to

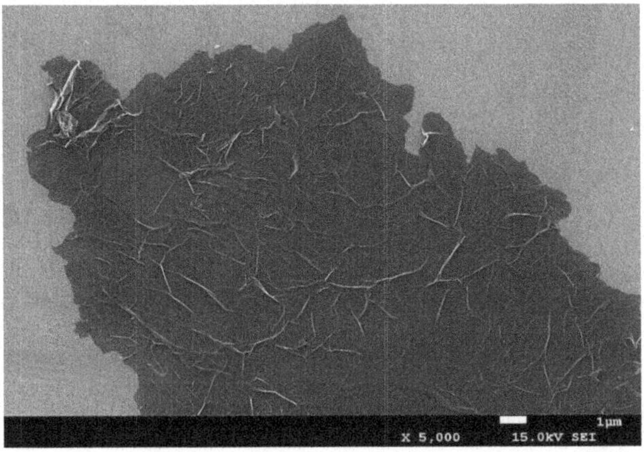

Fig. 3.7 Scanning electron microscopy image of a reduced graphene oxide layer (dark area)

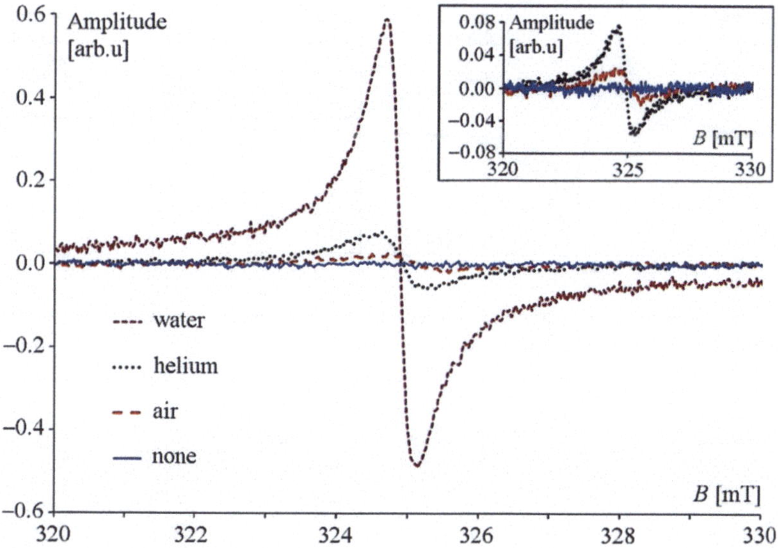

Fig. 3.8 EPR spectra of RGO sample in different surroundings at 10 K. The inset shows the magnification of the three low-amplitude signals

air (stage 2) caused the EPR signal of RGO to appear, but only in the low temperature range. Adsorption of guest molecules at the surface of graphene layers hindered the charge carrier transport by creating potential barriers for hopping. Thus, in low temperatures, where the thermal excitations were low, we got localized spins in the system. Introduction of helium resulted in the stronger EPR signal than for stage 2, most probably because much more helium was adsorbed on the RGO. Saturating the sample with heavy water (stage 6) resulted in the further increase of the EPR signal.

The EPR spectra of GO are observed in a whole temperature range at every stage of the sample treatment procedure. Low temperature behaviour, presented in Fig. 3.9a, was similar to RGO: signal intensity increased according to the sequence: pure–air–helium–water. Striking change appeared in high temperatures (Fig. 3.9b), where signal amplitudes of the air- and helium-filled sample equated due to the lack of the 'sorption pumping' effect in high temperatures. The above-mentioned observations are clear evidence that electronic properties of graphene-based systems strongly depend on the amount of adsorbed molecules.

The EPR spectra of RGO and GO at 10 K for stages 1 and 6 are shown in Fig. 3.10. To understand the difference between the spectra, note that graphene edges and defects (which are chemically active due to existence of the so-called dangling bonds and show some *sp*3 hybridization) have significantly different chemical and physical properties than the non-defect layers with *sp*2 hybridization. Therefore, the paramagnetic centres should also show different behaviour depending on whether they arise from edges (defects) or *sp*2 planes.

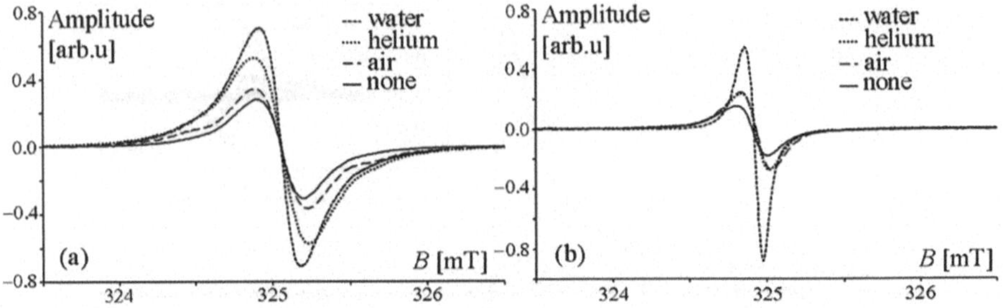

Fig. 3.9 EPR spectra of GO sample immersed in various media, recorded at 10 K (**a**) and room temperature (**b**)

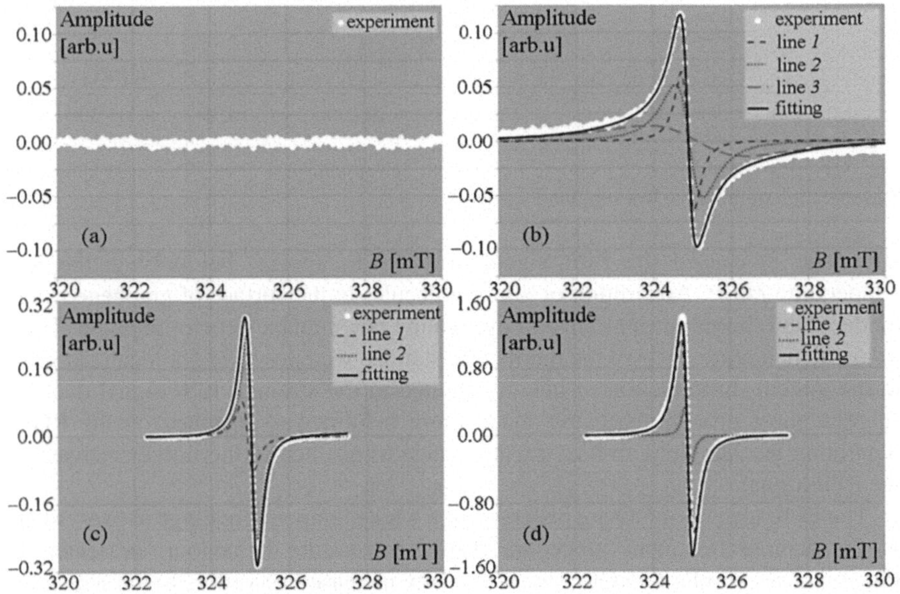

Fig. 3.10 EPR spectra of pure RGO (**a**), RGO + D_2O (**b**), pure GO (**c**) and GO saturated with heavy water (**d**). All spectra were recorded at 10 K

Purified RGO showed no EPR signal at all due to the strong delocalization of charge carriers. However, the signal appeared after adsorption of water, which is possible at the graphene edges with attached functional groups—hydrophilic adsorption sites. Such behaviour could be interpreted as a transition from the 'conducting state' (pure RGO), with lots of percolation paths, to the 'insulating state' (RGO + guest molecules), where thermal excitations were needed to transport the charge across the potential barriers formed at the graphene edges due to the host–guest interactions (hopping transport).

EPR signal of GO occurred in the whole temperature range at both sample treatment stages: purified, and saturated with D_2O. The existence of EPR signal resulted from the fact that edges and defects in the graphene layers were terminated with oxygen functional groups and there were no significant areas of well-conducting sp^2-graphene.

3.4 Summary and Conclusions

The effects of uniaxial tensile strain and different spatial configurations of adsorptions on electron density of states (DOS) and transport properties of graphene were studied. Spectral gap is sensitive to directions of the tensile strain. The presence of randomly distributed point defects does not avoid the minimum threshold zigzag deformations needed for the band-gap formation. Increase in point-defect concentrations acts against the band-gap opening for all defects considered herein, but their impact is different. However, spatially ordered impurities contribute to the band-gap manifestation and can reopen the gap that is normally suppressed by the randomly positioned dopants. Band gap varies non-monotonically with strain if zigzag deformation and impurity ordering act simultaneously.

For random adatomic distribution on hollow (H), bridge (B) or top (T) sites, the conductivity σ depends on their type: ${\sigma^H}_{\text{rnd}} > {\sigma^B}_{\text{rnd}} > {\sigma^T}_{\text{rnd}}$. If adatoms are correlated, σ is dependent on whether they act as interstitial or substitutional atoms: ${\sigma^H}_{\text{cor}} \approx {\sigma^B}_{\text{cor}} > {\sigma^T}_{\text{cor}}$. If adatoms form ordered superlattices with equal periods, σ is practically independent of the adsorption type: ${\sigma^H}_{\text{ord}} \approx {\sigma^B}_{\text{ord}} \approx {\sigma^T}_{\text{ord}}$. The conductivity for correlated and ordered adatoms is found to be enhanced dozens of times as compared to the cases of their random positions. The correlation and ordering effects manifest stronger for adatoms acting as substitutional atoms and weaker for those acting as interstitial atoms.

Lack of EPR signal of the purified RGO in the broad temperature range points out that there are not localized spins in the material, even if it is defective (sp^3 contribution) and there is some amount of oxygen functional groups attached to the graphene layers. RGO is a good conductor and has no localized spins in pure form. However, adsorption of atoms (molecules) followed by cooling of the system below 100 K resulted in the trapping of charge carriers in the localized states and the appearance of the EPR signal. This behaviour could be interpreted as the adsorption-driven metal–insulator transition. Nevertheless, further research is needed to prove it.

The existence of EPR signal of purified GO is due to the termination of most edges and defects in the graphene layers with oxygen functional groups. The electrical transport was suppressed, making GO an electrical insulator, where localized charge carriers existed even at high temperatures. In this case, adsorption of guest molecules also enhanced localization, with the biggest effect observed for water.

Results for both RGO and GO samples showed that amount of localized charge carriers (spins) correlated with the amount of adsorbed molecules responsible for the formation of potential barriers and, in turn, for the localization effects.

The localization phenomena in graphene-based systems depend heavily on the state of the layer edges, their functionalization and presence of 'foreign' molecules. Both factors can be controlled during and after the material synthesis, which allows for tuning the properties of graphene according to the type of application.

Acknowledgements The chapter generalizes results obtained within the framework of Polish–Ukrainian joint research project under the agreement on scientific cooperation between the Polish Academy of Sciences and the National Academy of Sciences of Ukraine for 2015–2017.

References

1. Katsnelson MI (2012) Graphene: carbon in two dimensions. Cambridge University Press, New York
2. Banhart F, Kotakoski J, Krasheninnikov AV (2010) Structural defects in graphene. ACS Nano 5(1):26–41. https://doi.org/10.1021/nn102598m
3. Elias D, Nair RR, Mohiuddin TMG, Morozov SV, Blake P, Halsall MP, Ferrari AC, Boukhvalov DW, Katsnelson MI, Geim AK, Novoselov KS (2009) Control of graphene's properties by reversible hydrogenation: evidence for graphane. Science 323:610–613. https://doi.org/10.1126/science.1167130
4. Ouyang F, Peng S, Liu Z, Zhongfan L, Zhirong L (2011) Bandgap opening in graphene antidot lattices: the missing half. ACS Nano 5:4023–4030. https://doi.org/10.1021/nn200580w
5. Radchenko TM, Tatarenko VA (2009) Statistical thermodynamics and kinetics of long-range order in metal-doped graphene. Solid State Phen 150:43–72. https://doi.org/10.4028/www.scientific.net/SSP.150.43
6. Radchenko TM, Tatarenko VA (2010) Kinetics of atomic ordering in metal-doped graphene. Solid State Sci 12(2):204–209. https://doi.org/10.1016/j.solidstatesciences.2009.05.027
7. Radchenko TM, Tatarenko VA (2010) A statistical-thermodynamic analysis of stably ordered substitutional structures in graphene. Phys E 42(8):2047–2054. https://doi.org/10.1016/j.physe.2010.03.024
8. Radchenko TM, Tatarenko VA, Sagalianov IY, Prylutskyy YI (2014) Effects of nitrogen-doping configurations with vacancies on conductivity in graphene. Phys Lett A 378 (30–31):2270–2274. https://doi.org/10.1016/j.physleta.2014.05.022
9. Radchenko TM, Tatarenko VA, Sagalianov IY, Prylutskyy YI (2014) Configurations of structural defects in graphene and their effects on its transport properties. In: Edwards BT (ed) Graphene: mechanical properties, potential applications and electrochemical performance. Nova Science Publishers, New York, pp 219–259
10. Bradley D (2012) Graphene straintronics: carbon. Mat Today 15(5):185. https://doi.org/10.1016/S1369-7021(12)70079-2
11. Amorim B, Cortijo A, de Juan F, Grushin AG, Guinea F, Gutierrez-Rubio A, Ochoa H, Parente V, Roldan R, San-Jose P, Schiefele J, Sturla M, Vozmediano MAH (2016) Novel effects of strains in graphene and other two dimensional materials. Phys Rep 617:1–54. https://doi.org/10.1016/j.physrep.2015.12.006
12. Si C, Sun Z, Liu F (2016) Strain engineering of graphene: a review. Nanoscale 8:3207–3217. https://doi.org/10.1039/C5NR07755A

13. Novoselov KS, Castro Neto AH (2012) Two-dimensional crystals-based heterostructures: materials with tailored properties. Phys Scr T146:014006-1–6. https://doi.org/10.1088/0031-8949/2012/T146/014006
14. Roche S, Leconte N, Ortmann F, Lherbier A, Soriano D (2012) Charlier J-Ch (2012) quantum transport in disordered graphene: a theoretical perspective. Solid State Com 153:1404–1410. https://doi.org/10.1016/j.ssc.2012.04.030
15. Botello-Méndez AR, Lherbier A, Charlier J-C (2013) Modeling electronic properties and quantum transport in doped and defective graphene. Solid State Com 175–176:90–100. https://doi.org/10.1016/j.ssc.2013.08.029
16. Kempiński W, Markowski D, Kempiński M, Śliwińska-Bartkowiak M (2013) Charge carrier transport control in activated carbon fibers. Carbon 57:533–536. https://doi.org/10.1016/j.carbon.2013.01.087
17. Náfrádi B, Choucair M, Southon PD, Kepert CJ, Forró L (2015) Strong interplay between the electron spin lifetime in chemically synthesized graphene multilayers and surface-bound oxygen. Chem Eur J 21(2):770–777. https://doi.org/10.1002/chem.201404309
18. Náfrádi B, Choucair M, Dinse KP, Forró L (2016) Room temperature manipulation of long lifetime spins in metallic-like carbon nanospheres. Nat Commun 7:12232. https://doi.org/10.1038/ncomms12232
19. Kempiński M, Kempiński W, Kaszyński J, Śliwińska-Bartkowiak M (2006) Model of spin localization in activated carbon fibers. App Phys Lett 88:143103-1–3. https://doi.org/10.1063/1.2187416
20. Radchenko TM, Tatarenko VA, Sagalianov IY, Prylutskyy YI, Szroeder P, Biniak S (2016) On adatomic-configuration-mediated correlation between electrotransport and electrochemical properties of graphene. Carbon 101:37–48. https://doi.org/10.1016/j.carbon.2016.01.067
21. Sagalianov IY, Radchenko TM, Prylutskyy YI, Tatarenko VA, Szroeder P (2017) Mutual influence of uniaxial tensile strain and point defect pattern on electronic states in graphene. Eur Phys J B 90(6):112-1–9. doi:10.1140/epjb/e2017-80091-x, 10.1140/epjb/e2017-80351-9
22. Kempiński M, Florczak P, Jurga S, Śliwińska-Bartkowiak M, Kempiński W (2017) The impact of adsorption on the localization of spins in graphene oxide and reduced graphene oxide, observed with electron paramagnetic resonance. Appl Phys Lett 111:084102-1–5. https://doi.org/10.1063/1.4996914
23. Radchenko TM, Shylau AA, Zozoulenko IV (2012) Influence of correlated impurities on conductivity of graphene sheets: time-dependent real-space Kubo approach. Phys Rev B 86(3):035418-1–13. https://doi.org/10.1103/PhysRevB.86.035418
24. Radchenko TM, Shylau AA, Zozoulenko IV, Ferreira A (2013) Effect of charged line defects on conductivity in graphene: numerical Kubo and analytical Boltzmann approaches. Phys Rev B 87(19):195448-1–14. https://doi.org/10.1103/PhysRevB.87.195448
25. Radchenko TM, Shylau AA, Zozoulenko IV (2014) Conductivity of epitaxial and CVD graphene with correlated line defects. Solid State Com 195:88–94. https://doi.org/10.1016/j.ssc.2014.07.012
26. Fung AWP, Dresselhaus MS, Endo M (1993) Transport properties near the metal–insulator transition in heat-treated activatedrotect carbon fibers. Phys Rev B 48(20):14953–14962. https://doi.org/10.1103/PhysRevB.48.14953
27. Lugo-Solis A, Vasiliev I (2007) Ab initio study of K adsorption on graphene and carbon nanotubes: role of long-range ionic forces. Phys Rev B 76:235431-1–8. https://doi.org/10.1103/PhysRevB.76.235431
28. Chan KT, Neaton JB, Cohen ML (2008) First-principles study of metal adatom adsorption on graphene. Phys Rev B 77:235430-1–12. https://doi.org/10.1103/PhysRevB.77.235430
29. Cao C, Wu M, Jiang J, Cheng H-P (2010) Transition metal adatom and dimer adsorbed on graphene: induced magnetization and electronic structures. Phys Rev B 81:205424-1–9. https://doi.org/10.1103/PhysRevB.81.205424
30. Nakada K, Ishii A (2011) Migration of adatom adsorption on graphene using DFT calculation. Solid State Com 151:13–16. https://doi.org/10.1016/j.ssc.2010.10.036

31. Ni ZH, Yu T, Lu YH, Wang YY, Feng YP, Shen ZX (2008) Uniaxial strain on graphene: Raman spectroscopy study and band-gap opening. ACS Nano 2(11):2301–2305. https://doi.org/10.1021/nn800459e
32. Ni ZH, Yu T, Lu YH, Wang YY, Feng YP, Shen ZX (2009) Uniaxial strain on graphene: Raman spectroscopy study and band-gap opening (correction). ACS Nano 3:483–483. https://doi.org/10.1021/nn8008323
33. Gui G, Li J, Zhong J (2008) Band structure engineering of graphene by strain: first-principles calculations. Phys Rev B 78:075435-1–6. https://doi.org/10.1103/PhysRevB.78.075435
34. Pereira VM, Castro Neto AH (2009) Strain engineering of graphene's electronic structure. Phys Rev Lett 103(4):046801-1–4. https://doi.org/10.1103/PhysRevLett.103.046801
35. Pereira VM, Castro Neto AH, Peres NMR (2009) Tight-binding approach to uniaxial strain in graphene. Phys Rev B 80(4):045401-1–8. https://doi.org/10.1103/PhysRevB.80.045401
36. Ribeiro RM, Pereira VM, Peres NMR, Briddon PR, Castro Neto AH (2009) Strained graphene: tight-binding and density functional calculations. New J Phys 11:115002-1–10. https://doi.org/10.1088/1367-2630/11/11/115002
37. Cocco G, Cadelano E, Colombo L (2010) Gap opening in graphene by shear strain. Phys Rev B 81:241412(R)-1–4. https://doi.org/10.1103/PhysRevB.81.241412
38. Guinea F, Katsnelson MI, Geim AK (2010) Energy gaps and a zero-field quantum hall effect in graphene by strain engineering. Nature Phys 6:30–33. https://doi.org/10.1038/NPHYS142
39. Low T, Guinea F, Katsnelson MI (2011) Gaps tunable by electrostatic gates in strained graphene. Phys Rev B 83:195436-1–7. https://doi.org/10.1103/PhysRevB.83.195436
40. Naumov II, Bratkovsky AM (2011) Gap opening in graphene by simple periodic inhomoheneous strain. Phys Rev B 84:245444-1–6. https://doi.org/10.1103/PhysRevB.84.245444
41. He X, Gao L, Tang N, Duan J, Xu F, Wang X, Yang X, Ge W, Shen B (2014) Shear strain induced modulation to the transport properties of graphene. Appl Phys Lett 105:083108-1–4. https://doi.org/10.1063/1.4894082
42. He X, Gao L, Tang N, Duan J, Mei F, Meng H, Lu F, Xu F, Wang X, Yang X, Ge W, Shen B (2014) Electronic properties of polycrystalline graphene under large local strain. Appl Phys Lett 104:243108-1–4. https://doi.org/10.1063/1.4883866
43. Yuan S, De Raedt H, Katsnelson MI (2010) Modeling electronic structure and transport properties of graphene with resonant scattering centers. Phys Rev B 82(11):115448-1–16. https://doi.org/10.1103/PhysRevB.82.115448
44. Yuan S, De Raedt H, Katsnelson MI (2010) Electronic transport in disordered bilayer and trilayer graphene. Phys Rev B 82(23):235409-1–13. https://doi.org/10.1103/PhysRevB.82.235409
45. Dresselhaus MS, Dresselhaus G (1981) Intercalation compounds of graphite. Adv Phys 30 (2):139–326. https://doi.org/10.1080/00018738100101367
46. McClure JW (1957) Band structure of graphite and de Haas-van Alphen effect. Phys Rev 108 (3):612–618. https://doi.org/10.1103/PhysRev.108.612
47. Slonczewski JC, Weiss PR (1958) Band structure of graphite. Phys Rev 109(2):272–279. https://doi.org/10.1103/PhysRev.109.272
48. Leconte N, Ferreira A, Jung J (2016) Efficient multiscale lattice simulations of strained and disordered graphene. Semiconductors and Semimetals 95:35–99. https://doi.org/10.1016/bs.semsem.2016.04.002
49. Castro Neto AH, Guinea F (2007) Electron-phonon coupling and Raman spectroscopy in graphene. Phys Rev B 75(4):045404-1–8. https://doi.org/10.1103/PhysRevB.75.045404
50. Blakslee L, Proctor DG, Seldin EJ, Stence GB, Wen T (1970) Elastic constants of compression-annealed pyrolytic graphite. J Appl Phys 41(8):3373–3382. https://doi.org/10.1063/1.1659428
51. Farjam M, Rafii-Tabar H (2009) Comment on "band structure engineering of graphene by strain: first-principles calculations". Phys Rev B 80(16):167401-1–3. https://doi.org/10.1103/PhysRevB.80.167401

52. Wehling TO, Yuan S, Lichtenstein AI, Geim AK, Katsnelson MI (2010) Resonant scattering by realistic impurities in graphene. Phys Rev Lett 105(5):056802-1–4. https://doi.org/10.1103/PhysRevLett.105.056802
53. Adessi C, Roche S, Blasé X (2006) Reduced backscattering in potassium-doped nanotubes: ab initio and semiempirical simulations. Phys Rev B 73:125414-1–5. https://doi.org/10.1103/PhysRevB.73.125414
54. Liu F, Ming P, Li J (2007) Ab initio calculation of ideal strength and phonon instability of graphene under tension. Phys Rev B 76(6):064120-1–7. https://doi.org/10.1103/PhysRevB.76.064120
55. Hummers WS Jr, Offeman RE (1958) Preparation of graphitic oxide. J Am Chem Soc 80(6):1339–1339. https://doi.org/10.1021/ja01539a017
56. Pei S, Cheng H-M (2012) The reduction of graphene oxide. Carbon 50(9):3210–3228. https://doi.org/10.1016/j.carbon.2011.11.010

Chapter 4
The Theory of Absorption and Emission of Light by Free Electrons in Ferromagnetic Semiconductors

Oleksandr Semchuk

4.1 Introduction

Until recently, there was no general approach to describing the phenomena of absorption and emission of light in ferromagnetic semiconductors (FMSC), unique materials that combine both semiconductor and magnetic properties simultaneously. The presence of strong s-d-exchange interaction between the electronic and magnetic subsystems of FMSC gives an opportunity to observe a series of unique effects in them: a metal-dielectric transition, a giant magnetic resonance, a strong shift of the optical absorption edge, photo-induced magnetic effects, etc. [1], which makes FMSC key materials for modern micro- and nano-electronics. Thus, new features in the processes of absorption and emission of light by FMSC should be expected, caused by the interaction between the electronic and magnetic subsystems. It is known that in order to place the act of absorption or radiation of a photon (a aquantum of light) by a free electron, it is necessary to add a "third component", which upholds the laws of conservation of energy and momentum in the collision process. This "third body" can be a phonon, a magnon, impurities, and the like. This explains the influence of the mechanisms of scattering of electrons on the processes of absorption and radiation.

On using the kinetic equation, which takes into account the influence of the external electromagnetic field on the mechanism of scattering of free carriers (the so-called quantum kinetic equation), the most common method is based on studying the absorption and scattering of light by free carriers in semiconductors [2, 3]. The convenience lies in the fact that in the single approach one can obtain an expression

O. Semchuk (✉)
Chyiko Institute of Surface Chemistry, National Academy of Sciences of Ukraine, Kyiv, Ukraine
e-mail: aleksandr1950@meta.ua

O. Fesenko, L. Yatsenko (eds.), *Nanooptics, Nanophotonics, Nanostructures, and Their Applications*, Springer Proceedings in Physics 210,
https://doi.org/10.1007/978-3-319-91083-3_4

for the absorption of light by free electrons in both classical and quantum cases. The same applies to the process of spontaneous radiation.

In this paper, the theory of absorption and radiation of light by free carriers in FMSC is constructed, and general expressions for the light emission coefficient of free carriers and the intensity of the spontaneous emission of light by hot electrons in FMSC are obtained. The main mechanism of carrier scattering considered is electron-two-magnon (bimagnon) scattering. Classical and quantum cases are considered. It is shown that the intensity of spontaneous radiation does not depend on the frequency of light in the classical frequency range and exponentially falls in the quantum region of frequencies.

4.2 Integral of Collision Electrons with Quasiparticles in the Presence of an Electromagnetic Wave

Consider the quantum-mechanical system: a conduction electron in a semiconductor + a diffuser (phonons, magnons, etc.) in a high-frequency electromagnetic field of laser radiation which vector-potential is given by the following expression $\vec{A}\,(t) = \vec{A}_0 \cos \omega t$ and their frequency ω satisfies the condition $\omega\tau < < 1$ (τ is the time of the free passage of conduction electrons between the collisions).

The complete Hamiltonian of such a system can be written as [4]

$$\begin{aligned} \widehat{H}(t) &= \widehat{H}_0 + \widehat{H}_{\text{int}}(t), \\ \widehat{H}_0 &= \frac{1}{2m}\sum_{\vec{p}}\left(\vec{P} - \frac{e}{c}\vec{A}\,(t)\right)a^+_{\vec{p}}a_{\vec{p}} + \sum_{\vec{k}} c_{\vec{k}}\left(b^+_{\vec{k}}b_{\vec{k}} + \frac{1}{2}\right), \end{aligned} \tag{4.1}$$

where $\widehat{H}_0$ is Hamiltonian of the undisturbed system, and $\widehat{H}_{\text{int}}$ is the operator of the quasiparticle interaction (electron–phonon, electron–magnon, etc.), which will be considered as a small perturbation formally dependent on time.

The evolution of such a system will be described by the Schrödinger time equation with Hamiltonian (4.1)

$$\imath\hbar\frac{\partial\Psi(\vec{r}\,,t)}{\partial t} = \widehat{H}(t)\Psi(\vec{r}\,,t). \tag{4.2}$$

Let us expand the solution $\Psi\left(\vec{r}, t\right)$ of Eq. (4.2) to the complete system of eigenfunctions $\psi_n\left(\vec{r}, t\right)$ of the undisturbed Hamiltonian $\widehat{H}_0$ (4.1)

$$\Psi(\vec{r},t) = \sum_n a_n(t)\psi_n(\vec{r},t), \tag{4.3}$$

at this,

$$\widehat{H}_0 u_n = \varepsilon_n u_n, \tag{4.4}$$

where ε_n is the energy of the system in the n-th quasi-stationary state.

Assuming that

$$a_n(t) = a_n^{(0)} + a_n^{(1)}(t) + a_n^{(2)}(t) + \cdots, \tag{4.5}$$

where $a_n^{(0)}$ is the undisturbed (initial) value of the coefficient $a_n(t)$, and $a_n^{(1)}(t)$, $a_n^{(2)}(t)$ are the first- and second-order corrections of smallness by $\widehat{H}_{\text{int}}(t)$, it can be shown that [4]

$$\frac{da_k^{(1)}}{dt} = \frac{1}{i\hbar}\sum_n H_{kn}^{\text{int}} a_n^{(0)}, \qquad \frac{da_k^{(2)}}{dt} = \frac{1}{i\hbar}\sum_n H_{kn}^{\text{int}} a_n^{(1)}, \tag{4.6}$$

where

$$H_{kn}^{\text{int}} = \int \psi_k^* \widetilde{H}_{\text{int}} \psi_n d\tau. \tag{4.7}$$

is the matrix element of the interaction operator (perturbation), and the symbol $\sum_n$ means summation discrete states and integration continuous states of the unbroken system.

Now let at the initial time ($t = 0$) the system be in the i-th quantum state, then $a_i^{(0)} = 1$, and all others $a_n^{(0)} = 0 \ (n \neq i)$. It is of interest that the amplitude $a_f^{(1)}(t)$ of the finite state f up to the time t when the perturbation $\widehat{H}_{\text{int}}$ "switches on" at the moment $t = 0$. Obviously, $a_f^{(1)}(0) = 0$; therefore, from (4.6) we have

$$a_f^{(1)}(t) = \frac{1}{i\hbar}\int_0^t H_{fi}^{\text{int}}(t')\, dt'. \tag{4.8}$$

The magnitude $\left|a_f^{(1)}(t)\right|^2$ is the probability that the quantum system is in a state $\psi_f(\vec{r},t)$ at the moment of time t. At $t = 0$ the system was in a state $\psi_f(\vec{r},0)$. Using (4.8), we obtain for the probability $\left|a_f^{(1)}(t)\right|^2$ [4]:

$$\left|a_f^{(1)}(t)\right|^2 = \frac{1}{\hbar^2}\left|\int_0^t H_{fi}^{\text{int}}(t')\ dt'\right|^2. \tag{4.9}$$

This value can be considered as the probability of transition of a quantum system from state $\Psi_i(\vec{r},t)$ to state $\Psi_f(\vec{r},t)$ under the action of perturbation (interaction between quasiparticles) $\widehat{H}_{\text{int}}(t)$ over time t. In addition to this value, the probability of the transition at the unit of time is introduced $W(i,f)$, which is connected with the value $\left|a_f^{(1)}(t)\right|^2$ by the following relation:

$$W(i,f) = \frac{d}{dt}\left|a_f^{(1)}(t)\right|^2 = \frac{1}{\hbar^2}\frac{d}{dt}\left|\int_0^t H_{if}^{\text{int}}(t')dt'\right|^2. \tag{4.10}$$

Passing into (4.10) to the impulse representation and writing the matrix element of the interaction operator in expanded form, we obtain from (4.10),

$$W(\vec{p}',\vec{p}) = \frac{1}{\hbar^2}\frac{d}{dt}\left|\int_0^t d\tau < \Psi_{\vec{p}'}^*(\tau)\left|\widehat{H}_{\text{int}}\right|\Psi_{\vec{p}}^*(\tau) >\right|^2. \tag{4.11}$$

For $t = 0$, the interaction between electrons and diffusers is included in the quantum-mechanical system: electrons in the conduction band and dispersers in the high-frequency field of laser radiation (phonons, magnons, ionized impurities, etc.). As a result of the interaction of an electron with a diffuser, it moves from one quasi-stationary state in which it has quasi-energy $E_{\vec{p}}$, quasi-momentum $\vec{p}$ and is described by a wave function $\Psi_{\vec{p}}(\vec{r},t)$ to another state described by a wave function $\Psi_{\vec{p}'}(\vec{r},t)$. In this state, its quasi-impulse becomes meaningful $\vec{p}'$, the quasi-energy becomes equal to $E_{\vec{p}'} \pm \varepsilon_{\vec{p}'\vec{p}}$ (here $\varepsilon_{\vec{p}'\vec{p}}$ is the quantity characterizing the energy change in the system of electron-diffuser as a result of interaction). Consequently, the problem is reduced to the calculation of a matrix element of the interaction operator (4.11). It can be calculated using the electron wave function found in the laser radiation field [5], which is now convenient to write, limited to the dipole approximation, in simplified form:

$$\Psi_{\vec{p}}(\vec{r},t) = \Psi(\vec{r}) \times \exp\left\{i\frac{e\gamma_{\vec{p}\vec{p}\prime}}{mc\hbar\omega}\sin\omega t\right\}\exp\left\{-\frac{i}{\hbar}\left(\left(\frac{p^2}{2m}+\frac{e^2A_0^2}{4mc^2}\right)t\right)\right\}. \quad (4.12)$$

In (4.12), the wave function is given as a product of two functions, one of which depends only on the coordinate ($\Psi(\vec{r})$), and the second depends on the time (exp $\{\ldots\}t$). Substituting in (4.11) the wave function of the conduction electron in the form (4.12), we obtain:

$$W(\vec{p},p) = \frac{1}{\hbar^2}\frac{d}{dt}\left|\int_0^t d\tau\left\langle\Psi^*_{\vec{p}'}(\vec{r})\middle|\widehat{H}_{\text{int}}\middle|\Psi_{\vec{p}}(\vec{r})\right\rangle\exp\left\{\frac{i}{\hbar}\left(\begin{matrix}\varepsilon_{\vec{p}'}-\varepsilon_{\vec{p}}\pm\\ \varepsilon_{\vec{p}'\vec{p}}\end{matrix}\right)\tau - i\frac{e\gamma_{\vec{p}'\vec{p}}}{mc\hbar\omega}\sin\omega\tau\right\}\right|^2. \quad (4.13)$$

In expressions (4.12) and (4.13), the notation $\gamma^2_{\vec{p}\vec{p}} = \sum'_{j,j'}(\vec{p}'-\vec{p})\vec{A}_j(\vec{p}'-\vec{p})\vec{A}_{\bar{j}}$ $\cos[\phi_j-\phi_{\bar{j}}]$. is introduced.

Extracting the matrix element of the operator of interaction $\left\langle\Psi^*_{\vec{p}'}(\vec{r})\middle|\widehat{H}_{\text{int}}\middle|\Psi_{\vec{p}}(\vec{r})\right\rangle$ of an electron with a diffuser from the integral sign and using the Jacobi–Anger formula [6],

$$\exp\{\pm iz\sin\phi\} = \sum_{l=-\infty}^{\infty} J_n(z)\exp\{\pm il\phi\},$$

($J_n(z)$ is the Bessel function of the real argument), we obtain from (4.13):

$$W(\vec{p},p) = \frac{1}{\hbar^2}\sum_{l=-\infty}^{\infty} J_l^2\left(\frac{e\gamma_{\vec{p}'\vec{p}}}{mc\hbar\omega}\right)\left|\left\langle\Psi^*_{\vec{p}'}(\vec{r})\middle|\widehat{H}_{\text{int}}\middle|\Psi_{\vec{p}}(\vec{r})\right\rangle\right|^2\frac{d}{dt}\left|\int_0^t d\tau\exp\left\{\frac{i}{\hbar}\left(\begin{matrix}\varepsilon_{\vec{p}'}-\varepsilon_{\vec{p}}\pm\varepsilon_{\vec{p}'\vec{p}}-\\ -l\hbar\omega\end{matrix}\right)\tau\right\}\right|^2. \quad (4.14)$$

Having obtained integration by $d\tau$ in expression (4.14), we get the result

$$W(\vec{p},p) = 2\sum_{l=-\infty}^{\infty} J_l^2\left(\frac{e\gamma_{\vec{p}'\vec{p}}}{mc\hbar\omega}\right)\left|\left\langle\Psi^*_{\vec{p}'}(\vec{r})\middle|\widehat{H}_{\text{int}}\middle|\Psi_{\vec{p}}(\vec{r})\right\rangle\right|^2\sin\frac{\omega_{\vec{p}'\vec{p}}t}{\omega_{\vec{p}'\vec{p}}}. \quad (4.15)$$

In expression (4.15) $\omega_{\vec{p}'\vec{p}} = \hbar^{-1}\left(\varepsilon_{\vec{p}'}-\varepsilon_{\vec{p}}\pm\varepsilon_{\vec{p}'\vec{p}}-l\hbar\omega\right)$.

Analyze now the result obtained. Consider the value $F = \sin \frac{\omega_{\vec{p}'\vec{p}} t}{\omega_{\vec{p}'\vec{p}}}$ as a function of $\omega_{\vec{p}'p'}$ and calculate the integral of this function depending on frequencies [6]:

$$\int_{-\infty}^{+\infty} \frac{\sin \omega_{\vec{p}'\vec{p}} t}{\omega_{\vec{p}'\vec{p}}} d\omega_{\vec{p}'\vec{p}} = \pi. \tag{4.16}$$

So, at large interaction time ($t \to \infty$) between an electron and a diffuser it comes from (4.16):

$$F = \frac{\sin \omega_{\vec{p}'\vec{p}} t}{\omega_{\vec{p}'\vec{p}}} \approx \frac{\pi}{\hbar} \delta\left(\varepsilon_{\vec{p}'} - \varepsilon_{\vec{p}} - \varepsilon_{\vec{p}'\vec{p}} - l\hbar\omega\right). \tag{4.17}$$

Thus, in the case when the interaction time of an electron with a diffuser is sufficiently large, the probability of a transition of an electron from a state where it is described by a quasi-momentum $\vec{p}'$ to another state where it will have a quasi-momentum $\vec{p}$ as a result of interaction with the disperser can be written as

$$W(\vec{p}', \vec{p}) = \frac{2\pi}{\hbar} \sum_{n=-\infty}^{+\infty} J_l^2 \left(\frac{e\gamma_{\vec{p}'\vec{p}}}{mc\hbar\omega} \right) \left| \left\langle n_{\vec{p}'}, N_{\vec{q}'} \middle| \widehat{H}_{\text{int}} \middle| n_{\vec{p}}, N_{\vec{q}} \right\rangle \right|^2 \delta\left(\varepsilon_{\vec{p}'} - \varepsilon_{\vec{p}} - \varepsilon_{\vec{p}'\vec{p}} - l\hbar\omega\right). \tag{4.18}$$

Here $\varepsilon_{\vec{p}'\vec{p}}$ is the change in the electron energy in the process of interaction with the disperser.

From (4.18) it follows that the probability of the transition $W(\vec{p}', \vec{p})$ is proportional to the square of the module of the matrix element of the interaction operator

$$\left| \left\langle \Psi^*_{\vec{p}'}(\vec{r}) \middle| \widehat{H}_{\text{int}} \middle| \Psi_{\vec{p}}(\vec{r}) \right\rangle \right|^2 = \left| \left\langle n_{\vec{p}'}, N_{\vec{q}} \middle| \widehat{H}_{\text{int}} \middle| n_{\vec{p}}, N_{\vec{q}} \right\rangle \right|^2 \tag{4.19}$$

and is different from zero only when the quanta of the external electromagnetic field-photons [5] are involved in the process of the interaction "electron-diffuser."

Note that in (4.18) we have written the matrix element of the interaction "electron-diffuser" through the numbers of electron $n_{\vec{p}}$ and diffuser $N_{\vec{q}}$ filling. Using (4.18), we can now write in general terms a quantum kinetic equation for conduction electrons in a semiconductor that interacts with a diffuser in a high-frequency nonuniform electromagnetic field of laser radiation $\vec{A}(t) = \vec{A}_0 \cos \omega t$.

$$\frac{\partial f_{\vec{p}}}{\partial t} + e\vec{F}_0 \frac{\partial f_{\vec{p}}}{\partial \vec{p}} + \frac{\vec{p}}{m}\frac{\partial f_{\vec{p}}}{\partial \vec{r}} = \frac{2\pi}{\hbar}\sum_{\vec{p}'}\sum_{l=-\infty}^{\infty} J_l^2\left(\frac{e\gamma_{\vec{p}'\vec{p}}}{mc\hbar\omega}\right)\delta\left(\varepsilon_{\vec{p}'} - \varepsilon_{\vec{p}} - \varepsilon_{\vec{p}'\vec{p}} - l\hbar\omega\right)\times$$
$$\left\{\left|\left\langle n_{\vec{p}'}, N_{\vec{q}'}\right|H_{\text{int}}\left|n_{\vec{p}}, N_{\vec{q}}\right\rangle\right|^2 f_{\vec{p}'}\left(1 - f_{\vec{p}}\right) - \left|\left\langle n_{\vec{p}}, N_{\vec{q}}\right|H_{\text{int}}\left|n_{\vec{p}'}, N_{\vec{q}'}\right\rangle\right|^2 f_{\vec{p}}\left(1 - f_{\vec{p}'}\right)\right\}. \tag{4.20}$$

In expressions (4.20), $f_{\vec{p}}$ is a function of the distribution of electrons according to pulses $\vec{p}$, $\vec{F}_0$ is the external constant electric field.

Now, in order to obtain an explicit form of the integral of electrons collision (right side of exp. (4.20)) for a particular scattering mechanism, it is necessary to set the Hamiltonian of interaction of the electron with the diffuser and to calculate the corresponding matrix element of the scattering operator.

4.3 Absorption of Light by Free Electrons in Ferromagnetic Semiconductors

Let us consider that electrons in ferromagnetic semiconductors (FMSC) relax on magnons both in energy and in momentum, and give the energy to acoustic phonons. Consider the case of electron–magnon interaction. For simplicity, we confine ourselves to one sub-band approximation, that is, we confine ourselves to considering the sub-band of conductivity with the spin "up." In addition, we assume that the electrons are relaxed on the magnons, and the magnons on the electrons both in energy and momentum. In the process of electron–magnon interaction we take into account both two-magnon processes in perturbation theory of the first order and one-magnon of the second order. For this case, the Hamiltonian of the electron–magnon interaction in the representation of the secondary quantization can be written as [1]

$$\tilde{H}_{2em} = \sum_{\vec{p},\vec{q},\vec{r}}\left\{C^{\uparrow}_{\vec{p}\vec{q}\vec{r}} a^{+}_{\vec{p}\uparrow} a_{\vec{p}-\vec{r}\uparrow} - C^{\downarrow}_{\vec{p}\vec{q}\vec{r}} a^{+}_{\vec{p}\downarrow} a_{\vec{p}-\vec{r}\downarrow}\right\} b^{+}_{\vec{q}} b_{\vec{q}+\vec{r}}, \quad C(\vec{p},\vec{k}) = \frac{J}{4N}\left[\frac{\varepsilon_{\vec{p}} - \varepsilon_{\vec{k}}}{JS + \varepsilon_{\vec{p}} - \varepsilon_{\vec{k}}}\right].$$
$$C^{\uparrow}_{\vec{p}\vec{q}\vec{r}} = C(\vec{p} + \vec{q}, \vec{k}) + C(\vec{p} + \vec{q}, \vec{p} - \vec{r}), \quad C^{\downarrow}_{\vec{p}\vec{q}\vec{r}} = C(\vec{p} - \vec{r}, \vec{p} - \vec{q} - \vec{r}). \tag{4.21}$$

Here J is the energy of the s-d-exchange interaction, N is the number of magnetic ions, S is the spin of the magnetic ion.

Let us make a series of simplifying approximations. Let us consider that the external constant electric field $\vec{F}_0$ is not too large; so, the average electron energy $\bar{\varepsilon} << JS$, and the electron energy in the field of laser radiation $\varepsilon_{\vec{P}} << JS$ ($\vec{P} = \vec{p} + \frac{e}{c}\vec{A}(\vec{r}, t)$ is the kinematic momentum of the electron). These inequalities allow us to limit ourselves to examining the sub-band with the $\sigma = \uparrow$, so further on spin index can be omitted.

Now, using (4.21), we calculate the square of the matrix element of the electron–bimagnon interaction $\left|\left\langle f_{\vec{p}'}, N_{\vec{q}}' \middle| \widehat{H}_{\text{int}} \middle| f_{\vec{p}}, N_{\vec{q}} \right\rangle\right|^2$. The calculations carried out by the method described in [5] yield the following results:

$$\begin{aligned}\left|\left\langle f_{\vec{p}'}, N_{\vec{q}}' \middle| \widehat{H}_{\text{int}} \middle| f_{\vec{p}}, N_{\vec{q}} \right\rangle\right|^2 &= \left|C_{\vec{p}'\vec{q}\vec{q}'}\right|^2 N_{\vec{q}'}(\vec{r},t)\left(1 + N_{\vec{q}}(\vec{r},t)\right)\delta_{\vec{p}+\vec{q}}^{\vec{p}'+\vec{q}'},\\ \left|\left\langle f_{\vec{p}}, N_{\vec{q}} \middle| \widehat{H}_{\text{int}} \middle| f_{\vec{p}'}, N_{\vec{q}}' \right\rangle\right|^2 &= \left|C_{\vec{p}'\vec{q}\vec{q}'}\right|^2 N_{\vec{q}}(\vec{r},t)\left(1 + N_{\vec{q}'}(\vec{r},t)\right)\delta_{\vec{p}+\vec{q}}^{\vec{p}'+\vec{q}'}.\end{aligned} \tag{4.22}$$

Given that for electron-double-magnon scattering the magnitude $\varepsilon_{\vec{p}'\vec{p}} = \omega_{\vec{q}'} - \omega_{\vec{q}}$ (magnon energy difference), we can obtain the following expression for the integral of electrons' collisions with magnons FMSC, which is in the external heating electric field, and a heterogeneous high-frequency electromagnetic field of laser radiation, in the form

$$\left(\frac{\partial f}{\partial t}\right)_{sc} = \frac{2\pi}{\hbar}\sum_{p',q',\vec{q}}\sum_{l=-\infty}^{\infty}\left|C_{\vec{p}'\vec{q}\vec{q}'}\right|^2 J_l^2(x)\delta_{\vec{p}+\vec{q}}^{\vec{p}'+\vec{q}'}\left\{\begin{matrix} f_{\vec{p}'}(1-f_{\vec{p}})N_{\vec{q}'}(1+N_{\vec{q}}) - \\ f_{\vec{p}}(1-f_{\vec{p}'})N_{\vec{q}}(1+N_{\vec{q}'}) \end{matrix}\right\}$$
$$\delta\begin{pmatrix} \varepsilon_{\vec{P}'} - \varepsilon_{\vec{P}} + \omega_{\vec{q}} \\ -\omega_{\vec{q}'} - l\hbar\omega \end{pmatrix}. \tag{4.23}$$

In (4.23) the symbol $x = e\gamma_{\vec{p}\vec{P}'}/(mc\hbar\omega)$, $\omega_{\vec{q}}$ is the energy of a magnon with a quasi-momentum $\vec{q}$.

The energy absorbed by electrons per unit time in the process of electron–magnon scattering in the presence of an electromagnetic wave is determined by the following expression [2, 3]:

$$P = \int \varepsilon_{\vec{p}}\left(\frac{\partial f}{\partial \vec{p}}\right)_{sc} d\,\vec{p}\,. \tag{4.24}$$

Substituting (4.23) in (4.24) and taking into account that $f_{\vec{p}} << 1$, $\hbar\omega >> \varepsilon_{\vec{p}}, \omega_{\vec{q}}$, considering that the electron–bimagnon scattering is a quasi-elastic ($N_{\vec{q}} \approx N_{\vec{q}} + 1$), it can be observed that

$$P = \frac{2\pi}{\hbar}\frac{V}{(2\pi\hbar)^3}\sum_{\vec{q}} W(\vec{q})\sum_{l=-\infty}^{\infty}\int d\vec{p}\, d\vec{p}' J_l^2\left(\frac{e\gamma_{\vec{p}\vec{p}'}}{mc\hbar\omega}\right)\varepsilon_{\vec{p}}(f_{\vec{p}'} - f_{\vec{p}})\delta(\varepsilon_{\vec{p}'} - \varepsilon_{\vec{p}} - l\hbar\omega). \tag{4.25}$$

In (4.25), the following notation is introduced:

$$W(\vec{q}) = \frac{J^2 q^4 N_q^2}{\left(q^2 + q_0^2\right)^2}, \quad q_0^2 = 2mJS, \tag{4.26}$$

where $V = a^3$ is the volume of the elementary cell, a is the constant of a crystal lattice.

If in the (4.25) that contains function $f_{\vec{p}'}$ substitution is made $\vec{p}' \to \vec{p}$, $l \to -l$ and $\varepsilon_{\vec{p}'}$ is expressed in terms of $\varepsilon_{\vec{p}}$, after using the δ-function, the expression (4.27) it is obtained:

$$P = -\frac{2\pi}{\hbar}\frac{V}{(2\pi\hbar)^3}\sum_{\vec{q}} W(\vec{q})\sum_{l=-\infty}^{\infty}\hbar\omega l\int d\vec{p}\, d\vec{p}' J_l^2\left(\frac{e\gamma_{\vec{p}\vec{p}'}}{mc\hbar\omega}\right)f_{\vec{p}}\delta(\varepsilon_{\vec{p}'} - \varepsilon_{\vec{p}} - l\hbar\omega). \tag{4.27}$$

In the future, we confine ourselves only to single-photon transitions; that is, we suppose that $l = \pm 1$. In this approximation,

$$P = P(+) + P(-), \tag{4.28}$$

where

$$P(\pm) = \pm 2\pi\omega\frac{V}{(2\pi\hbar)^3}\sum_{\vec{q}} W(\vec{q})\int d\vec{p}\, d\vec{p}' J_1^2\left(\frac{e\gamma_{\vec{p}\vec{p}'}}{mc\hbar\omega}\right)f_{\vec{p}}\delta(\varepsilon_{\vec{p}'} - \varepsilon_{\vec{p}} \pm \hbar\omega). \tag{4.29}$$

The sign plus means an increase in the energy of the electronic subsystem due to the absorption of the photon, and the minus sign means the decrease of this energy due to photon emission.

The argument of the Bessel function J_1^2 in (4.29), as the estimates show, for all frequencies of the optical range, is much smaller than the unit. Therefore, in (4.29) we can confine ourselves to the first term in the Taylor series (4.30).

$$J_1^2(x) \approx \frac{x^2}{4} = \frac{e^2\left[\vec{A}_0(\vec{p} - \vec{p}')\right]^2}{4m^2c^2\hbar^2\omega^2}. \tag{4.30}$$

It is easy to hold the summation over $\vec{q}$ in (4.29), considering that the electron–bimagnon scattering is quasi-elastic ($N_{\vec{q}} \approx T_m/\omega_{\vec{q}} = 2m_s T_m/q^2$; T_m is the magnons temperature in the energy units, $\omega_{\vec{q}} = q^2/2m_s$, m_s is the mass of the magnon):

$$\frac{V}{(2\pi\hbar)^3}\sum_{\vec{q}} W(\vec{q}) = \frac{V^2}{(2\pi\hbar)^6}\int W(\vec{q})d\vec{q} = \frac{a^6 J^2 m_s^2 T_m^2}{2^7 \pi^4 \hbar^6 \sqrt{2mJS}}, \tag{4.31}$$

and, as a result, obtain

$$P(\pm) = \pm\frac{e^2 a^6 J^2 m_s^2 T_m^2}{2^8 \pi^3 m^2 c^2 \omega \hbar^8 \sqrt{2mJS}}\int d\vec{p}\, d\vec{p}'\left[\vec{A}_0(\vec{p}-\vec{p}')\right]^2 f_{\vec{p}}\delta(\varepsilon_{\vec{p}'} - \varepsilon_{\vec{p}} \pm \hbar\omega). \tag{4.32}$$

Also, we assume that

$$f_{\vec{p}} = \frac{n}{(2\pi m T_e)^{\frac{3}{2}}}\exp\left(-\frac{\varepsilon_{\vec{p}}}{T_e}\right). \tag{4.33}$$

(n, m, T_e is the concentration, effective mass, and temperature of conduction electrons, respectively).

Let us make a replacement $\vec{p} - \vec{p}' = \vec{k}$ and write in the new variables (4.32):

$$P(\pm) = \pm\frac{e^2 a^6 J^2 m_s^2 T_m^2}{2^8 \pi^3 m^2 c^2 \omega \hbar^8 \sqrt{2mJS}}\int d\vec{p}\, d\vec{k}\left[\vec{A}_0\vec{k}\right]^2 f_{\vec{p}}\delta\left(\frac{k^2}{2m} - \frac{pk\cos\theta}{m} \pm \hbar\omega\right), \tag{4.34}$$

where θ is the angle between the vectors $\vec{p}$ and $\vec{k}$.

Now consider that $d\vec{p} \rightarrow p^2 dp d\Omega \rightarrow p^2 dp \sin\theta d\theta d\varphi$. The integral over φ is to be taken simply, since nothing depends on it, but the integral over θ can be taken using the δ -function:

$$\int_0^\pi d\theta \sin\theta\delta\left(\frac{k^2}{2m} - \frac{pk\cos\theta}{m} \pm \hbar\omega\right) = \frac{m}{pk} \tag{4.35}$$

Equality (4.35) is performed on condition

$$|\cos\theta| = \left|\pm\hbar\omega - \frac{k^2}{2m}\right| \Big/ \frac{pk}{2m} \le 1. \tag{4.36}$$

Condition (4.36) means that for a given k argument of the δ-function can be zero. In other words, inequality (4.36) defines the limits of integration by k. In case of absorption it follows from (4.36),

$$k_{\max} = p + \sqrt{p^2 + 2m\hbar\omega}, \quad k_{\min} = -p + \sqrt{p^2 + 2m\hbar\omega}. \tag{4.37}$$

After calculating the integrals over the angles of the vector $\vec{p}$, it can be obtained from (4.34) taking into account (4.33):

$$P(+) = \frac{e^2 a^6 J^2 m_s^2 T_m^2 n A_0^2 2\pi}{2^8 \pi^3 m^2 c^2 \omega \hbar^8 \sqrt{2mJS}(2\pi m T_e)^{3/2}} \int_0^\infty \exp(-\varepsilon_p/T_e)\ p^2 dp \int_{k_{\min}}^{k_{\max}} k^3 dk \int_0^\pi \sin\vartheta \cos^2\vartheta d\vartheta, \tag{4.38}$$

where ϑ is the angle between the vectors $\vec{A}$ and $\vec{k}$.

Integrals over the corners and k can be easily taken:

$$\int_0^\pi \sin\vartheta\cos^2\vartheta d\vartheta = \frac{1}{3}, \quad \int_{k_{\min}}^{k_{\max}} k^3 dk = \frac{1}{4}\left(k_{\max}^4 - k_{\min}^4\right) = m\left[\begin{array}{l}\left(\sqrt{\varepsilon} + \sqrt{\hbar\omega + \varepsilon}\right)^4 - \\ \left(-\sqrt{\varepsilon} + \sqrt{\hbar\omega + \varepsilon}\right)^4\end{array}\right]. \tag{4.39}$$

The remaining integral over p has the form

$$\int_0^\infty p^2 dp \exp(-\varepsilon_p/T_e)\sqrt{\hbar\omega + \varepsilon}(\hbar\omega + 2\varepsilon) = \frac{1}{2}\left(\frac{T_e}{\pi}\right)^{3/2} \int_0^\infty dx e^{-x}\{x(\alpha + x)\}^{1/2}(\alpha + 2x), \tag{4.40}$$

where $\alpha = \hbar\omega/2T$.

The integrals of type (4.40) can be expressed through the Bessel function of the imaginary argument $K_g(\alpha)$ using the recurrence formula [6]

$$\int_0^\infty dx e^{-x} x^n (x^2 + 2\alpha x)^{g-\frac{1}{2}} = (-1)^n \frac{2^g}{\sqrt{\pi}} \Gamma\left(g + \frac{1}{2}\right) \frac{d^n}{d\alpha^n}\left(\alpha^{-g} e^{\alpha} K_g(\alpha)\right), \tag{4.41}$$

($\Gamma(\alpha)$ is a gamma function) and, as a result, it is obtained from (4.38)

$$P(+) = \frac{2e^2 n A_0^2 T_e^{3/2}}{3\sqrt{\pi} m \tau_p^{em} c^2 \hbar\omega} \left\{ e^{\alpha} \alpha^3 \frac{d}{d\alpha} \left(\frac{K_1(\alpha)}{\alpha} \right) \right\}. \tag{4.42}$$

Here, τ_p^{em} is the time of the relaxation of electrons by momentum on magnons taken at $\varepsilon = T$ [7].

The Bessel function $K_1(\alpha)$ has an asymptotic behavior:

$$K_1(\alpha) = \begin{cases} 1/\alpha, & \alpha \to 0 \\ \sqrt{\pi/2\alpha}\exp(-\alpha), & \alpha \to \infty. \end{cases} \tag{4.43}$$

As for $P(-)$ magnitude, in this case, in contrast to (4.38), the integration over ε must be carried out from $\hbar\omega$ to ∞ (since only electrons with energy $\varepsilon \geq \hbar\omega$ can emit photons). Now, to make in the expression for $P(-)$ a shift $\varepsilon \to \varepsilon - \hbar\omega$, it follows that

$$P(-) = -\exp\left(-\frac{\hbar\omega}{T_e} \right) P(+). \tag{4.44}$$

In the experiment, the absorption coefficient is measured, which in our notation has the form

$$K = \frac{P(+) + P(-)}{\Pi_0}. \tag{4.45}$$

In expression (4.45), Π_0 is the electromagnetic flux falling on the surface of the semiconductor:

$$\Pi_0 = \frac{c\sqrt{\varepsilon_0}}{4\pi} \left(\frac{\omega}{c} \right)^2 \frac{A_0^2}{2}, \tag{4.46}$$

where ε_0 is the static dielectric constant, A_0 is the amplitude of the vector-potential of the field of laser radiation.

Now, taking into account (4.42) and (4.44), the general expression for the absorption coefficient of light by free carriers in ferromagnetic semiconductors can be expressed as

$$\begin{aligned} K &= \frac{P(+) + P(-)}{\Pi_0} \\ &= -\frac{16\sqrt{\pi} e^2 n T_e}{3\sqrt{\varepsilon} c \hbar \omega^3 m \tau_p^{em}} \sqrt{\frac{T_e}{T}} \left(1 - \exp\left(-\frac{\hbar\omega}{T_e} \right) \right) \left\{ e^{\alpha} \alpha^3 \frac{d}{d\alpha} \left(\frac{K_1(\alpha)}{\alpha} \right) \right\}. \end{aligned} \tag{4.47}$$

The general expression for the absorption coefficient for electron–bimagnon scattering (4.47) can be greatly simplified in classical ($\hbar\omega/T_e \ll 1$) and quantum ($\hbar\omega/T_e \gg 1$) cases. So, let us consider these cases.

1. The area of classical absorption ($\hbar\omega/T_e < < 1$).

In this case, $\exp(-\hbar\omega/T_e) \approx 1 - \hbar\omega/T_e$, $K_1(\alpha) \approx 1/\alpha$ and it follows from (4.47),

$$K = \frac{32\sqrt{\pi}e^2 n}{3\sqrt{\varepsilon}c\omega^2 m\tau_p^{em}}\sqrt{\frac{T_e}{T}}. \tag{4.48}$$

If there is no heating of the electrons ($T_e = T$) then expression (4.48) differs from the classical Drude's formula [8] only for the factor $8/3\sqrt{\varepsilon}$. This is due to the fact that in the classical Drude theory, the energy dependence of relaxation time is not taken into account.

2. Area of quantum absorption ($\hbar\omega/T_e > > 1$).

In this case, $\exp(-\hbar\omega/T_e) \to 0$, $K_1(\alpha) \approx \sqrt{\pi/2\alpha}\exp(-\alpha)$ and from (4.47) it follows that

$$K = \frac{4\pi e^2 n}{3\sqrt{\varepsilon}c\omega^2 m\tau_p^{em}}\sqrt{\frac{T_e}{T}}\left(\frac{\hbar\omega}{T_e}\right)^{1/2}. \tag{4.49}$$

If the notation is introduced: $\tau(\hbar\omega) = \tau_p^{em}(\hbar\omega)$ (the time of the electrons relaxation by the momentum on the magnons at $\varepsilon = \hbar\omega$), then (4.49) can be rewritten in the form

$$K = \frac{4\pi ne^2}{3mc\sqrt{\varepsilon}\omega^2 m}\frac{1}{\tau(\hbar\omega)}. \tag{4.50}$$

4.4 Polarizing Effect in the Emission of Light by Free Electrons

If the electron gas is warmed up, then the effect reversible to the absorption of Drude will be observed, when free carriers will emit light. This spontaneous emission can be valued using the previously obtained expression for the radiation field induced $P(-)$.To do this, the vector- potential of the laser wave field ($\vec{A}_0$) must first be normalized in such a way that N_{ph} photons are to be in the volume V, that is, use the condition

$$\frac{1}{V}N_{ph}\hbar\omega = \frac{E^2}{4\pi} = \frac{1}{8\pi} = \frac{1}{8\pi}\left(\frac{\omega}{c}\right)A_0^2. \tag{4.51}$$

From this,

$$A_0 = 2c\left(\frac{2\pi\hbar}{V\omega N_{ph}}\right)^{1/2}. \tag{4.52}$$

Further, substituting expression (4.51) in the formula for $P(-)$, putting in advance $N_{ph} = 1$, and multiplying obtained expression on the density of finite states of the field in the unit frequency range and the cortical angle $d\Omega$:

$$d\rho(\omega) = \frac{V}{(2\pi c)^3}\omega^2 d\Omega.$$

The expression will be obtained for the energy $W(-)$ emitted by free electrons in the FMSC per unit time in a cortical angle $d\Omega$:

$$W(-) = e^2 nT \frac{(\hbar\omega)^{3/2}}{3\pi^2 c^3 m\tau_p^{em}(T_e)\sqrt{T_e}} \exp\left\{-\frac{\hbar\omega}{T_e}\right\} d\Omega. \tag{4.53}$$

From (4.53), for the classical frequency range ($\hbar\omega < < T_e$) the following can be obtained:

$$W(-) = \frac{4e^2 nT_e}{3\pi^{5/2} c^2 m\tau_p^{em}(T_e)} d\Omega. \tag{4.54}$$

Also, a simple expression can also be obtained from (4.53) for the quantum frequency range ($\hbar\omega > > T_e$):

$$W(-) = e^2 nT \frac{(\hbar\omega)^{3/2}}{3\pi^2 c^3 m\tau_p^{em}(T_e)\sqrt{T_e}} \exp\left\{-\frac{\hbar\omega}{T_e}\right\} d\Omega. \tag{4.55}$$

Expressions (4.54) and (4.55) show that the intensity of the radiation of hot electrons in the FMSC does not depend on the frequency of light in the classical frequency range and exponentially falls in the quantum frequency region.

4.5 Conclusions

The theory of absorption and emission of light by free carriers in the FMSC is constructed and general expressions for absorption coefficient are obtained. In the presence of hot electrons, the results also obtained for the intensity of light emission by free electrons in the FMSC for the case when the electron–bimagnon scattering predominates. In all cases, the dependence on the concentration of electrons and their temperature is taken into account. It is shown that the intensity of radiation does not depend on the frequency of light in the classical frequency range and exponentially falls in the quantum region of frequencies.

References

1. Nagaev EL (1979) Physics of magnetic semiconductors (in Russian). Nauka, Moscow
2. Tomchuk PM (2004) Peculiarities of the light absorption and emission by free electrons of multivalley stmiconductors. Ukr Fiz Jor 49:682
3. Bondar VM, Sarbey OG, Tomchuk PM (2002) Polarizability dependence of hot eelectrons. Fiz Tverd Tela 44:1540
4. Landau LD, Lifshitz EM (1965) Quantum mechanics, A course of theoretical physics, vol 2. Pergamon Press, London
5. Semchuk OY, Willander M, Karlsteen M (2004) A wave function and quasi-energy of conduction electrons in a field of coherent light beams. Phys Status Solidi 241:2549. https://doi.org/10.1002/pssb.200302024
6. Abramowitz M, Stegun IA (1964) Handbook of mathematical functions with formulas, graphs, and mathematical tables. Courier Corporation, New York
7. Semchuk OY, Rozhkov SS (1981) Heating of electrons and magnons in ferromagnetic semiconductors. Sov Phys Solid State (US) 23:1118
8. Seeger K (2004) Semiconductor physics. Springer, Berlin/Heidelberg

Chapter 5
Mechanism of Active Electron Transfer in a Protein-Like Nanowire Under Real Conditions

L. V. Shmeleva and A. D. Suprun

5.1 Introduction

Donor–acceptor transfer of electrons between polypeptide fragments of protein molecules (cell organelles) is already studied in sufficient details [1, 2]. Study of mechanisms of their transfer in these fragments has just begun. Such study is based on the general principles of energy or charge transfer in polypeptide fragments of protein molecules [3], in particular, taking into account their actual structure, for example, the finite length of the spiral section of the protein molecule [4].

The current arising as a result of charge transfer is determined on the basis of the concept of an injected electron as a free quasiparticle of the classical type in the conduction band of the primary structure of the protein molecule [5, 6, 7].

The question of the model application, based on previous studies [8, 9, 10], was considered for electron transfer along a fragment similar to the polypeptide section of the protein molecule. The physical justification of the model was that under any temperature conditions the current in the fragment under consideration did not arise in the absence of external fields or other factors disturbing the electrostatic equilibrium of the electronic subsystem.

The model was additionally generalized by the inclusion of a factor named in the previous studies by an effective electrostatic field. This factor is due to the presence of amino acid heterogeneity. In its presence, an effective residual electrostatic field appears in the fragment under consideration. The value of the energy of this field is estimated and it is shown that it is small in relation to other characteristic energies of the considered object. The states of conductivity were calculated for this field. Their dependence on the field was analyzed. It is shown that with the increase of the value

L. V. Shmeleva (✉) · A. D. Suprun
Taras Shevchenko National University of Kyiv, Kyiv, Ukraine
e-mail: LShmel@univ.kiev.ua

O. Fesenko, L. Yatsenko (eds.), *Nanooptics, Nanophotonics, Nanostructures, and Their Applications*, Springer Proceedings in Physics 210,
https://doi.org/10.1007/978-3-319-91083-3_5

of the field, the conductivity states are shifted toward a decrease of energy. The current created by one electron carried by this field was estimated. The performed calculations show that the current increases continuously in the physiologically relevant range 33–41 °C (306–314 °K) with the temperature increase. In this case, for polypeptides (the number of amino acid residues is 100), it grows from 201.85 pA to 204.90 pA, and for the longest protein-like chains (the number of amino acid residues is 300), increases from 22.43 pA to 22.77 pA. Such increases in currents in the range (310–314 °K) (sickness) is useful in view of the possibility of increasing the synthesis of ATP in mitochondria of cells. ATP synthesis increase facilitates the energy of the body struggling against the sickness. At the same time, the current reduction in the range 37–33 °C (310–306 °K) (overcooling of the body) negatively affects the process of ATP synthesis, reducing it and, consequently, reducing the protective energy resources of the body. It is also shown that the current decreases with an increase in the length of the protein-like nanowire from 100 (polypeptides) to 300 (proteins) of the amino acid residues. But in more detail [10], taking into account the effective electrostatic field, the number of amino acid residues can affect its value: this may slightly offset the decrease in current with an increase in the number of amino acid residues.

5.2 Methods

5.2.1 *Some Properties of an Effective Residual Electrostatic Field*

In previous studies [8, 9, 10] it was found that the primary structure of polypeptide fragments of protein-like polymers has an expressed electronic structure inherent in semiconductors. It was also found that due to the amino acid inhomogeneity in this protein-like structure there is an uncompensated electrostatic field with energy of the general form

$$W_{\mathbf{n}} \equiv -\langle \varphi(\mathbf{r}) | \sum_{\mathbf{m}(\neq \mathbf{n})} \frac{ze^2}{13|\mathbf{r}+\mathbf{n}-\mathbf{m}|} | \varphi(\mathbf{r}) \rangle . \tag{5.1}$$

Hereinafter, the summation is carried out for all atoms, and the angle brackets mean quantum energy averaging: $-\sum_{\mathbf{m}(\neq \mathbf{n})} \frac{ze^2}{13|\mathbf{r}+\mathbf{n}-\mathbf{m}|}$, by variable $\mathbf{r}$ in wave functions $\varphi(\mathbf{r})$ for the conductivity state of a one-electron ion. In the polypeptide fragments under consideration of the protein-like polymers, the conductivity states are associated with the 2p-free oxygen states. In this case, the charge number is $z = 7$. Vector variables $\mathbf{n}$ and $\mathbf{m}$ describe the spatial positions of molecular groups of the amino acid residues type.

This energy causes the transfer of the electron from the acceptor to the donor amino groups of the protein-like polymer in the absence of any external fields and, as established in previous studies, it is included in the energy functional of the general form

$$E(\{a\}) = \sum_{\mathbf{n}\mathbf{m}}{}^{\prime} w\mathbf{nm} + \sum_{\mathbf{n}} (W_{\mathbf{n}} + D_{\mathbf{n}}) \cdot |a_{\mathbf{n}}|^2 + \sum_{\mathbf{n}\mathbf{m}}{}^{\prime} M\mathbf{nm} \left(a_{\mathbf{n}}^* a_{\mathbf{m}} + a_{\mathbf{m}}^* a_{\mathbf{n}}\right). \tag{5.2}$$

In this functional, $a_{\mathbf{n}}$ is the required part of the wave function of the inoculated electron, $w_{\mathbf{nm}}$ is the energy of interatomic interaction, and $M_{\mathbf{nm}}$ is so-called energy of the exchange resonance interaction, which describes the dynamics of excitation, if considered as a quasiparticle, that is, as an object of the classical type. These two energies were discussed in detail in [8, 9]. Therefore, we do not stop here, because in future their explicit form will not be used. Energy $D_{\mathbf{n}}$ was also discussed in detail in [8, 9], but here it will be needed for comparison with energy $W_{\mathbf{n}}$. Therefore, let us recall its explicit form. Accurate to the terms that determine the interelectron interaction (these terms are almost by order of magnitude smaller), it is determined by the ratio:

$$D_{\mathbf{n}} \equiv - \sum_{\mathbf{m}(\neq \mathbf{n})} \left[\langle \varphi(\mathbf{r}) | \frac{ze^2}{|\mathbf{r} + \mathbf{n} - \mathbf{m}|} | \varphi(\mathbf{r}) \rangle + \cdots \right]. \tag{5.3}$$

As it is seen, the main remaining contribution almost coincides with the definition (5.1) for the field. Such a comparison is necessary, as in previous studies with a matrix element $D_{\mathbf{n}}$ certain transformations were performed, which cannot be repeated now, but simply transferred to the field term.

5.2.2 Basic System of Equations Taking into Account the Effective Residual Electrostatic Field

In [8] it was shown that taking into account the real structure of the amino acid residue and in the approximation of the closest neighbors, the functional (5.2) can be modeled as follows:

$$E(\{a\}) = N_a w + \sum_{n=1}^{N_0} \left[\sum_{\alpha=0}^{4} (\tilde{W}_n^{\alpha} + P_{\alpha}) |a_{n\alpha}|^2 + M \left\{ \sum_{\alpha=0}^{1} (a_{n\alpha}^* a_{n,\alpha+1} + a_{n,\alpha+1}^* a_{n\alpha}) + \sum_{\alpha=2}^{3} (a_{n,\alpha-1}^* a_{n,\alpha+1} + a_{n,\alpha+1}^* a_{n,\alpha-1}) + (a_{n2}^* a_{n+1,0} + a_{n+1,0}^* a_{n2}) \right\} \right]. \tag{5.4}$$

In this functional, the variable n searches all the amino acid residues, and a variable α searches five values—by the effective number of atoms in each such residue. Energy $\tilde{W}_n^\alpha$ is the total energy, which consists of energy (5.1) and energy of the external field [9]. In the absence of external fields, the energy $\tilde{W}_n^\alpha$ is reduced only to expression (5.1), which, taking into account the actual structure of the amino acid residue and in the approximation of its closest neighbors, loses the index n. Here the functional (5.4) takes the following form:

$$E(\{a\}) = N_a w + \sum_{n=1}^{N_0}\left[\sum_{\alpha=0}^{4}(W_\alpha + P_\alpha)|a_{n\alpha}|^2 + M\left\{\sum_{\alpha=0}^{1}(a^*_{n\alpha}a_{n,\alpha+1} + a^*_{n,\alpha+1}a_{n\alpha}) + \right.\right.$$
$$\left.\left. + \sum_{\alpha=2}^{3}(a^*_{n,\alpha-1}a_{n,\alpha+1} + a^*_{n,\alpha+1}a_{n,\alpha-1}) + (a^*_{n2}a_{n+1,0} + a^*_{n+1,0}a_{n2})\right\}\right], \tag{5.5}$$

where the energy W_α, due to the properties (5.2), (5.3), completely takes over the symmetric properties of the parameter P_α. Namely, if P_α, as shown in [8, 9], have such properties as $P_0 \equiv D$; $P_1 = P_2 = 3D/2$; $P_3 = P_4 = D/2$ (here $D = -\langle\varphi(\mathbf{r})|\frac{ze^2}{|\mathbf{r}+\mathbf{b}|}|\varphi(\mathbf{r})\rangle + \cdots$, and $\mathbf{b}$ – vector of lattice), then obviously the field factor W_α has similar properties: $W_0 \equiv W$; $W_1 = W_2 = 3W/2$; $W_3 = W_4 = W/2$. Moreover, we can even write down the definitions: $W = -\langle\varphi(\mathbf{r})|\frac{ze^2}{13|\mathbf{r}+\mathbf{b}|}|\varphi(\mathbf{r})\rangle$. But the expression (5.1) is a very significant approximation. In fact, according to [10], its value is determined by the real amino acid composition of the polypeptide fragment of the protein-like polymer. Therefore, in the future we will allow for the possibility of changing this parameter in some numerical range.

As before [9], in a result of variation (5.5), but already provided $W_\alpha \neq 0$ as well as $a_{n\alpha} = \mathrm{A}_\alpha \exp(ikRn)$, where k is one-dimensional wave vector, and R is intermolecular distances, we can obtain the following system of five equations:

$$\begin{gathered}
(x + 1 + \mathrm{w})\mathrm{A}_0 + v\mathrm{A}_1 + ve^{-ikR}\mathrm{A}_2 = 0;\\
v\mathrm{A}_0 + (x + 3(1 + \mathrm{w})/2)\mathrm{A}_1 + v\mathrm{A}_2 + v\mathrm{A}_3 = 0;\\
ve^{ikR}\mathrm{A}_0 + v\mathrm{A}_1 + (x + 3(1 + \mathrm{w})/2)\mathrm{A}_2 + v\mathrm{A}_4 = 0;\\
v\mathrm{A}_1 + (x + (1 + \mathrm{w})/2)\mathrm{A}_3 = 0;\\
v\mathrm{A}_2 + (x + (1 + \mathrm{w})/2)\mathrm{A}_4 = 0.
\end{gathered} \tag{5.6}$$

It is a homogeneous system for coefficients A_α and own value determination:

$$x \equiv \varepsilon/|D|. \tag{5.7}$$

Now in this system, in addition to free parameter, which is a wave vector k, and parameter $v \equiv |M|/|D|$ [9], an additional parameter occurs: $w \equiv |W|/|D|$, related to an effective field W. Here signs of all parameters are taken into account explicitly. By the definitions of parameters W and D that have just been discussed, $w < 1/13$ is obvious. In other words, the dimensionless effective field is much less than unity and therefore it will be more convenient to work with the factor $f = 1 + w$, especially since the field in such form of structure is included in the system (5.6). This factor will be called field factor.

System compatibility condition (5.6), subject to replacement $f = 1 + w$, results, as in previous studies [8, 9], in the fifth degree equation relative to x:

$$x^5 + 5fx^4 + \frac{1}{2}(19f^2 - 10v^2)x^3 + \frac{1}{2}(17f^3 - 26v^2f + 4v^3\cos(kR))x^2 + \\ + \frac{1}{16}(57f^4 - 164v^2f^2 + 32v^3f\cos(kR) + 48v^4)x + \\ + \frac{1}{16}f(9f^4 - 40v^2f^2 + 8v^3f\cos(kR) + 32v^4) = 0. \quad (5.8)$$

As it can be seen from this equation, the convenience of using the field factor $f = 1 + w$ is connected also with the fact that in the absence of a field ($w = 0$), this factor becomes equal to unity and Eq. (5.8) passes into already known form [8, 9].

5.3 Results and Discussion

5.3.1 *Graphical–Numerical Analysis of the Compatibility Condition for Conductivity States Detection in the Presence of an Effective Electrostatic Field*

In the absence of a field when $f = 1$, Eq. (5.8) has five roots that depend on two parameters: the wave vector k and factor v. The wave vector is a free parameter and determines the dispersion of physical quantities, in particular the electron velocity and the current corresponding to it. Factor v, as it seen from Eq. (5.8), determines the width of the corresponding sub-zone for each of the five solutions of Eq. (5.8). It is fixed for a particular material and a given temperature. Since the material is a protein-like macromolecule, this parameter varies only depending on the temperature. Dependence of conductivity states on these two parameters in the absence of external fields ($f = 1$) was studied in detail in [8, 9]. Now our interest will be the dependence of the conductivity states, velocity, and current on the field factor $f > 1$.

Equation (5.8) can be solved only by graphical–numerical methods. The dependence of the roots of equation (5.8) (conduction states) on the field factor f is determined at fixed parameters k and v. To have an idea of the effect of the field factor f on conductivity states, calculations were made for three values: $k = 0$, $|k| R = \pi/2$, and $|k|R = \pi$ within the entire range of wave vector k. For parameter v, the typical value $v = 0.4$ was chosen. This value will be used here as a similar analysis in previous studies, for $w = 0$ (or $f = 1$), was conducted exactly for this value, and it is important for the accuracy of the current analysis.

The results of the graphical–numerical analysis of the solutions for Eq. (5.8) are shown in the Fig. 5.1.

The graphs in Fig. 5.1 show that with field increase in the studied range $1 \leq f \leq 1.1$, all conductivity bands are shifted toward energy decrease. Energy decrease in the presence of an effective electrostatic field can be physically interpreted as the effect of electron "pulling" into a conductivity band through an acceptor center.

But for finding the velocity of the injected electron and the current corresponding to it, numerical calculations are not convenient. Therefore, we will find satisfactory analytic approximations of the dependence of conductivity states on an effective

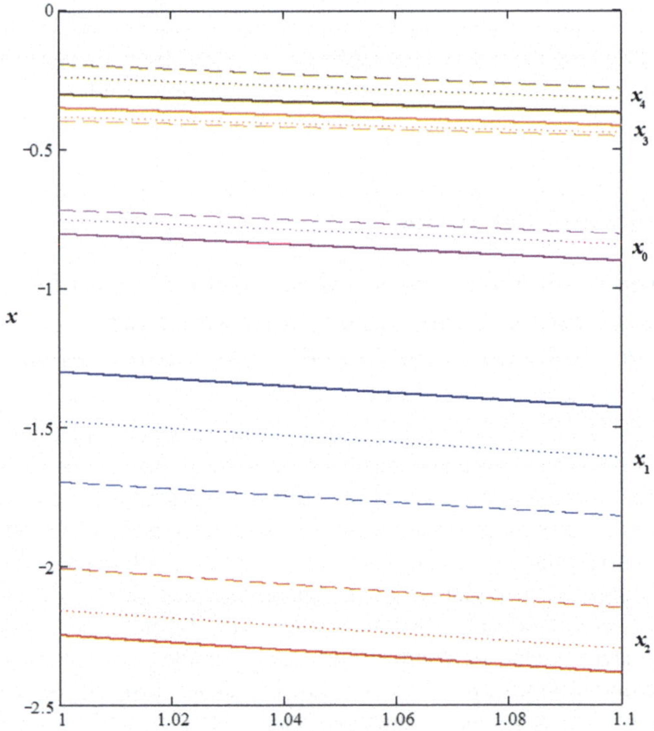

Fig. 5.1 Direct numerical solution of compatibility conditions (5.8) within the range $1 \leq f \leq 1.1$ for values: $k = 0$ (solid curves), $|k|R = \pi/2$ (dotted line), $|k|R = \pi$ (dashed curves), $v = 0.4$. The distance between solid curves and dashed curves corresponds to the width of the corresponding conductivity sub-band. In total, there are five fairly well-separated sub-zones

electrostatic field. To do this, we use the analytic representation for the case of the field absence w = 0 (or $f = 1$) [9]:

$$x_s(v,k) = -\begin{pmatrix} 1.0 \\ 3/2 \\ 3/2 \\ 1/2 \\ 1/2 \end{pmatrix} + \begin{pmatrix} v^2 + 3.15\,v^4 \\ v - 2.49v^2 \\ -v - 1.42v^2 \\ v^2 - 1.23v^4 \\ v^2 + 3.72v^4 \end{pmatrix} - v^3 \begin{pmatrix} 0.68 \\ -3.10 \\ 1.90 \\ -0.35 \\ 0.87 \end{pmatrix} \cos(kR)\,, \tag{5.9}$$

where $s = \{0,\ 1,\ 2,\ 3,\ 4\}$. In the case of an effective electrostatic field presence, this expression was modified by comparing the modified expression (5.9) with the direct numerical solutions of Eq. (5.8) given in Fig. 5.1.

As a result of this comparison, an improved expression for conductivity states was obtained in the presence of an effective electrostatic field:

$$x_s(f,v,k) = -\begin{pmatrix} 1.0 \\ 3/2 \\ 3/2 \\ 1/2 \\ 1/2 \end{pmatrix} f + \begin{pmatrix} v^2 + 3.15\,v^4 \\ v - 2.49v^2 \\ -v - 1.42v^2 \\ v^2 - 1.23v^4 \\ v^2 + 3.72v^4 \end{pmatrix} - v^3 \begin{pmatrix} 0.68 \\ -3.10 \\ 1.90 \\ -0.35 \\ 0.87 \end{pmatrix} \cos(kR) +$$

$$+(1 - \theta_{-}(1-f)) \begin{pmatrix} 0.0021 - 0.006\ \cos(kR) \\ 0.0063 - 0.004\ \cos(kR) \\ 0.0030 + 0.003\ \cos(kR) \\ 0.0120 - 0.004\ \cos(kR) \\ 0.0066 + 0.009\ \cos(kR) \end{pmatrix} f\,. \tag{5.10}$$

As is shown, this analytical solution differs from (5.9) by the change of the right-hand side of equality. Namely, the first term is changed and a new fourth term is added. Both of these terms are linearly dependent on the field factor f. In addition, the new term has an additional dispersion dependence on the wave vector k. Multiplier $(1 - \theta_{-}(1 - f))$ ensures the absence of the last term in the absence of a field when w = 0 and $f = 1$, due to the definition [11]: $\theta_{-}(x) = \left\{ \begin{array}{l} 0, npu\ \ x < 0 \\ 1, npu\ \ x \geq 0 \end{array} \right|$. The analytic approximation (5.10) for solutions of Eq. (5.8) reproduces the spectrum shown in Fig. 5.1 with a fairly good accuracy. To illustrate this, Fig. 5.2 shows a similar spectrum as that of Fig. 5.1, but already calculated by the formula (5.10).

In specified scale, the graphics in Figs. 5.1 and 5.2 are almost similar. This fact alone means that the approximation of a direct numerical solution (Fig. 5.1) in the

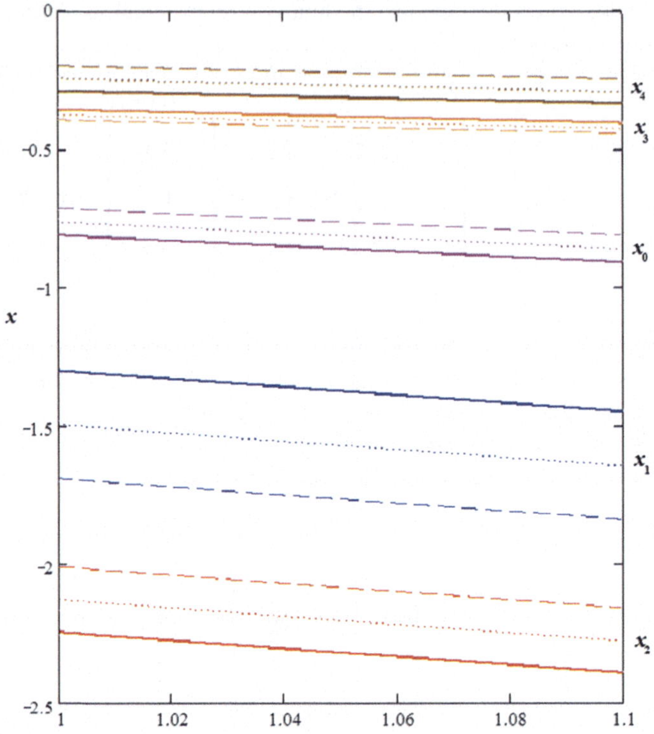

Fig. 5.2 Calculation of the conductivity states spectrum by the formula (5.10) for the same values of parameters as in the Fig. 5.1

field factor f by linear approximation (Fig. 5.2) with the help of formula (5.10) is a fairly good approximation. But in the linear field-factor approximation used here, it is virtually impossible to ensure a high degree of correlation between exact and approximate solutions. Linear approximation allows such correlation only with an average error of 3%. In this case, the error of the three lowest energy states: x_0, x_1, x_2, does not exceed 1% (respectively, 1%, 0.7%, and 0.5%). Two states with highest energy, x_3 and x_4, are approximated with errors, respectively, 4% and 9%. But such a significant error for these two states is not due to the quality of approximation, but to the small (in comparison with the other three states) depth of their location along the axis x (for dimensionless energy). In absolute values, the deviations are the same as for states x_0, x_1, x_2. Such accuracy of approximation in a linear approximation (5.10) for the field factor f allows its use for the further analysis of the current and the influence of temperature on it.

5.3.2 Active electron transfer over the primary structure of the protein molecule. Estimates of electric current

Effective electrostatic field $w \equiv |W|/|D|$ creates a constantly acting electrostatically nonequilibrium state in the primary structure of a protein molecule. This provides for the active transfer of the electron through its primary structure from the acceptor centers to the donor centers of the molecule. In this case the current occurs, which is, if not the only one, the quantitative characteristic of the process. In this sense, estimates of such a current and the influence of temperature on it are quite important.

The current generated by an electron directly depends on its velocity. To determine it, we should use the following definition [6, 7, 12, 13]:

$$V = \frac{1}{\hbar}\frac{dE(k)}{dk}. \tag{5.11}$$

In this definition, $E(k)$ is a dispersion (the dependence of the energy of the injected electron on the wave vector). In this case, it represents the electron energy in the conductivity band. Formula (5.10) shows its dimensionless form: $x_s(f, v, k)$. Since the definition (5.11) includes a dimensional dispersion, it is also necessary to use its dimensional form. According to definition (5.7), the following can be obtained for such a representation: $\varepsilon_s(f, v, k) = |D| \cdot x_s(f, v, k)$. Substituting energy $\varepsilon_s(f, v, k)$ in the definition (5.11) instead of $E(k)$, we can obtain the following:

$$V_s(f, v, k) = \frac{1}{\hbar}|D|R\left\{ v^3\begin{pmatrix} 0.68 \\ -3.10 \\ 1.90 \\ -0.35 \\ 0.87 \end{pmatrix} + (1-\theta_-(1-f))\begin{pmatrix} 0.006 \\ 0.004 \\ -0.003 \\ 0.004 \\ -0.009 \end{pmatrix} f \right\} \sin(kR).$$

With one of the simplest (among the series [14]) values of current density: $J = enV$, we can obtain the following:

$$J_s = \frac{e|D|R}{\hbar NV}\left\{ v^3\begin{pmatrix} 0.68 \\ -3.10 \\ 1.90 \\ -0.35 \\ 0.87 \end{pmatrix} + (1-\theta_-(1-f))\begin{pmatrix} 0.006 \\ 0.004 \\ -0.003 \\ 0.004 \\ -0.009 \end{pmatrix} f \right\} \sin(kR), \tag{5.12}$$

where it is taken into account that $n \equiv 1/NV$ for one injected electron. Here N is the number of amino acid residues, and V its effective mean volume. Let us take into account that the effective mean volume of the amino acid residue is $V = RS$, where

R is an effective distance between the amino acid residues, which actually coincides with their mean longitudinal linear size, and S is effective cross-sectional area of the primary structure of the polypeptide fragment of the protein-like molecule. In (5.12) we substitute the product of RS in place of V. Next, Eq. (5.12) multiply by the area S and, taking into account that $I = JS$, for the current we will obtain

$$I_s = \frac{e|D|}{\hbar N}\left\{ v^3 \begin{pmatrix} 0.68 \\ -3.10 \\ 1.90 \\ -0.35 \\ 0.87 \end{pmatrix} + (1 - \theta_-(1-f)) \begin{pmatrix} 0.006 \\ 0.004 \\ -0.003 \\ 0.004 \\ -0.009 \end{pmatrix} f \right\} \sin(kR) \quad (5.13)$$

To determine the full current I we need to find the sum [9]

$$I = \sum_{s=0}^{4} I_s. \quad (5.14)$$

This algorithm for determining the total current is due to two factors. In the classical sense, it corresponds to the parallel connection of conduction channels. And in the quantum sense it corresponds to their equal probability.

Substituting (5.13) into the sum (5.14) and taking into account that the sum of the coefficients of the first term in the curly brackets of expression (5.13) is equal to zero, for full current I we will obtain the following:

$$I = 0.002\frac{e|D|}{\hbar N} f \sin(kR)(1 - \theta_-(1-f)). \quad (5.15)$$

In order to estimate the current created by the electron injected into the primary structure of the polypeptide fragments of protein-like molecules, some simplifications are required in formula (5.15). First, let us consider that $f > 1$ (w > 0). Then the factor $1 - \theta_-(1-f)$ equals to unity: $1 - \theta_-(1-f) = 1$. Second, let us consider the smallest (typical) values of the wave vector k. Then $\sin(kR) = 2\pi/N$ and formula (5.15) takes the form $I = 0.004\frac{\pi e|D|}{\hbar N^2} f$. Field factor f will be considered, though larger than unity, but so close to the unity that the current can be presented as $I = \frac{\pi e|D|}{250\,\hbar N^2}$. Now, with this definition, we can not only evaluate the current, but also analyze the impact of temperature on it. In [8], the total energy dependence $|D|$ on temperature was found: $|D| = |D_0| + \mathrm{sign}(D_0)\mathrm{sign}(D_1)\ |D_1|\frac{227.45^\circ}{T}$, where parameters $|D_0|$ and $|D_1|$ satisfy the condition: $|D_0| >> |D_1|\frac{227.45^\circ}{T}$. Here we will use the most typical of its representations for the considered system: $|D| = |D_0|\left(1 - \frac{114^\circ}{T}\right)$, where $|D_0| \approx 1$ eV. Consequently, the final current is determined by equality:

Table 5.1 Dependence of electric current on temperature and number of amino acid residues

	t				
	33 °C	35 °C	37 °C	39 °C	41 °C
N	(306 °K)	(308 °K)	(310 °K)	(312 °K)	(314 °K)
100	201.85 pA	202.63 pA	203.40 pA	204.16 pA	204.90 pA
150	89.71 pA	90.06 pA	90.40 pA	90.74 pA	91.07 pA
200	50.46 pA	50.66 pA	50.85 pA	51.04 pA	51.23 pA
250	32.30 pA	32.42 pA	32.54 pA	32.66 pA	32.78 pA
300	22.43 pA	22.51 pA	22.60 pA	22.68 pA	22.77 pA

$$I = \frac{I_0}{N^2}\left(1 - \frac{114^{\circ}}{T}\right), \tag{5.16}$$

where $I_0 \equiv \frac{\pi e |D_0|}{250\,\hbar}$. From (5.16) it follows that the current depends on two parameters—the number of amino acid residues N and temperature T. Typical range of residues N is from 100 (polypeptides) to 300 (proteins). The physiologically significant temperature range is 33–41 °C (306–314 °K). Using the value $|D_0| = 1.6 \cdot 10^{-19}$J, $e = 1.6 \cdot 10^{-19}$C, $\hbar = 10^{-34}$ J c in an amplitude multiplier I_0 of expression (5.16), we can obtain: $I_0 = 3.217$ μA. Given this value, the dependence of the current on the temperature and the number of amino acid residues in accordance with (5.16) is given in Table 5.1.

As it shown in the table, for very weak fields ($f \sim 1$, but $f > 1$) with the increase in the length of the protein-like nanowire from 100 (polypeptides) to 300 (proteins) of the amino acid residues, the current decreases. But, if we are not limited to very weak fields for estimates, then the definition (5.16) would look as follows: $I = \frac{I_0}{N^2}\left(1 - \frac{114^{\circ}}{T}\right)f$, and would grow with field increase. Generally speaking, in more detail [10] consideration of the effective electrostatic field, the number of amino acid residues will affect its value, which can slightly compensate the decrease of the current with increase of N.

As for the influence of temperature on the current, in the physiologically relevant range 33–41 °C (306–314 °K) the current always increases with temperature increase. In this case, for polypeptides (N = 100), it grows from 201.85 pA to 204.90 pA, while for the longest protein-like chains it grows from 22.43 pA to 22.77 pA. Such increases in currents in the range 37–41 °C (310–314 °K) (sickness) is useful in view of the possibility of increasing the synthesis of ATP in mitochondria of cells (ATP synthesis increase facilitates the energy of the body struggling against the sickness). At the same time, the current reduction in the range 37–33 °C (310–306 °K) (overcooling of the body) negatively affects the process of ATP synthesis, reducing it and, consequently, reducing the protective energy resources of the body.

5.4 Conclusions

This work analyzes one of the mechanisms of an electron metabolic transfer in the primary structure of a protein-like molecule. This structure is a nanowire with semiconductor properties.

The main purpose of the study was to calculate the current from the injected electron in the presence of a disturbance of the electrostatic equilibrium of the system, which is called an effective electrostatic field. The resulting current was determined on the basis concept of an injected electron as a free quasiparticle of the classical type in the conduction band of the primary structure of the protein molecule.

The physical cause of the residual effective electrostatic field occurrence is due to the amino acid heterogeneity of the protein-like molecule. The value of the energy of this field is estimated and it is shown that it is small in relation to other characteristic energies of the considered object. Given this field, the conductivity states were calculated. Their dependence on the effective electrostatic field was analyzed. It is shown that with the increase of the field value, the conductivity states are shifted toward the energy decrease. That is, the so-called inclination of conductivity bands occurs, as it should be.

Current calculations were performed in the presence of temperature not equal to zero. It has been shown that an electron transferred by an effective electrostatic field produces a microcurrent in the range of 22.77–201.85 pA, depending on the length of the protein-like nanowire (corresponding to 300–100 amino acid residues) and the temperature in the physiologically relevant range: 33–41 °C (306–314 °K). It is shown that with an increase of the nanowire length, the current decreases at any temperature. This means that the short protein-like molecules more effectively perform the function of the metabolic transfer of the electron. However, with a more detailed consideration of an effective electrostatic field, the number of amino acid residues can affect its value, which can slightly compensate the decrease in current with the growth of the number of amino acid residues. At the same time, for any length of a protein-like nanowire, the current increases, which is characteristic of semiconductors. But such growth can have both positive effects (in the case of the sickness) and negative consequences (in the case of overcooling).

References

1. Petrov EG, Shevchenko YV, May V (2003) On the length dependence of bridge-mediated electron transfer reactions. Chem Phys 288(2):269–279. http://www.sciencedirect.com/science/article/pii/S0301010403000326
2. Petrov EG, May V, Hänggi P (2002) Controlling electron transfer processes through short molecular wires. Chem Phys 281(2):211–224. http://www.sciencedirect.com/science/article/pii/S0301010402003403

3. Brizhik L, Eremko A, Piette B, Zakrzewski W (2006) Charge and energy transfer by solitons in low-dimensional nanosystems with helical structure. Chem Phys 324(1):259–266. http://www.sciencedirect.com/science/article/pii/S0301010406000863
4. Suprun AD, Shmeleva LV (2014) Alpha-helical regions of the protein molecule as organic nanotubes. Nanoscale Res Lett 9(1):200. https://nanoscalereslett.springeropen.com/articles/10.1186/1556-276X-9-200
5. Davydov AS (1982) Biology & quantum mechanics, vol. 109. Pergamon Press. 229 p. https://books.google.com.ua/books/about/Biology_quantum_mechanics.html?id=P5ApAQAAMAAJ&redir_esc=y
6. Ishizaki A, Calhoun TR, Schlau-Cohen GS, Fleming GR (2010) Quantum coherence and its interplay with protein environments in photosynthetic electronic energy transfer. Phys Chem Chem Phys 12:7319–7337. http://pubs.rsc.org/en/content/articlelanding/2010/cp/c003389h#!divAbstract
7. Suprun AD, Shmeleva LV (2012) Degeneracy effect of dynamical properties of quasi-particles of electronic origin in semiconductor materials. Funct Mater 19(4):508–519. http://functmaterials.org.ua/contents/19-4/
8. Suprun AD, Shmeleva LV (2017) Chapter 13: temperature effect on the basis states for charge transfer through a polypeptide fragments of proteins and on the nanocurrent in it. In Fesenko O, Yatsenko L (eds) Nanophysics, nanomaterials, surface studies, and applications, vol. 195, Springer proceedings in physics. Springer International Publishing AG, pp 175–186. https://link.springer.com/chapter/10.1007/978-3-319-56422-7_13
9. Suprun AD, Shmeleva LV (2016) Current in the protein nanowires: quantum calculations of the base states. Nanoscale Res Lett 11(1):74. https://doi.org/10.1186/s11671-016-1269-0
10. Suprun AD, Shmeleva LV (2015) Primary structure of proteins as a nanowire for metabolic electronic transport. Nanoscale Res Lett 10(1):121. https://doi.org/10.1186/s11671-015-0763-0
11. Korn GA, Korn TM (2000) Mathematical handbook for scientists and engineers: definitions, theorems, and formulas for reference and review. Courier Corporation, New York. 1097 p. https://books.google.com.ua/books?hl=ru&lr=&id=xUQc0RZhQnAC&oi=fnd&pg=PP1&dq=Korn&ots=QDWUbJ9xI6&sig=N__AgdlGQ-soUjn6QfDIjM_7bE8&redir_esc=y#v=onepage&q=Korn&f=false
12. Davydov AS (2013) Theory of molecular excitons. Springer. 312 p. https://books.google.com.ua/books?hl=ru&lr=&id=wPcHCAAAQBAJ&oi=fnd&pg=PA1&ots=_G4n_XXxJx&sig=EkDayI4mWsJaX5FKwr0UyryOEnc&redir_esc=y#v=onepage&q&f=false
13. Suprun AD, Shmeleva LV (2014) The centrally-symmetric solutions of electronic excitations of semiconductors in the conditions of relativistic like degeneracy of dynamical properties. Funct Mater 21(1):69–79. https://doi.org/10.15407/fm21.01.069
14. Martin RM (2004) Electronic structure: basic theory and practical methods. Cambridge University Press. 625 p. https://books.google.com.ua/books?hl=ru&lr=&id=v1YhAwAAQBAJ&oi=fnd&pg=PR17&dq=Martin,+R.+M.+(2004).+Electronic+structure:+basic+theory+and+practical+methods.+Cambridge+university+press.&ots=8OCb8x9gbg&sig=hu_VycdapoJ6UmUE5tqwroKydK0&redir_esc=y#v=onepage&q&f=false

Chapter 6
Electron Correlation Effects in Theoretical Model of Doped Fullerides

Yu. Skorenkyy, O. Kramar, L. Didukh, and Yu. Dovhopyaty

6.1 Introduction

The diversity of physical properties of doped fullerides has not been explained so far at a microscopic level despite the intensive experimental and theoretical studies conducted in recent decades. A variety of fullerene-based organic compounds have been synthesized with metals [1], hydrogen [2, 3], halogens [4–6], and benzene [7] that form a new class of organic conductors and semiconductors with tunable parameters. In polycrystalline C_{60} doped with alkali metals, at temperatures under 33 K superconductivity has been observed [8–13] with critical temperatures varying from 2.5 K for Na_2KC_{60} to 33 K for $RbCs_2C_{60}$. Along with the phonon mechanism of Cooper pairing [14, 15], purely electronic pairing mechanism has been proposed [16]. To date, superconductivity of molecular conductors remains an open problem. According to the theoretical band structure calculations (see [17] for a review), fullerides with integer band-filling parameter n should be Mott–Hubbard insulators because all of them possess large enough intra-atomic Coulomb correlation. At the same time, the doped systems A_3C_{60} (where A = K, Rb, Cs) turn out to be metallic at low temperatures [1]. It has been noted in papers [18, 19] that for a proper description of the metallic behavior of A_3C_{60} (with $x = 3$ corresponding to the half filling of conduction band), the orbital degeneracy of the energy band is to be taken into account.

Adding to fullerene C_{60} the radicals containing metals of platinum group creates fullerene-based ferromagnetic materials [1]. Another example of ferromagnetic system in which neither component per se is ferromagnetic, are stacked Pd/C_{60} bilayers [20]. A purely organic compound TDAE-C_{60} (TDAE stands for tetrakis

Y. Skorenkyy (✉) · O. Kramar · L. Didukh · Y. Dovhopyaty
Ternopil Ivan Puluj National Technical University, Ternopil, Ukraine
e-mail: skorenkyy@tntu.edu.ua

O. Fesenko, L. Yatsenko (eds.), *Nanooptics, Nanophotonics, Nanostructures, and Their Applications*, Springer Proceedings in Physics 210,
https://doi.org/10.1007/978-3-319-91083-3_6

(dimethylamino)ethylene) represents a pronounced example of ferromagnetic behavior [21–23] of unclear nature. Studies of polymerized fullerenes [24–26] are promising as well. Distinct from polycrystalline fulleride, where fullerene molecules are bound weakly by van der Waals forces, in such polymers a chemical binding is realized. In temperature interval from 300 K to 500 K the polymerized fullerene C_{60} has the features of semiconductor with 2.1 eV energy gap [27].

Solid-state fullerenes (fullerides) are molecular crystals with intra-molecular interaction much stronger than intermolecular one. In a close packing structure each fullerene molecule has 12 nearest neighbors. Such structures are of two types, namely, base-centered cubic and hexagonal ones [28–30]. At low temperatures, the cubic O_h crystal lattice symmetry does not correspond to the icosahedral Y symmetry of individual C_{60} molecules. There are four C_{60} molecules per unit cell of fulleride lattice, arranged in tetrahedra so that molecules' orientations in every tetrahedron are identical. The tetrahedra form a simple cubic lattice. At ambient temperatures, solid fullerides have one of closely packed lattices [31, 32] and are semiconductors with band gap of 1.5–1.95 eV for C_{60} [33, 34], 1.91 eV for C_{70} [35], 0.5–1.7 eV for C_{78} [36], 1.2–1.7 eV for C_{84} [37–39]. Electrical resistivity of polycrystalline C_{60} [37, 40, 41] decreases monotonically with temperature and energy gap depends on the external pressure. Experimental studies of fulleride films [42] have shown nonexponential dependences with characteristic relaxation times of $\tau \sim 5 \cdot 10^{-8}$ s. The absence of temperature dependence in the temperature region 150–400 K favors the carrier localization and the recombination mechanism related to electron tunneling between the localized states. Transition from electronic to hole conductivity is proven by change of Hall coefficient sign. Such a transition is inherent, for example, to half-filled conduction band of K_3C_{60}.

In single-particle approximation, neglecting electron correlations, the following spectrum has been calculated [43]: 50 of 60 p_z electrons of a neutral molecule fill all orbitals up to $L = 4$. The lowest $L = 0,1,2$ orbitals correspond to icosahedral states a_g, t_{1u}, h_g. All states with greater L values undergo the icosahedral-field splitting. There are ten electrons in partially filled $L = 5$ state. Icosahedral splitting ($L = 5 \rightarrow h_u + t_{1u} + t_{2u}$) of these 11-fold degenerate orbitals leads to the electronic configuration shown in Fig. 6.1.

Microscopic calculations and experimental data show that the completely filled highest occupied molecular orbital is of h_u symmetry, and LUMO (threefold degenerate) has t_{1u} symmetry. The HOMO–LUMO gap is caused by icosahedral perturbation in $L = 5$ shell, with experimentally found value of about 1 eV [45]. t_{1g} (LUMO+1) state formed by $L = 6$ shell is 1 eV above the t_{1u} LUMO. Different phases of the alkali fullerides are formed at changes of temperature, alkali metal concentration, or lattice structure. In particular, metallic or insulating phases occur at different fillings n of LUMO in C_{60} (n can take values from 0 to 6).

For a detailed theoretical study of electrical and magnetic properties of doped fullerides a model is to be formulated, which takes into account orbital degeneracy of energy levels, Coulomb correlation, as well as correlated hopping of electrons in narrow energy bands. The proper treatment of these interactions is important for a consistent description of a competition between on-site Coulomb correlation

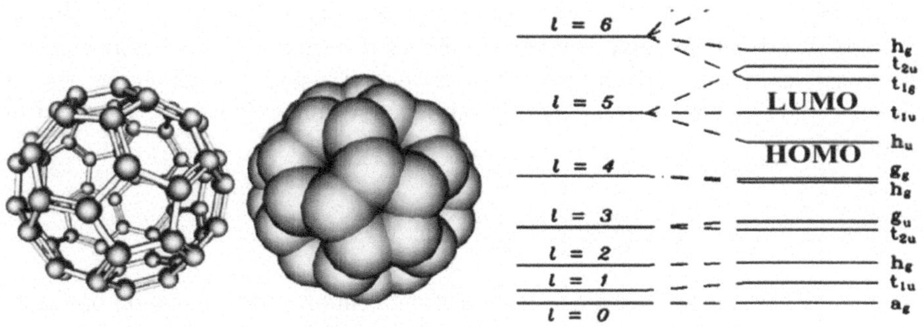

Fig. 6.1 Single-electron energy levels of fullerene C_{60} (from paper [44]). The highest occupied molecular orbital (HOMO) and the lowest unoccupied molecular orbital (LUMO) are of particular importance

(characterized by Hubbard parameter U) and delocalization processes (translational motion of electrons is determined by bare bandwidth and energy levels' degeneracy). Section 6.2 is devoted to the formulation of such model. Energy spectrum of electronic subsystem and the ground-state energy have been calculated within the Green function approach in Sect. 6.3. A generalization of the magnetization and Curie temperature calculations [46, 47] has been developed, which allows us to extend the phase diagram of the model and discuss driving forces for ferromagnetic state stabilization observed in TDAE-doped fullerides and polymerized fullerides. The competition of itinerant behavior enhanced by the external pressure application and localization due to the correlation effects is discussed.

6.2 Theoretical Model of Doped Fulleride Electronic Subsystem

Within the second quantization formalism, the Hamiltonian of interacting electrons (with spin-independent interaction $V^{ee}(r-r')$ in crystal field $V^{\text{ion}}(r)$) may be written as

$$H = H_0 + H_{int}, \tag{6.1}$$
$$H_0 = \sum_{\sigma} \int d^3 r c_{\sigma}^{+}(r) \left[-\frac{\hbar}{2m}\Delta + V^{\text{ion}}(r) \right] c_{\sigma}(r),$$
$$H_{int} = \sum_{\sigma,\sigma'} \int d^3 r \int d^3 r' V^{ee}(r-r') n_{\sigma}(r) n_{\sigma'}(r').$$

Here $c_{\sigma}^{+}(r)$, $c_{\sigma}(r)$ are field operators of electron with spin σ creation and annihilation, respectively, $n_{\sigma}(r) = c_{\sigma}^{+}(r)c_{\sigma}(r)$. Interaction term is diagonal with respect to spatial coordinates r, r'; therefore, it depends only on the electron fillings of the sites interacting, with energy $V^{ee}(r-r')$. The lattice potential $V^{ion}(r)$ causes the splitting of initial band into multiple sub-bands numbered by index λ. For noninteracting electrons description, Bloch wave functions $\psi_{\lambda k}(r)$ and band energies $\epsilon_{\lambda k}$ of corresponding states are used. Let us introduce Wannier functions, localized at position R_i:

$$\varphi_{\lambda i}(r) = \frac{1}{\sqrt{N}} \sum_k e^{-ikR_i} \psi_{\lambda k},$$

where N is the number of lattice sites. Electron creation and annihilation operators $a^+_{i\lambda\sigma}$, $a_{i\lambda\sigma}$ on lattice site i in band λ can be introduced:

$$a^+_{i\lambda\sigma} = \int d^3r \varphi_{\lambda i}(r) \widehat{c}^+_\sigma(r),$$
$$a_{i\lambda\sigma} = \int d^3r \varphi_{\lambda i}(r) \widehat{c}_\sigma(r),$$

with the inverse transform

$$\widehat{c}^+_\sigma(r) = \sum_{i\sigma} \varphi^*_{\lambda i}(r) a^+_{i\lambda\sigma},$$
$$\widehat{c}_\sigma(r) = \sum_{i\sigma} \varphi^*_{\lambda i}(r) a_{i\lambda\sigma}.$$

In this way a general Hamiltonian can be rewritten in Wannier (site) representation as

$$H = \sum_{\lambda ij\sigma} t_{ij\lambda} a^+_{i\lambda\sigma} a_{i\lambda\sigma} + \frac{1}{2} \sum_{\alpha\beta\gamma\delta} \sum_{ijkl} \sum_{\sigma\sigma'} J^{\alpha\beta\gamma\delta}_{ijkl} a^+_{i\alpha\sigma} a^+_{j\beta\sigma'} a_{l\delta\sigma'} a_{k\gamma\sigma}, \tag{6.2}$$

where the matrix elements are defined by formulae

$$t_{ij\lambda} = \int d^3r \varphi^*_{\lambda i}(r-R_i) \left[-\frac{\hbar}{2m} \Delta + V^{ion}(r) \right] \varphi_{\lambda j}(r-R_j), \tag{6.3}$$

$$J^{\alpha\beta\gamma\delta}_{ijkl} = \int d^3r \int d^3r' V^{ee}(r-r') \phi^*_{i\alpha}(r-R_i) \phi^*_{j\beta}(r'-R_j) \phi_{l\delta}(r'-R_l) \phi_{k\gamma}(r-R_k) \tag{6.4}$$

Note that the Hamiltonian in Wannier representation is nondiagonal, which is essential feature of strongly correlated electron systems and requires specific theoretical approaches for energy spectrum calculation. By analogy to the nondegenerate model [48, 49], we obtain the following Hubbard-type Hamiltonian for orbitally degenerate band with matrix elements of electron interactions describing correlated electron hoppings:

$$
\begin{aligned}
H = & -\mu \sum_{i\lambda\sigma} a^{+}_{i\lambda\sigma} a_{i\lambda\sigma} + \sum_{ij\lambda\sigma}{}' a^{+}_{i\lambda\sigma} \Big(t_{ij} + \sum_{k'\lambda'} J(i\lambda k'\lambda' j\lambda k'\lambda') n_{k'\lambda'} \Big) a_{j\lambda\sigma} + \\
& + U \sum_{i\lambda} n_{i\lambda\uparrow} n_{i\lambda\downarrow} + \frac{U'}{2} \sum_{i\sigma\lambda\lambda'} n_{i\lambda\sigma} n_{i\lambda'\bar{\sigma}} + \frac{(U' - J_0)}{2} \sum_{i\sigma\lambda\lambda'} n_{i\lambda\sigma} n_{i\lambda'\sigma} + \\
& + \frac{1}{2} J \sum_{ij\lambda\lambda'\sigma\sigma'} a^{+}_{i\lambda\sigma} a^{+}_{j\lambda'\sigma'} a_{i\lambda\sigma'} a_{j\lambda'\sigma},
\end{aligned}
\tag{6.5}
$$

where μ is chemical potential, $n_{i\lambda\sigma} = a^{+}_{i\lambda\sigma} a_{i\lambda\sigma}$ is number operator of electrons of spin σ in orbital λ of site i, $\bar{\sigma}$ denotes spin projection opposite to σ, $n_{i\lambda} = n_{i\lambda\uparrow} n_{i\lambda\downarrow}$; t_{ij} is electron hopping integral from site j to site i (interorbital hoppings are neglected),

$$
J(i\lambda k\lambda' j\lambda k\lambda') = \int\int \varphi^{*}_{\lambda}(r - R_i)\varphi_{\lambda}(r - R_j)\frac{e^2}{|r - r'|}|\varphi_{\lambda'}(r' - R_k)|^2 dr dr' \tag{6.6}
$$

(ϕ_{λ} are Wannier functions),

$$
U = \int\int |\varphi_{\lambda}(r - R_i)|^2 \frac{e^2}{|r - r'|}|\varphi_{\lambda}(r' - R_i)|^2 dr dr' \tag{6.7}
$$

is on-site Coulomb correlation, assumed to have the same magnitude for all orbitals,

$$
J_0 = \int\int \varphi^{*}_{\lambda}(r - R_i)\varphi_{\lambda'}(r - R_i)\frac{e^2}{|r - r'|}\varphi^{*}_{\lambda'}(r' - R_i)\varphi_{\lambda}(r' - R_i) dr dr' \tag{6.8}
$$

is on-site Hund's rule exchange integral, which stabilizes states $|\lambda \uparrow \lambda' \uparrow\rangle$ and $|\lambda \downarrow \lambda' \downarrow\rangle$, forming the atomic moments. Values of U, U', and J_0 are related by condition [50]

$$
U' = U - 2J_0.
$$

Intersite exchange coupling is parameterized as

$$
J = J(i\lambda j\lambda' j\lambda i\lambda') = \int\int \varphi^{*}_{\lambda}(r - R_i)\varphi_{\lambda}(r - R_j)\frac{e^2}{|r - r'|}\varphi^{*}_{\lambda}(r - R_j)\varphi_{\lambda}(r - R_i) dr dr'. \tag{6.9}
$$

Intra-site Coulomb repulsion U and intersite exchange J are two principal energy parameters of the model. In fullerides, the competition between the Coulomb repulsion and delocalization processes (translation motion of electrons) determines the metallic or insulating state realization [51].

For fullerides, magnitude of U may be estimated from different methods. Within the local density approximation the Coulomb repulsion of 3.0 eV was obtained [52, 53]. From experimental data of paper [54] based on the electron affinity to ion C^{-}_{60} the value 2.7 eV has been obtained. In solid state closely spaced C_{60} molecules cause screening, which leads to the repulsion energy reduction to 0.8–1.3 eV [52, 53]. Auger spectroscopy and photo-emission spectroscopy gave values in the

1.4–1.6 eV range [44, 55]. It is worthwhile to note that electrons of the same spin projection spend less energy to sit on the same site than those of antiparallel spins; thus, orbitally degenerated levels are filled in accordance to Hund's rule. Experiments [55] give for singlet-triplet splitting a value of 0.2 eV $\pm$ 0.1 eV; in paper [45], U_2 is taken to be 0.05 eV.

We reduce the term $\sum'_{ijk\lambda\lambda'\sigma} J(i\lambda k\lambda' j\lambda k\lambda')\, a^{+}_{i\lambda\sigma} n_{k\lambda'} a_{j\lambda'\sigma}$ in Hamiltonian (6.5) to

$$\sum_{ij\lambda\sigma}{}'\left(J(i\lambda i\lambda j\lambda i\lambda)a^{+}_{i\lambda\sigma}a_{j\lambda\sigma}n_{i\bar{\sigma}} + h.c.\right) + \sum_{ij\lambda\bar{\lambda}\sigma}{}'\left(J(i\lambda i\bar{\lambda} j\lambda i\bar{\lambda})a^{+}_{i\lambda\sigma}a_{j\lambda\sigma}n_{i\bar{\lambda}} + h.c.\right)$$
$$+ \sum_{ij\lambda\lambda'\sigma}{}' \sum_{\substack{k\neq i\\ k\neq j}} \left(J(i\lambda k\lambda j\lambda k\lambda')a^{+}_{i\lambda\sigma}a_{j\lambda\sigma}n_{k\lambda'}\right)$$

(here $\bar{\lambda}$ denotes the orbital other than λ). The first and the third sums in this expression generalize the correlated hopping, introduced for nondegenerate model (see, e.g., [49]). The second sum describes the correlated hopping type, which is a peculiarity of orbitally degenerated systems. Among such processes one can distinguish three distinct types of hopping, of which the first and the second are influenced by the occupancies of sites between which hopping takes place and the third one depends on neighboring sites' filling. The latter can be taken into account in a mean-field type approximation:

$$\sum_{ij\lambda\lambda'\sigma}{}' \sum_{\substack{k\neq i\\ k\neq j}} J(i\lambda k\lambda' j\lambda' k\lambda')a^{+}_{i\lambda\sigma}a_{j\lambda\sigma}n_{k\lambda'} \cong n \sum_{ij\lambda\sigma}{}' T_1(ij)a^{+}_{i\lambda\sigma}a_{j\lambda\sigma},$$

where $n = \langle n_{i\alpha} + n_{i\beta} + n_{i\gamma} \rangle$ is mean number of electrons per site,

$$T_1(ij) = \sum_{\substack{k\neq i\\ k\neq j}} J(i\lambda k\lambda' j\lambda k\lambda')$$

and we assume that $J(i\lambda\, k\alpha\, j\lambda\, k\alpha) = J(i\lambda\, k\beta\, j\lambda\, k\beta) = J(i\lambda\, k\gamma\, j\lambda\, k\gamma)$ and $T_1(ij)$ is the same for all orbitals. If α-, β-, and γ-orbital states are equivalent, one can take:

$$J(i\lambda i\bar{\lambda}\, j\lambda i\bar{\lambda}) = t'_{\alpha\alpha}(ij) = t'_{\beta\beta}(ij) = t'_{\gamma\gamma}(ij) = t'_{ij},$$
$$J(i\lambda i\lambda\, j\lambda i\lambda) = t''_{\alpha\alpha}(ij) = t''_{\beta\beta}(ij) = t''_{\gamma\gamma}(ij) = t''_{ij}.$$

The Hamiltonian takes its final form

$$
\begin{aligned}
H = & -\mu \sum_{i\lambda\sigma} a^{+}_{i\lambda\sigma} a_{i\lambda\sigma} + U \sum_{i\lambda} n_{i\lambda\uparrow} n_{i\lambda\downarrow} + \frac{U'}{2} \sum_{i\lambda\sigma} n_{i\lambda\sigma} n_{i\lambda'\bar{\sigma}} + \\
& + \frac{(U' - J_0)}{2} \sum_{i\lambda\lambda'\sigma} n_{i\lambda\sigma} n_{i\lambda'\sigma} + \sum_{ij\lambda\sigma}{}' t_{ij}(n) a^{+}_{i\lambda\sigma} a_{j\lambda\sigma} + \\
& + \sum_{ij\lambda\bar{\lambda}\sigma}{}' \left(t'_{i\,j} a^{+}_{i\lambda\sigma} a_{j\lambda\sigma} n_{i\bar{\lambda}} + h.c. \right) + \sum_{ij\lambda\sigma}{}' \left(t''_{i\,j} a^{+}_{i\lambda\sigma} a_{j\lambda\sigma} n_{i\lambda\bar{\sigma}} + h.c. \right) \\
& + \frac{1}{2} \sum_{ij\lambda\sigma\sigma'} J(ij) a^{+}_{i\lambda\sigma} a^{+}_{j\lambda\sigma'} a_{i\lambda\sigma'} a_{j\lambda\sigma},
\end{aligned}
\tag{6.10}
$$

with effective concentration-dependent hopping integral $t_{ij}(n) = t_{ij} + nT_1(ij)$. Estimations of bare half bandwidth ($w = z|t_{pj}|$, z being the number of nearest neighbors to a site) are 0.5–0.6 eV from data of paper [17] and 0.6 eV from paper [45].

6.3 Results and Discussion

For calculation of single-particle electron spectrum we apply the Green function method. The equation for single-particle Green function is

$$
E\left\langle\left\langle a_{p\alpha\uparrow} | a^{+}_{p'\alpha\uparrow} \right\rangle\right\rangle = \frac{\delta_{pp'}}{2\pi} + \left\langle\left\langle \left[a_{p\alpha\uparrow}, H \right] | a^{+}_{p'\alpha\uparrow} \right\rangle\right\rangle. \tag{6.11}
$$

Taking into account the Hamiltonian (6.10) structure, in the commutator from the above equation we approximate nondiagonal terms in a mean-field manner as

$$
\begin{aligned}
& \sum_{ij\lambda\sigma\lambda'} t'_{ij}(n) \left(\left[a_{p\alpha\uparrow}, a^{+}_{i\lambda\sigma} a_{j\lambda\sigma} n_{i\lambda'} \right] + \left[a_{p\alpha\uparrow}, a^{+}_{i\lambda\sigma} a_{j\lambda\sigma} n_{j\lambda'} \right] \right) = \\
& = \sum_{j} t'_{pj} \left(\langle n_{p\beta} \rangle + \langle n_{p\gamma} \rangle \right) a_{j\alpha\uparrow} + \sum_{j\sigma} t'_{pj} \left(\left\langle a^{+}_{p\beta\sigma} a_{j\beta\sigma} \right\rangle + \left\langle a^{+}_{p\gamma\sigma} a_{j\gamma\sigma} \right\rangle \right) a_{p\alpha\uparrow} + \\
& + \sum_{j} t'_{pj} \left(\langle n_{j\beta} \rangle + \langle n_{j\gamma} \rangle \right) a_{j\alpha\uparrow} + \sum_{j\sigma} t'_{pj} \left(\left\langle a^{+}_{j\beta\sigma} a_{p\beta\sigma} \right\rangle + \left\langle a^{+}_{j\gamma\sigma} a_{p\gamma\sigma} \right\rangle \right) a_{p\alpha\uparrow}.
\end{aligned}
\tag{6.12}
$$

In a general case, averaged value of electron number operator depends on orbital $\langle n_{i\lambda} \rangle = n_{\lambda}$,

$$
\beta'_{\lambda} = \sum_{j\sigma,\, \lambda' \neq \lambda} t'_{pj} \left\langle a^{+}_{p\lambda'\sigma} a_{j\lambda'\sigma} + h.c. \right\rangle. \tag{6.13}
$$

According to the terminology of work [56], we classify the quantity β'_{λ} as orbital-dependent shift of the band center.

Analogously, we process terms

$$\sum_{ij\lambda\sigma} t''_{ij}\left(\left[a_{p\alpha\uparrow}, a^{+}_{i\lambda\sigma}a_{j\lambda\sigma}n_{i\lambda\bar{\sigma}}\right] + \left[a_{p\alpha\uparrow}, a^{+}_{i\lambda\sigma}a_{j\lambda\sigma}n_{j\lambda\bar{\sigma}}\right]\right) = \\ = 2\left(\sum_{j} t''_{pj}\langle n_{p\alpha\downarrow}\rangle a_{j\alpha\uparrow} + \sum_{j} t''_{pj}\left\langle a^{+}_{p\alpha\downarrow}a_{j\alpha\downarrow}\right\rangle a_{p\alpha\uparrow}\right) \tag{6.14}$$

and introduce notations $n_{\lambda\sigma} = \langle n_{j\lambda\sigma}\rangle$ and

$$\beta''_{\lambda\sigma} = 2\sum_{j} t''_{pj}\left\langle a^{+}_{p\lambda\bar{\sigma}}a_{j\lambda\bar{\sigma}}\right\rangle. \tag{6.15}$$

According to the terminology of paper [56], we classify the quantity $\beta''_{\lambda\sigma}$ as spin-dependent shift of the sub-band center.

Then the commutator (6.14) can be represented in the form

$$2n_{\lambda\sigma}\sum_{j} t''_{pj}a_{j\alpha\uparrow} + \beta''_{\lambda\sigma}.$$

For the exchange interaction we have

$$\frac{1}{2}J\sum_{ij\lambda\lambda'\sigma\sigma'}\left[a^{+}_{p\alpha\uparrow}, a^{+}_{i\lambda\sigma}a^{+}_{j\lambda'\sigma'}a_{i\lambda\sigma'}a_{j\lambda'\sigma}\right] = J\sum_{j\lambda'\sigma'} a^{+}_{j\lambda'\sigma'}a_{p\alpha\sigma'}a_{j\lambda'\uparrow}. \tag{6.16}$$

Averaging this expression, one has to take into account that $\left\langle a^{+}_{j\lambda\sigma}a_{p\bar{\lambda}\sigma}\right\rangle = 0$ because the electron transfer between different orbitals is excluded $\left\langle a^{+}_{j\lambda\sigma}a_{j\lambda\bar{\sigma}}\right\rangle = \left\langle S^{\pm}_{j\lambda}\right\rangle = 0$, then we have

$$J\sum_{\sigma}\left\langle a^{+}_{j\alpha\sigma}a_{p\alpha\sigma}\right\rangle\sum_{j} a_{j\alpha\uparrow} - zJ\left(n_{\alpha\uparrow} + n_{\beta\uparrow} + n_{\gamma\uparrow}\right)a_{p\alpha\uparrow}.$$

Other commutators in Eq. (6.11) are trivial. Hence, the equation of motion in the mean-field approximation takes the form

$$\left\{E + \mu - \beta'_{\alpha} - \beta''_{\alpha\uparrow} - Un_{\alpha\downarrow} - 2U'n_{\beta\downarrow} - 2(U' - J_0)n_{\beta\uparrow} + zJ\left(n_{\alpha\uparrow} + n_{\beta\uparrow} + n_{\gamma\uparrow}\right)\right\}\left\langle\left\langle a_{p\alpha\uparrow}|a^{+}_{p'\alpha\uparrow}\right\rangle\right\rangle = \\ = \frac{\delta_{pp'}}{2\pi} + \sum_{j}\left\{t_{pj}(n) - 4t'_{pj}n_{\bar{\lambda}} - 2t''_{pj}n_{\lambda\downarrow} + J\sum_{\sigma}\left\langle a^{+}_{j\alpha\sigma}a_{p\alpha\sigma}\right\rangle\right\}\left\langle\left\langle a_{j\alpha\uparrow}|a^{+}_{p'\alpha\uparrow}\right\rangle\right\rangle. \tag{6.17}$$

Let us introduce notations

$$\widetilde{\mu}_{\lambda\sigma} = \mu - \beta'_{\lambda} - \beta''_{\lambda\sigma} - Un_{\lambda\bar{\sigma}} - 2U'n_{\bar{\lambda}\bar{\sigma}} - 2(U' - J_0)n_{\bar{\lambda}\sigma} + zJ\sum_{\lambda} n_{\lambda\sigma}, \tag{6.18}$$

for the renormalized chemical potential and

$$\alpha^{\lambda\sigma} = 1 - \tau_1 n - 4\tau' n_{\bar{\lambda}} - 2\tau'' n_{\lambda\bar{\sigma}} - \frac{zJ}{w} \sum_{j\sigma} \left\langle a^{+}_{j\alpha\sigma} a_{p\alpha\sigma} \right\rangle. \tag{6.19}$$

for the correlation band narrowing factor.

Here, dimensionless parameters of correlated hopping $\tau' = t'_{pj}/|t_{pj}|$; $\tau'' = t''_{pj}/|t_{pj}|$ are introduced. In absence of the correlated hopping $\tau_1 = 0$, $\tau' = \tau'' = 0$.

After the Fourier-transform, we obtain for the Green function

$$\left\langle\left\langle a_{p\lambda\sigma} | a^{+}_{p'\lambda\sigma} \right\rangle\right\rangle_{\vec{k}} = \frac{1}{2\pi} \frac{1}{E - E^{\lambda\sigma}(\vec{k})}, \tag{6.20}$$

where the energy spectrum is

$$E^{\lambda\sigma}(\vec{k}) = -\widetilde{\mu}_{\lambda\sigma} + \alpha^{\lambda\sigma} t(\vec{k}). \tag{6.21}$$

The chemical potential is to be calculated from the equation

$$n_{\lambda\sigma} = \frac{1}{N} \sum_{\vec{k}} \int_{-\infty}^{\infty} J_{\vec{k}}^{\lambda\sigma}(E) dE, \tag{6.22}$$

where the spectral density of the Green function is defined by the expression

$$J_{\vec{k}}^{\lambda\sigma} = \delta\left(E - E^{\lambda\sigma}(\vec{k})\right) \left[1 + \exp((E - \widetilde{\mu}_{\lambda\sigma})/\theta)\right]^{-1}. \tag{6.23}$$

where $\theta = kT$. In case of arbitrary density of electronic states $\rho(\varepsilon)$ for nonzero temperatures one has

$$n_{\lambda\sigma} = \int_{-w}^{w} \rho(\varepsilon) \left[1 + \exp\left(\left(E^{\lambda\sigma}(\vec{k}) - \widetilde{\mu}_{\lambda\sigma}\right)/\theta\right)\right]^{-1} d\varepsilon. \tag{6.24}$$

The ground-state energy can be calculated by the method of work in [57] as

$$\begin{aligned} E_0 &= \frac{1}{2N} \sum_{\vec{k}\lambda\sigma} \int_{-\infty}^{\infty} \left(t_{\vec{k}}(n) + E^{\lambda\sigma}(\vec{k})\right) J_{\vec{k}}^{\lambda\sigma}(E) dE = \\ &= \frac{1}{2N} \sum_{\vec{k}\lambda\sigma} \left(t_{\vec{k}}(n) + E^{\lambda\sigma}(\vec{k})\right) \left[1 + \exp\left(\left(E^{\lambda\sigma}(\vec{k}) - \widetilde{\mu}_{\lambda\sigma}\right)/\theta\right)\right]^{-1}. \end{aligned} \tag{6.25}$$

For arbitrary density of states (at $T = 0$ K) one has

$$E_0 = \frac{1}{2}\sum_{\lambda\sigma}\int\limits_{-w}^{\varepsilon^{\lambda\sigma}} \rho(\varepsilon)\big((1-\tau_1 n)\varepsilon + E^{\lambda\sigma}(\varepsilon)\big)d\varepsilon. \tag{6.26}$$

Single-electron states occupation numbers are related by the constraints

$$\sum_{\lambda\sigma} n_{\lambda\sigma} = n, \quad \sum_{\lambda}\left(n_{\lambda\uparrow} - n_{\lambda\downarrow}\right) = m,$$

in the case of no orbital ordering, we have $n_{\lambda\uparrow} = (n + m)/3$, $n_{\lambda\downarrow} = (n - m)/3$.

In the case of rectangular density of states, the ground-state energy can be calculated analytically:

$$E_0 = -\sum_{\lambda\sigma}\left\{(1-\tau_1 n + \alpha^{\lambda\sigma})(1-n_{\lambda\sigma})n_{\lambda\sigma}\frac{w}{2} + \widetilde{\mu}_{\lambda\sigma}\frac{n_{\lambda\sigma}}{2}\right\}, \tag{6.27}$$

where

$$\widetilde{\mu}_{\lambda\sigma} = \mu - \beta'_\lambda - \beta''_{\lambda\sigma} - (U+2U')\frac{n-\eta_\sigma m}{3} - 2(U'-J_0)\frac{n+\eta_\sigma m}{3} + zJ(n+\eta_\sigma m),$$
$$\alpha^{\lambda\sigma} = 1 - \tau_1 n - \frac{8}{3}\tau' n - \frac{2}{3}\tau''(n-\eta_\sigma m) - \frac{2}{3}\frac{zJ}{w}\left(n - \frac{1}{3}(n^2+m^2)\right),$$

$\eta_\sigma = 1$ for spin-up electrons and -1 otherwise. In these calculations we take into account that

$$\frac{zJ}{w}\sum_{j\sigma}\left\langle a^+_{j\alpha\sigma}a_{p\alpha\sigma}\right\rangle = -\frac{zJ}{w}\sum_{j\sigma}\frac{t_{pj}}{|t_{pj}|}\left\langle a^+_{j\alpha\sigma}a_{p\alpha\sigma}\right\rangle = \frac{zJ}{w}\sum_{\sigma} n_{\alpha\sigma}(1-n_{\lambda\sigma}) \tag{6.28}$$

and

$$\beta'_\lambda = 4\sum_{j\sigma} t'_{pj}\left\langle a^+_{j\bar{\lambda}\sigma}a_{p\bar{\lambda}\sigma}\right\rangle = -4\tau'\sum_{\sigma}\int\limits_{-w}^{w}\rho(\varepsilon)\varepsilon d\varepsilon. \tag{6.29}$$

For the rectangular density of states

$$\beta'_\lambda = -\frac{8}{3}\tau' w\left(n - \frac{1}{3}(n^2+m^2)\right). \tag{6.30}$$

Analogously,

$$\beta''_{\lambda\uparrow} = \frac{2}{9}\tau'' w\big(n(3-n) - \eta_\sigma m(3-2n) - m^2\big). \tag{6.31}$$

Thus, the model is parameterized and dependences of the ground-state energy of the considered model E_0 on energy parameters U, J_H, zJ, correlated hopping parameters $\tau_1, \tau^{'}, \tau^{''}$, electron concentration n, and magnetization m can be studied numerically. Using the expression (6.27), one can obtain the equilibrium value of system magnetization m_{GS}, which is a zeroth-order approximation in magnetization calculation at nonzero temperature. In case of the second-order transition, one obtains equation for the Curie temperature at arbitrary density of states

$$\theta_c = 6 \int_{-w}^{w} \rho(\varepsilon)(A + B\varepsilon)\exp(E^*/\theta_c)(\exp(E^*/\theta_c) + 1)^{-2} d\varepsilon, \tag{6.32}$$

where E^* is paramagnetic spectrum. The chemical potential is determined by the condition

$$n = 3 \int_{-w}^{w} \rho(\varepsilon)(\exp(E^*/\theta_c) + 1)^{-1} d\varepsilon. \tag{6.33}$$

Equations (6.32) and (6.33) generalize corresponding results for nondegenerate case [47] on triple orbital degeneracy of energy levels and allow modeling of the Curie temperature at various densities of electronic states in a wide range of the model energy parameters values for electron concentrations $0 < n < 3$ (which corresponds to doped fullerides A_nC_{60} and TDAE-C_{60}). Behavior of Curie temperature appears to be closely related to the ground-state magnetization concentrational dependence. The correlated hopping reduces the obtained values of the Curie temperature considerably.

In Fig. 6.2, the concentration dependence of Curie temperature is shown in different scenarios, corresponding to different acting mechanisms of correlated hopping of electrons, both second-type correlated hopping parameters considered to have the same magnitude $\tau_2 = \tau^{'} = \tau^{''}$. One can see that the correlated hopping, which is known [49] to suppress conductance, enhances ferromagnetic tendencies greatly. Applicability of the particular scenario to a given fulleride requires further studies and will be considered elsewhere. On a qualitative level, taking into account the correlated hopping allows to obtain reasonable estimates for Curie temperature within the considered model of triply degenerate band with intersite exchange small enough to be characteristic for polymerized fullerenes [58] and αTDAE-C_{60} [59]. We note that the system remains semiconducting at Coulomb interaction energies greater than the bandwidth at integer average occupation number of a site in this model [19].

Thus, both ferromagnetic ordering and semiconducting behavior [21] can be observed. Simultaneous taking into account of triple orbital degeneracy of energy levels and correlated hopping of electrons makes standard methods of theoretical treatment complicated to apply, though it gives a clue for description of

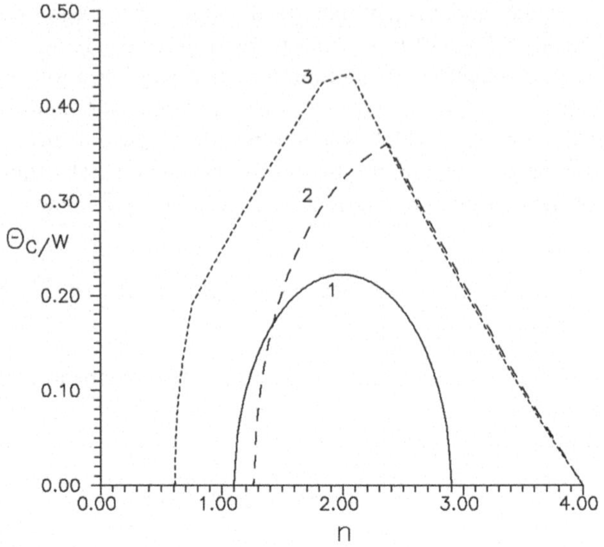

Fig. 6.2 Curie temperature dependence of band filling at $U/w = 1.2$, $J_0/w = 0/35$. Curve 1 corresponds to $zJ/w = 0.25$ and $\tau_1 = \tau_2 = 0$; 2: $zJ/w = 0.1$ and $\tau_1 = \tau_2 = 0.05$; 3: $zJ/w = 0.1$; and $\tau_1 = 0.05$, $\tau_2 = 0.07$

ferromagnetic behavior of doped fullerides. At the same time, the approach used in this investigation has its own shortcomings [47], namely, it overestimates numerical values for Curie temperature. This is rewarded by the natural description of the electron–hole asymmetry, which is clearly seen from curves 2 and 3 in Fig. 6.2. As expected, Hund's rule coupling affects the Curie temperature considerably (see Fig. 6.3); however, in the absence of the correlated hopping, the values of J_0/w required for a ferromagnetic solution are too large to be the case of the doped fullerides.

In Fig. 6.3, there is a region of sharp critical temperature increase at increasing J_0/w (this region corresponds to a partial spin polarization of the system) and a region of linear proportionality between Curie temperature and J_0/w (this region corresponds to saturated ferromagnetic state). The competition of itinerant behavior and localization due to the correlation effects can be enhanced by the external pressure application renormalizing the effective band width. This effect is particularly important near the critical value of J_0/w parameter. In our opinion, based on estimations of papers [17, 45, 55], curve 3 can be used as a reasonable model for the pressure-driven transition to ferromagnetic state.

From Fig. 6.4 one can see that the change of electron concentration qualitatively changes tendencies in the present model, which is yet another example of electron–hole asymmetry. In the present approach, the difference between ferromagnetic order stabilization by the correlated hopping for scenarios 1,2 and destabilization for 3,4,5 can be attributed to the different consequences of sub-bands' narrowing caused by the correlated hopping in energy band picture at electron concentrations $n = 1$ and $n = 3$.

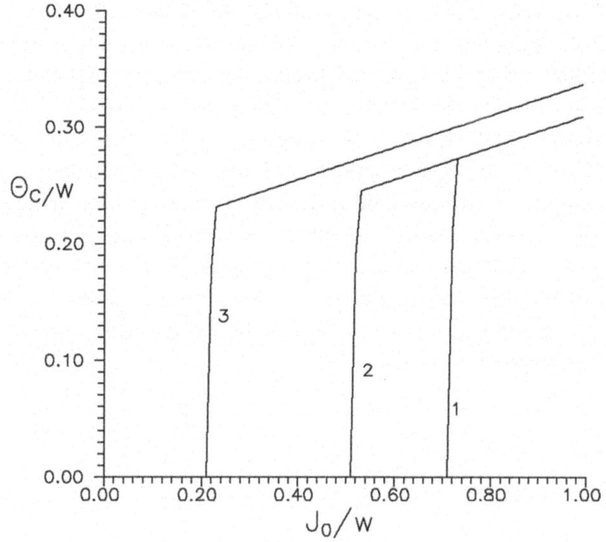

Fig. 6.3 Curie temperature dependence on J_0/w for quarter-filling of the band at $U/w = 1.2$ and $zJ/w = 0.05$. Curve 1 corresponds to the absence of correlated hopping; 2: $\tau_1 = 0.1$, $\tau_2 = 0$; 3: $\tau_1 = 0.1$, $\tau_2 = 0.1$

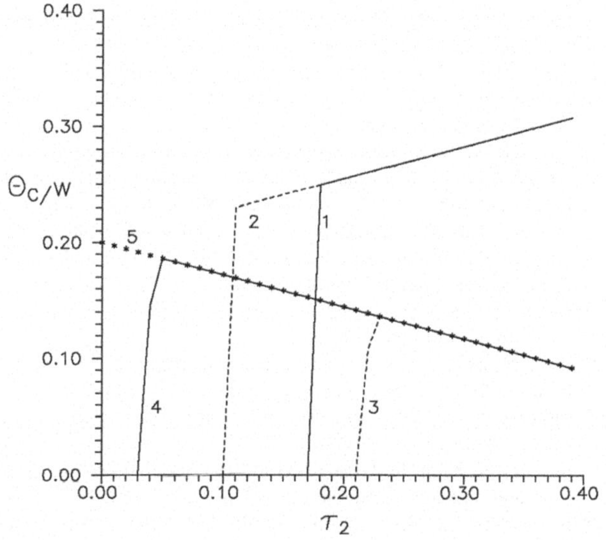

Fig. 6.4 Curie temperature dependence on the correlated hopping parameter τ_2 at $U/\mathrm{w} = 1$, $zJ/\mathrm{w} = 0.05$, $J_0/w = 0.4$. Curve 1 corresponds to $n = 1$, $\tau_1 = 0$; curve 2: $n = 1$, $\tau_1 = 0{,}1$; curve 3: $n = 3$, $\tau_1 = 0{,}05$; curve 4: $n = 3$, $\tau_1 = 0{,}08$; curve 5: $n = 3$, $\tau_1 = 0{,}1$

6.4 Conclusions

The general model for electronic subsystem of doped fullerides can describe both semiconducting behavior and magnetic order onset, if Coulomb correlation, intersite exchange interaction, correlated hopping of electrons, and orbital degeneracy of energy levels are all taken into account. In such a model, the magnetization and

Curie temperature calculations allow us to extend the model phase diagram and discuss driving forces for ferromagnetic state stabilization observed in polymerized fullerenes and tetrakis(diethylamino)ethylene-fullerene. Curie temperature dependence on integer electron concentration at realistic values of the Coulomb correlation strength and Hund's rule coupling, associated with orbital degeneracy of energy levels, is strongly asymmetrical with respect to half-filling. Hund's rule coupling stabilizes ferromagnetic ordering in quarter-filled band. There is a region of sharp critical temperature increase at increasing Hund's rule coupling parameter corresponding to a partial spin polarization of the system and a region of saturated ferromagnet state. The balance of itinerant behavior and localization due to the correlation effects can be shifted by the external pressure application renormalizing the effective band width.

References

1. Gunnarsson O (2004) Alkali-doped fullerides: narrow-band solids with unusual properties. World Scientific Publishing Co, Singapore
2. Diederich F (1991) The higher fullerenes: isolation and characterization of C_{76}, C_{84}, C_{90}, C_{94}, and $C_{70}O$, an oxide of D5h-C_{70}. Science 252:548–551
3. Jin C (1994) Direct solid-phase hydrogenation of fullerenes. J Phys Chem 98:4215–4217
4. Fischer JE, Heyney PA, Smith AB (1992) Solid-state chemistry of fullerene-based materials. Acc Chem Res 25:112–118
5. Selig GH (1991) Fluorinated fullerenes. J Am Chem Soc 113:5475–5476
6. Olah GA (1991) Chlorination and bromination of fullerenes. J Am Chem Soc 113:9385–9387
7. Lundgren MP, Khan S, Baytak AK, Khan A (2016) Fullerene-benzene purple and yellow clusters: theoretical and experimental studies. J Mol Struct 1123:75–79
8. Haddon RC, Hebard AF, Rosseinsky MJ (1991) Conducting films of C_{60} and C_{70} by alkali-metal doping. Nature 350:320–322
9. Hebard AF, Rosseinsky MJ, Haddon RC (1991) Superconductivity at 18 K in potassium-doped C_{60}. Nature 350:600–601
10. Zhou Q, Zhu JE (1992) Fischer compressibility of M_3C_{60} fullerene superconductors: relation between Tc and lattice parameter. Science 255:833–835
11. Fleming RM (1992) Relation of structure and superconducting transition temperatures in A_3C_{60}. Nature 352:787–788
12. Holczer K, Klein O, Huang SM (1991) Alkali-fulleride superconductors: synthesis, composition, and diamagnetic shielding. Science 252:1154–1157
13. Rosseinsky MJ, Ramirez AP, Glarum SH (1991) Superconductivity at 28 K in Rb_xC_{60}. Phys Rev Lett 66:2830–2832
14. Varma CM, Zaanen J, Raghavachari K (1991) Superconductivity in the fullerenes. Science 254:989–992
15. Zhang FC, Ogata M, Rice TM (1991) Attractive interaction and superconductivity for K_3C_{60}. Phys Rev Lett 67:3452–3455
16. Chakravarty S, Gelfand MP, Kivelson S (1991) Electronic correlation effects and superconductivity in doped fullerenes. Science 254:970–974
17. Manini N, Tosatti E (2006) Theoretical aspects of highly correlated fullerides: metal-insulator transition. Cornell University Library. Available via arxiv.org. https://arxiv.org/pdf/cond-mat/0602134.pdf

18. Jian Ping L (1994) Metal-insulator transitions in degenerate Hubbard models and A_xC_{60}. Phys Rev B 49:5687–5690
19. Yu D et al (2012) Mott-Hubbard localization in a model of the electronic subsystem of doped fullerides. Ukr J Phys 57:920–928
20. Ghosh S, Tongay S, Hebard AF, Sahin H, Peeters FM (2014) Ferromagnetism in stacked bilayers of Pd/C_{60}. J Magn Magn Mater 349:128–134
21. Makarova T (2003) Magnetism of carbon-based materials. In: Narlikar A (ed) Studies of high-Tc superconductivity, vol 45. Nova Science Publishers, New York, pp 107–169
22. Allemand P-M et al (1991) Organic molecular soft ferromagnetism in a fullerene C_{60}. Science 253:301–303
23. Schilder A, Bietsch W, Schwoerer M (1999) The role of TDAE for magnetism in [TDAE]C_{60}. New J Phys 1:5(1–11)
24. Wudl F (1992) The chemical properties of buckminsterfullerene (C_{60}) and the birth and infancy of fulleroids. Acc Chem Res 25:157–161
25. Wood RA et al (2002) Ferromagnetic fullerene. J Phys Condens Matter 14:L385–L391
26. Blundell SJ (2002) Magnetism in polymeric fullerenes: a new route to organic magnetism. J Phys Condens Matter 14:V1–V3
27. Takahashi N et al (1993) Plasma-polymerized C_{60}/C_{70} mixture films: electric conductivity and structure. J Appl Phys 74:5790–5798
28. Krätschmer W, Lamb LD, Fostiropoulos K, Huffman DR (1990) Solid C_{60}: a new form of carbon. Nature 347:354–358
29. Bethune DS et al (1990) The vibrational Raman spectra of purified solid films of C_{60} and C_{70}. Chem Phys Let 179:219–222
30. Meijer G, Bethune DS (1990) Laser deposition of carbon clusters on surfaces: a new approach to the study of fullerenes. J Chem Phys 93:7800–7802
31. Vaughan GBM, Heiey PA, Luzzi DE (1991) Orientational disorder in solvent-free solid C_{70}. Science 254:1350–1353
32. Chen T et al (1992) Scanning-tunneling-microscopy and spectroscopy studies of C_{70} thin films on gold substrates. Phys Rev B 45:14411–14414
33. Ajie H, Alvarez MM, Anz SJ (1990) Characterization of the soluble all-carbon molecules C_{60} and C_{70}. J Phys Chem 94:8630–8633
34. Saito S, Oshyama A (1991) Cohesive mechanism and energy bands of solid C_{60}. Phys Rev Lett 66:2637–2640
35. Achiba Y (1991) Visible, UV, and VUV absorption spectra of C_{60} thin films grown by the molecular-beam epitaxy (MBE) technique. Chem Lett 20:1233–1236
36. Wang XQ et al (1993) Structural and electronic properties of large fullerenes. Z Phys D Suppl 26:264–266
37. Kikuchi K (1991) Separation, detection, and UV/visible absorption spectra of fullerenes C_{76}, C_{78} and C_{84}. Chem Lett 9:1607–1610
38. Kuzuo R, Terauchi M, Tanaka M (1994) Electron-energy-loss spectra of crystalline C_{84}. Phys Rev B 49:5054–5057
39. Armbruster JF, Roth M, Romberg HA (1994) Electron energy-loss and photoemission studies of solid C_{84}. Phys Rev B 50:4933–4936
40. Duclos SJ et al (1991) Thiel effects of pressure and stress on C_{60} fullerite to 20 GPa. Nature 351:380–382
41. Regueiro MN (1991) Absence of a metallic phase at high pressures in C_{60}. Nature 354:289–291
42. Cheville RA, Halas NJ (1992) Time-resolved carrier relaxation in solid C_{60} thin films. Phys Rev B 45:4548–4550
43. Erwin SC (1993) Electronic structure of the alkali-intercalated fullerides, endohedral fullerenes, and metal-adsorbed fullerenes. In: Billups WE, Ciufolini MA (eds) Buckminsterfullerenes. VCH Publishers, New York, pp 217–255
44. Brühwiler PA et al (1993) Auger and photoelectron study of the Hubbard U in C_{60}, K_3C_{60} and K_6C_{60}. Phys Rev B 48:18296–18299

45. Martin RL, Ritchie JP (1993) Coulomb and exchange interactions in $C_{60}{}^{n-}$. Phys Rev B 48:4845–4849
46. Didukh L, Kramar O (2002) Metallic ferromagnetism in a generalized Hubbard model. Low Temp Phys 28:30–36
47. Didukh L, Kramar O, Skorenkyy Yu (2005) Metallic ferromagnetism in a generalized Hubbard model. In: Murray VN (ed) New developments in ferromagnetism research. Nova Science Publishers Inc, New York, pp 39–80
48. Skorenkyy Y et al (2007) Mott transition, ferromagnetism and conductivity in the generalized Hubbard model. Acta Phys Pol A 111:635–644
49. Didukh L, Skorenkyy Y, Kramar O (2008) Electron correlations in narrow energy bands: modified polar model approach. Condens Matter Phys 11:443–454
50. Lacroix-Lyon-Caen C, Cyrot M (1977) Alloy analogy of the doubly degenerate Hubbard model. Solid State Communs 21:837–840
51. Gunnarsson O, Koch E, Martin RM (1997) Mott-Hubbard insulators for systems with orbital degeneracy. Phys Rev B 56:1146–1152
52. Pederson MR, Quong AA (1992) Polarizabilities, charge states, and vibrational modes of isolated fullerene molecules. Phys Rev B 46:13584–13591
53. Antropov VP, Gunnarsson O, Jepsen O (1992) Coulomb integrals and model Hamiltonians for C_{60}. Phys Rev B 46:13647–13650
54. Hettich RL, Compton RN, Ritchie RH (1991) Doubly charged negative ions of carbon 60. Phys Rev Lett 67:1242–1245
55. Lof RW et al (1992) Band gap, excitons and Colomb interaction in solid C_{60}. Phys Rev Lett 68:3924–3927
56. Didukh L et al (2001) Ground state ferromagnetism in a doubly orbitally degenerate model. Phys Rev B 64:144428(1–10)
57. Roth LM (1969) Electron correlation in narrow energy bands. The two–pole approximation in a narrow s band. Phys Rev 184:451–459
58. Kvyatkovskii OE et al (2005) Spintransfer mechanism of ferromagnetism in polymerized fullerenes: ab initio calculations. Phys Rev B 72:214426(1–8)
59. Forró L, Mihály L (2001) Electronic properties of doped fullerenes. Rep Prog Phys 64:649–699

Chapter 7
The Possible Mechanisms of Conductivity in Polyene-Like Polymers and Types of Conductivity in Maximally Feeble External Fields

A. D. Suprun, S. V. Vasylyuk, and V. N. Yashchuk

Linear conjugated systems are the systems of connected p-orbitals with delocalized electrons in molecular compounds with the alternation of simple and multiple bonds. In general, the linear conjugated systems, which will be discussed here, can be represented by the formula $[R_1 - (CH)_m - R_2]$, where R1 and R2 are terminal group connected to the polymethine chain. These systems can be either electric neutral and charged positively or negatively, depending on what they injected – electron or hole. Methine carbon atom of (CH) in the chain is in sp2-hybridized state. To explain the properties of some organic compounds such as quasi-metallic conductivity, a significant change in the spectral properties of ionic dyes that absorb and emit light in the near infrared region of the spectrum [1–4] managed, including through the use of the concept of solitons. It is established that injection of electrons/holes in the conduction band leads to soliton level appearance inside the restricted area and this is accompanied by marked shift of the valence band top and the conduction band bottom, when electron injected – downward energy, in the case of hole injection – in the direction of energy increasing [1–4]. In this work we use the charge transfer model based on the concept of solitons, according to Davidov A. [4] which complements and develops other models [1–4].

1. *Mishra A.* Cyanine during 1990s: A review. Chem. Rev., 2000, **100**, p.1973–2011.
2. *F.Meyers, S.R.Marder, J.W.Perry*, Introducing to Nonlinear Optical Properties Organic Materials. In Chemistry Advanced Materials. An Overreviw. Chapt. **6**. Ed. L. V. Interrante, J.Hampden-Smith. 1998. Wiley-VCH. Inc. New York-Chicherster-Weinheim-Brisbane-Singapore-Toronto P.207–268.

A. D. Suprun
Taras Shevchenko National University of Kyiv, Kyiv, Ukraine

S. V. Vasylyuk (✉) · V. N. Yashchuk
Physics Faculty, Kyiv National Taras Shevchenko University, Kyiv, Ukraine

O. Fesenko, L. Yatsenko (eds.), *Nanooptics, Nanophotonics, Nanostructures, and Their Applications*, Springer Proceedings in Physics 210,
https://doi.org/10.1007/978-3-319-91083-3_7

3. *Bredas J.L., Street G.B.* Polaron, bipolaron and solitons in conductiong polymers. Acc.Chem.Res. 1985, **18**, 309–315.
4. *Davydov A. S.* (1973). "The theory of contraction of proteins under their excitation". Journal of Theoretical Biology **38** (**3**): 559–569. doi:10.1016/0022-5193(73)90256-7. PMID 4266326.

7.1 Introduction

The possibility to reveal quasi-metal conductivity is one of the most important and specific properties of the linear conjugated systems. This fact resulted in the review of the theoretical concepts, which were based on the representations concerning behavior of π-electrons in the conjugated molecules in the part, which concerns anion and cation polymethine dyes and on cations/anions radicals of polyenes, which stimulated development of a new theoretical concept of similar systems concerning peculiarities of the distribution of the injected charge [1–4]. Besides using of the conjugate systems [1–5], they find their application in the linear and nonlinear optics. So, ionic polymethine dyes can intensively absorb light quanta in visible and near IR spectral bands [5, 6]. It is established [1, 7, 8] that such unique spectral features are caused by the specific distribution of charge in the main and excited states.

According to the modern representations, transfer from the conjugated molecule in the neutral state to its ionic state, which is caused by the injection of the electron or hole in highly molecular π-electron system, is accompanied by the appearance of so-called energy level in the forbidden zone, similar to the impurity level (in fact, it is a very narrow energy band), and significant shift of both the bottom of the conductivity zone and top of the valence zone [2, 3]. The appearance of such a level is frequently associated with the excitation of soliton states. An injected charge (the electron or hole) is not delocalized uniformly along the whole system of conjunction but can autolocalize itself, according to O. S. Davydov's terminology [4], id est., to form the charge wave of the final size (soliton). According to calculations, the length of the soliton-like comprises 15–17 carbon-carbon bonds [2–7]; however, in the general case, it is determined by the parameters of the crystal lattice or its analogue [9]. In the case of molecular ions with comparatively long polymethine chain, when the length of π-system considerably exceeds sizes of the charge wave, soliton becomes mobile and can transfer charge without the change of the total energy of the molecule. Such displacement of the charge wave, for example, in cations of the polymethine dyes, results in a certain violation of the electron configuration, which manifests itself experimentally in the substantial change of the spectral band in absorption spectra of the IR dyes [10–13].

In the first part of this work [14], the general statement of work concerning the transfer of charge, injected into the conduction zone of the polymer chain of the polyene type, was fulfilled. A response of the "crystal lattice" of the chain is taken into account and a set of equations for description of the charge transfer is obtained.

Approximations, which reduce the problem to such, which corresponds to the transfer by the mechanism of the charge soliton and its solutions for the case of the extremely feeble fields, are found. It is shown that allowance for the structural features of the “lattice” generates two modes of excitations. In this, the second part of the work, on the basis of the results of the first part, the issues, connected with the possible types of the conductivity of the polymer chain of the polyene type are analyzed. An expression for the current, caused by the injected electron, is found. Particular cases, which correspond to different types of conductivity (metallic, semiconductor) are analyzed as well. An experimental inspection concerning a possibility of realization of this excitation is proposed [14].

In the work as a whole, an idealized situation of the infinite chain is considered as a basic one. However, introduction of the terminal radical groups, which will be fulfilled in the further investigations, can result in appearance of the effective field of stresses and, as a consequence, to displacement of charge solitons in these fields under conditions of the absence of the external electric field.

The purpose of this part of the work is a study of the manifestation of the charge soliton waves in the conductivity of the polymer chain of the polyene type under conditions of the accepted approximations of the maximally feeble external fields and under ideal conditions stated above.

7.1.1 General Definition of the Current of the Injected Electron

Using the results of the first part of this work [14], let us consider now in the approximation, zero by field (this approximation in the physical sense corresponds to the external field switched off), the conductivity caused by the transfer of the injected charge. From the following formulae of the first part as [14]:

$$\left(\Omega + \frac{1}{\mu} \pm \tau_{\parallel} \nu_{\perp} \left[\frac{1}{2} + \frac{1}{\mu}\right] \equiv \varepsilon_{\pm}\right) \tag{7.1}$$

$$\left(1 \pm \tau_{\parallel} \nu_{\perp} \equiv s_{\pm}\right) \tag{7.2}$$

$$\left(\kappa_{\pm} = \frac{g\mu}{2} \cdot \frac{J_{\pm}^2}{s_{\pm}}; C_{\pm} = J_{\pm}^2 \sqrt{\frac{g\mu}{4 s_{\pm}}}; \varepsilon_{\pm} = -\frac{g^2 \mu}{8} \cdot \frac{J_{\pm}^4}{s_{\pm}}\right) \tag{7.3}$$

one can obtain:

$$\Omega_{\pm} = -\frac{1}{\mu} \mp \tau_{\parallel} \nu_{\perp} \left[\frac{1}{2} + \frac{1}{\mu}\right] - \frac{g^2 \mu}{8\left(1 \pm \tau_{\parallel} \nu_{\perp}\right)} J_{\pm}^4$$

As it was noted in the first part [14], parameter Ω is a dimensionless analogue of the frequency constant ω in the phase of a plain wave: $(\mathbf{k} \cdot \mathbf{r} - \omega t)$. In the considered zero by field approximation and even in the case of the maximally feeble fields, this parameter is constant and has a direct sense of such frequency. So, within the accuracy of the factor $\hbar$, it is simultaneously an eigenvalue of the considered state of the conductivity in the quantum-mechanical sense.

Taking into account (14.37) [14] $\left(\mu(p) \equiv \frac{1}{\cos(p)}\right)$, his relationship can be written as:

$$\Omega_{\pm} = \mp\frac{\tau_{\|}v_{\perp}}{2} - \left(1 \pm \tau_{\|}v_{\perp}\right)\cos(p) - \frac{g^2 J_{\pm}^4}{8\left(1 \pm \tau_{\|}v_{\perp}\right)\cos(p)} \tag{7.4}$$

Then, taking into account (14.48) [14] $\left(\Omega \equiv \overset{\circ}{\gamma}\right)$, for the phase factor γ, we have

$$\gamma_{\pm}(\tau) = \int\limits_{0}^{\tau} \Omega_{\pm} d\tau' \tag{7.5}$$

For the wave function $a_{in}(\tau)$, which determines distribution of the electron, injected into the conductivity zone, in the polyene chain in the zero approximation by the field, taking into account relationships (14.37) [14], (7.2), (7.3), and (7.5) and also such relationships from the first part [14], as:

$$(a_{\alpha n}(t) = A_{\alpha n}(t)\exp[i(p(t)n - \gamma(t))]) \tag{7.6}$$

$$\left(\xi_{\pm} = x - x_c^{\pm}(\tau)\right) \tag{7.7}$$

$$\left(A_0^{\pm}(\xi_{\pm}) = \pm A_1^{\pm}(\xi_{\pm}) = \Phi_{\pm}(\xi_{\pm})\right) \tag{7.8}$$

$$\left(\overset{(0)}{\Phi}_{\pm}(\xi_{\pm}) = \frac{C_{\pm}}{ch(\kappa_{\pm}\xi_{\pm})}\right) \tag{7.9}$$

we have a complete soliton solution:

$$\overset{(0)}{a}_{\pm}(\tau, x) = \sqrt{\frac{g^2 J_{\pm}^4}{4\left(1 \pm \tau_{\|}v_{\perp}\right)\cos(p)}} \cdot \frac{\exp[i(px - \gamma_{\pm}(\tau))]}{ch\left[\frac{g^2 J_{\pm}^2}{2\left(1 \pm \tau_{\|}v_{\perp}\right)\cos(p)}\left(x - x_c^{\pm}(\tau)\right)\right]}$$

Let us now consider criteria of applicability of the continually soliton model. There are two of them. The first one pertains to the applicability of the continual model and is determined by the general relationship [14]: $\left|\frac{\overset{(0)}{\Phi'}_{\pm}(\xi_{\pm})}{\overset{(0)}{\Phi}_{\pm}(\xi_{\pm})}\right| \ll 1$. Substituting here $\overset{(0)}{\Phi}_{\pm}(\xi_{\pm})$ from (7.9), it is possible to obtain this criterion in entirely simple form: $|\kappa_{\pm}| \ll 1$. An explicit form of this criterion with regard for the first (left) definition in (7.3), as well as relationships (14.37) [14] and (7.2) reduces itself to the inequity:

$$\frac{g^2 J_\pm^2}{2\left|1 \pm \tau_\| \nu_\perp\right| \cdot \left|\cos(p)\right|} \ll 1 \qquad (7.10)$$

The second criterion pertains to the applicability of the soliton model [14]. For its formulation it is necessary, first of all, to determine the effective sizes of the soliton. These sizes are determined by the general relationship:

$$\left|\frac{\overset{(0)}{\Phi}_\pm(\xi_\pm)}{\overset{(0)}{\Phi}_\pm(0)}\right| = \delta \le \frac{1}{2},$$

where the parameter δ determines an accepted "level of zero." In particular, $\delta = \frac{1}{2}$ determines "half-width" of the soliton on the level of its "half-height."

Using definition (7.9) again, it is possible to obtain:

$$\left|\xi_\pm^\delta\right| = \frac{2\left|1 \pm \tau_\| \nu_\perp\right| \cdot \left|\cos(p)\right|}{g J_\pm^2} \ln\left(\frac{1}{\delta} + \sqrt{\frac{1}{\delta^2} - 1}\right)$$
$$\ge \frac{2\left|1 \pm \tau_\| \nu_\perp\right| \cdot \left|\cos(p)\right|}{g J_\pm^2} \ln\left(2 + \sqrt{3}\right)$$

The whole sizes of the area, which is covered by the soliton excitation, id est., the effective sizes of the soliton $\Delta\xi_\pm^\delta \equiv 2\left|\xi_\pm^\delta\right|$, are equal to:

$$\Delta\xi_\pm^\delta = \frac{4\left|1 \pm \tau_\| \nu_\perp\right| \cdot \left|\cos(p)\right|}{g J_\pm^2} \ln\left(\frac{1}{\delta} + \sqrt{\frac{1}{\delta^2} - 1}\right)$$
$$\ge \frac{4\left|1 \pm \tau_\| \nu_\perp\right| \cdot \left|\cos(p)\right|}{g J_\pm^2} \ln\left(2 + \sqrt{3}\right)$$

It is clear that for application of the soliton model, the area $\Delta\xi_\pm^\delta$ should be much less than the length of the polymer chain, or, in the dimensionless representation, much be less than the number of the carbon C_2H_2 groups [14]. Designating this number by N, we obtain $\Delta\xi_\pm^\delta \ll N$.

From here, in turn, it is possible to obtain $\frac{g J_\pm^2}{2\left|1 \pm \tau_\| \nu_\perp\right| \cdot \left|\cos(p)\right|} \gg \frac{2}{N} \ln\left(\frac{1}{\delta} + \sqrt{\frac{1}{\delta^2} - 1}\right)$.

Comparing it with (7.10), both criteria can be written by one inequity:

$$\frac{2}{N}\ln\left(\frac{1}{\delta}+\sqrt{\frac{1}{\delta^2}-1}\right) \ll \frac{gJ_{\pm}^2}{2\left|1\pm\tau_{\|}v_{\perp}\right|\cdot\left|\cos(p)\right|} \ll 1$$

In this case $\delta \leq \frac{1}{2}$. To determine the current, caused by the transfer of the injected electron under conditions of the absence of the external field, let us use a general relationship for the current density (in one-dimension representation):

$$j = en\nu$$

Adapting this relationship to the case of one electron, we determine its volume density n by the relationship:

$$n = \frac{1}{\sigma R_0 N},$$

where σ is an effective area of the cross-section of the polymer chain. As the current I is connected with the current density by the relationship $I = j\sigma$, then for the current we have:

$$I = \frac{e\nu}{R_0 N} \tag{7.11}$$

The velocity ν is determined by the standard [14, 15] for the theory of solids (and condensed systems of solid-state type) dispersion relationship:

$$\nu = \frac{1}{\hbar}\frac{dE}{dk},$$

where k is the wave vector of the considered quasi particle (here – the injected electron), and E is respective energy (energy zone). As the dimensionless impulse p is expressed here as $p = kR_0$, then

$$\nu = \frac{R_0}{\hbar}\frac{dE}{dp}. \tag{7.12}$$

Dimensionless frequency, (14.2) [14] which is determined in (7.4), can, after necessary transformations, be intended to the role of the energy E in the case of $|p| \to 0$. Proceeding from the definition (14.48) [14]

$$\Omega_{\pm} = \frac{d\gamma}{d\tau}$$

and definition of the dimensionless time $\tau = v_{\|}t$ for the frequency, one can obtain $\omega_{\pm} = v_{\|}\Omega_{\pm}$, and for the energy $E_{\pm}$, we shall have $E_{\pm} = \hbar v_{\|}\Omega_{\pm}$, respectively. Thus, from (7.12) we shall have:

$$\nu_{\pm} = R_0 v_{\parallel} \frac{d\Omega_{\pm}}{dp} \tag{7.13}$$

Using definitions (14.25) [14] in (7.13):

$$v_{\alpha} \equiv \frac{|M_{\alpha}|}{\hbar}; \quad v_{\alpha\beta}^{\text{out}} \equiv \frac{\left|M_{\alpha\beta}^{\text{out}}\right|}{\hbar}; \quad v_{\beta\alpha}^{\text{out}} \equiv \frac{\left|M_{\beta\alpha}^{\text{out}}\right|}{\hbar}; \quad v_{01}^{\text{in}} \equiv \frac{\left|M_{01}^{\text{in}}\right|}{\hbar}; \quad \tau_{\alpha} \equiv \frac{1}{v_{\alpha}};$$

$$\Pi_{\alpha} \equiv \frac{W}{|M_{\alpha}|}; \quad \Pi_x^{\alpha} \equiv \frac{W_x}{|M_{\alpha}|}; \quad \Pi_y^{\alpha} \equiv \frac{W_y}{|M_{\alpha}|}; \quad g_{\alpha\alpha}^{(n)} \equiv \frac{G_{\alpha\alpha}^{(n)}}{|M_{\alpha}|}; \quad g_{\alpha\beta}^{(n)} \equiv \frac{G_{\alpha\beta}^{(n)}}{|M_{\alpha}|},$$

$$v_{\alpha} \equiv \frac{|M_{\alpha}|}{\hbar}; \quad v_{\alpha\beta}^{\text{out}} \equiv \frac{\left|M_{\alpha\beta}^{\text{out}}\right|}{\hbar}; \quad v_{\beta\alpha}^{\text{out}} \equiv \frac{\left|M_{\beta\alpha}^{\text{out}}\right|}{\hbar}; \quad v_{01}^{\text{in}} \equiv \frac{\left|M_{01}^{\text{in}}\right|}{\hbar}; \quad \tau_{\alpha} \equiv \frac{1}{v_{\alpha}};$$

$$\Pi_{\alpha} \equiv \frac{W}{|M_{\alpha}|}; \quad \Pi_x^{\alpha} \equiv \frac{W_x}{|M_{\alpha}|}; \quad \Pi_y^{\alpha} \equiv \frac{W_y}{|M_{\alpha}|}; \quad g_{\alpha\alpha}^{(n)} \equiv \frac{G_{\alpha\alpha}^{(n)}}{|M_{\alpha}|}; \quad g_{\alpha\beta}^{(n)} \equiv \frac{G_{\alpha\beta}^{(n)}}{|M_{\alpha}|},$$

and (14.29) [14]:
($v_0 = v_1 = v_{\parallel}$), we shall further obtain for the velocity:

$$\nu_{\pm} = \frac{R_0 \left|M_{\parallel}\right|}{\hbar} \frac{d\Omega_{\pm}}{dp} \tag{7.14}$$

where, remind, $|M_{\parallel}| \equiv |M_0| \equiv |M_1|$ are the exchange resonance energies along the polymer chain. Substituting in the last expression factor $\Omega_{\pm}$ from (7.4) and then the expression obtained in (7.11), we shall obtain a final expression for the current:

$$I_{\pm} = \frac{e\left|M_{\parallel}\right|}{\hbar N} \left[\left(1 \pm \tau_{\parallel}\nu_{\perp}\right) \sin(p) - \frac{g^2 J_{\pm}^4 \sin(p)}{8\left(1 \pm \tau_{\parallel}\nu_{\perp}\right) \cos^2(p)} \right] \tag{7.15}$$

In this case velocities, as follows from (7.14), take the form $\nu_{\pm} = C_* \left[\left(1 \pm \tau_{\parallel}\nu_{\perp}\right) \sin(p) - \frac{g^2 J_{\pm}^4 \sin(p)}{8\left(1 \pm \tau_{\parallel}\nu_{\perp}\right) \cos^2(p)} \right]$. Here $C_* = \frac{R_0 \left|M_{\parallel}\right|}{\hbar}$.

It should be noted that the last expression for the velocity, taking into account expressions (14.36) [14] ($\beta(p) \equiv \sin(p)$) and (14.41) [14] $\left(\overset{\circ}{x}{}_c^{\pm} = \left(1 \pm \tau_{\parallel} v_{\perp}\right) \beta(p) \right)$, can be presented in the form:

$$\nu_{\pm} = C_* \overset{\circ}{x}{}_c^{\pm} \left[1 - \frac{g^2 J_{\pm}^4}{8} \cdot \frac{1}{\left(1 \pm \tau_{\parallel}\nu_{\perp}\right)^2 - \left(\overset{\circ}{x}{}_c^{\pm}\right)^2} \right]$$

This relationship is interesting by the fact that the velocity $\nu_{\pm}$ is mainly determined by the velocity of the center of the soliton localization $\overset{\circ}{x}{}_c^{\pm}$; however, a dissipative additive appears in it, which at $|p| \to 0$ is proportional to the value $-\frac{g^2 J_{\pm}^4}{8} \cdot \overset{\circ}{x}_c^{\pm}$ and is caused by the interaction of the soliton and the polymer chain [14].

From (7.15) for each mode of the $I_{\pm}$, we can, at last, write current

$$I_+ = \frac{e|M_{\parallel}|}{\hbar N}\left[(1 + \tau_{\parallel}\nu_{\perp})\sin(p) - \frac{g^2 \sin^4(\phi)\sin(p)}{8(1 + \tau_{\parallel}\nu_{\perp})\cos^2(p)}\right] \tag{7.16}$$

$$I_- = \frac{e|M_{\parallel}|}{\hbar N}\left[(1 - \tau_{\parallel}\nu_{\perp})\sin(p) - \frac{g^2 \cos^4(\phi)\sin(p)}{8(1 - \tau_{\parallel}\nu_{\perp})\cos^2(p)}\right] \tag{7.17}$$

7.1.2 General Analysis of the Conductivity of Molecular Structures of the Polyene-Like Type

From any physical considerations, of both classical and quantum ones (classical language is used to describe conductors, connected in parallel, and a quantum one describes the channels of conductivity with total probability), it is clear that the value of the total current in each local point of the polymer molecule should be determined by the sum:

$$I = I_+ + I_- \tag{7.18}$$

However, before using of the relationship (7.18), it should be taken into account that there can be two organizations of the "lattice" of the polyene chain, which are different in principle.

One of them corresponds the case, when the product $\tau_{\parallel}\nu_{\perp}$ substantially exceeds the one: $\tau_{\parallel}\nu_{\perp} \gg 1$. Another one describes the case, when the product $\tau_{\parallel}\nu_{\perp}$ has the same order of magnitude as one $\tau_{\parallel}\nu_{\perp} \sim 1$, or this product is substantially less than one $\tau_{\parallel}\nu_{\perp} \sim 1$.

The first possibility: $\tau_{\parallel}\nu_{\perp} \gg 1$, taking into account definitions (14.25) [14] and approximations (14.28) [14] $\left(\nu_{\alpha\beta}^{out} = \nu_{\beta\alpha}^{out} = \nu_{\perp}\right)$, (14.29) [14] $(\nu_0 = \nu_1 = \nu_{\parallel})$, (14.30) [14] $\left(\nu_{01}^{in} = \nu_{\perp}^{in}\right)$, and (14.38) [14] $\left(\nu_{\perp} = \nu_{\perp}^{in}\right)$ resolves itself to the inequity:

$$\frac{|M_{\perp}|}{|M_{\parallel}|} >> 1, \tag{7.19}$$

where designation $|M_{\parallel}|$, as after formula (7.14), means that $|M_{\parallel}| \equiv |M_0| \equiv |M_1|$, and $|M_{\perp}|$ is similarly determined as $|M_{\perp}| \equiv \left|M_{01}^{\text{out}}\right| \equiv \left|M_{10}^{\text{out}}\right| \equiv \left|M_{01}^{\text{in}}\right|$. Inequity (7.19) means the strong "suppression" of the processes of the resonance exchange along the chain with regard to the similar processes between the chains (this model is usually considered as a generally accepted one). This is seen especially well if to write the inequity (7.19) in the form $|M_{\|}| \ll |M_{\perp}|$. If the force of the inequity (7.19) is big enough to neglect by one with regard to the product $\tau_{\|}\nu_{\perp}$, then currents (7.16) and (7.17) take the form:

$$I_{+} = \frac{e|M_{\|}|}{\hbar N}\tau_{\|}\nu_{\perp}\sin(p)\left[1 - \frac{g^2\sin^4(\phi)}{8(\tau_{\|}\nu_{\perp})^2\cos^2(p)}\right]$$

$$I_{-} = -\frac{e|M_{\|}|}{\hbar N}\tau_{\|}\nu_{\perp}\sin(p)\left[1 - \frac{g^2\cos^4(\phi)}{8(\tau_{\|}\nu_{\perp})^2\cos^2(p)}\right]$$

Substituting these current modes into (7.18), we obtain:

$$I = \frac{e|M_{\perp}|\tilde{g}^2}{8\hbar N}\frac{\sin(p)}{\cos^2(p)}\cos(2\phi) \tag{7.20}$$

where parameter $\tilde{g}$ differs from the parameter g by the fact that it is reduced to the dimensionless form using the constant $|M_{\perp}|$, not $|M_{\|}|$.

That is the constant $|M_{\|}|$ falls out from the consideration at all, which allows you to perform the transition to the limit $|M_{\|}| \to 0$, which completely suppresses the exchange resonance processes along the chain.

It is seen that this current can be equal to zero (that is, natural conditions of the absence of the current at the absence of the external field) in two cases only: $p = 0$ and $\phi = \pi/4$. The first from these conditions ($p = 0$) at the absence of the dissipative losses at $\{\Pi\} \to 0$, according to the eq. (14.46) [14] $(\mathring{p} = \Pi)$, leads to the general conclusion that $p \to p_0 = \text{const}$. In this case the constant p_0 can have any value in the area:

$$0 \le p_0 < \frac{\pi}{2}. \tag{7.21}$$

An upper limit of this inequity is determined by the following considerations. If the free electron, which is injected, has the initial dimensionless mechanical impulse p_m, then after injection this impulse is transformed (if it takes place without the losses) into the impulse p_m according to the equation $p_m = \mu(p_0)\beta(p_0)$. Let us remind that here the dimensionless mass $\mu(p)$ and dimensionless velocity $\beta(p)$ are determined by the relationships (14.36) [14] and (14.37) [14] and lead to the generalized relationship of de Broglie relation [14, 15]

$$p_m = tg\,(p_0). \tag{7.22}$$

As p_m changes within the range from 0 to ∞, then from (7.22) follows the area (7.21) for the impulse p_0. In this sense condition $p = 0$ is rough enough. The second condition

$$\phi = \frac{\pi}{4} \tag{7.23}$$

is the condition of the physical equivalence of modes "+" and "–," thus, consequently, of the polyene "under-chains" "0" and "1." This follows from the fact that at $\phi = \frac{\pi}{4}$, the following equity is fulfilled:

$$\cos(\phi) = \sin(\phi) = J_{\pm} = \frac{1}{\sqrt{2}} \tag{7.24}$$

Under conditions of the ideal chain (it means that it is linearly drawn and is not under any mechanical influence), it is the only physically correct condition. Under fulfillment of the condition (7.23) for the considered case $|M_{\|}| \ll |M_{\perp}|$, the current in the considered chain is absent at any values of the initial impulse p_0, if an ideality is not violated in the system. In this case the condition (7.23) itself should be considered as a basic one – such, as resulted in (7.24), at least, in the case of the suppressed longitudinal resonance exchange ($|M_{\|}| \ll |M_{\perp}|$).

The second possibility $\tau_{\|}\nu_{\perp} \sim 1$ or $\tau_{\|}\nu_{\perp} \ll 1$ corresponds similar to (7.1–7.3) conditions:

$$\frac{|M_{\perp}|}{|M_{\|}|} \sim 1; \frac{|M_{\perp}|}{|M_{\|}|} \ll 1. \tag{7.25}$$

Thus, condition $|M_{\perp}| \sim |M_{\|}|$ allows that $|M_{\perp}|$ can be both more than $|M_{\|}|$ and less than it. In this case the general expressions (7.16) and (7.17) for the modes of the excited states should be substituted into the total current (7.18). As a result we have:

$$I = \frac{2e|M_{\|}|}{\hbar N} \sin(p) \left[1 - \frac{g^2}{32\cos^2(p)} \cdot \frac{1 + 2\tau_{\|}\nu_{\perp}\cos(2\phi) + \cos^2(2\phi)}{1 - (\tau_{\|}\nu_{\perp})^2} \right] \tag{7.26}$$

First of all, it is seen that there is a degraded situation here: $\tau_{\|}\nu_{\perp} = 1$ (i.e., $|M_{\perp}| = |M_{\|}|$), which requires a separate consideration almost from the very beginning. Secondly, at the absence of the field ($\Pi = 0$), this current, just the same as in the previous case ($|M_{\perp}| \gg |M_{\|}|$), can become zero at the value $p_0 = 0$ only. However, at $p_0 \neq 0$, this current does not become zero because of the condition (7.23) for the equivalence of the modes ("sub-chains") if the dissipative losses are not taken into considerations. In this case the current is determined by the final expression:

$$I = \frac{2e|M_{\|}|}{\hbar N} \sin(p_0) \left[1 - \frac{g^2}{32\cos^2(p_0)} \cdot \frac{1}{\left[1 - (\tau_{\|}\nu_{\perp})^2\right]} \right] \tag{7.27}$$

This result means that if the dissipative losses, (which lead to the decrease of p_0 to zero during the time of the charge flight), are absent or are not high enough, the effect similar to the superconductivity arises when the current different from zero at the absence of the external field takes place.

To be honest, from the expression (7.26), if we put there $p = p_0$ follows that in the case if the big bracket in that expression is equal to zero, the current becomes zero as well at any value of the impulse p_0.

However, this condition results in, first, the fact that phase parameter ϕ becomes different from $\frac{\pi}{4}$, because modes "+", "–" or "subchains" "0", "1" become non-equivalent (under ideal conditions), that is difficult to explain from the physical point of view, and, secondly, the phase parameter ϕ can be also the complex one that is physically inadmissible.

Thus, this type of organization of the "crystalline lattice" of the considered polymer chain of the polyene type, which is determined by the relation (7.25), is more likely can be referred to the metallic type, as the most physically substantiated value of the current there is expression (7.27), which allows the existence of the current also in the absence of the external field.

Alternatively, the previous type of the organization of the "crystal lattice" of the considered polymer chain of the polyene chain, which is determined by the condition (7.27) and, respectively, by the current (7.20) (which becomes zero under condition of the physical equivalence of modes), can be classified as the semiconductor or dielectric one, as the current without the field is absent. More than that, for such model of organization of the "crystal lattice" of the chain in the considered situation $\Pi \rightarrow 0$, the current, according to (7.20), will not appear even in the case of the presence of the external field, as $\phi = \frac{\pi}{4}$, and feeble fields do not violate this condition.

So, before the investigation of the issue on the influence of the field on the current (i.e., the issue on the features of volt-ampere characteristics), one should determine (e.g., experimentally) with the type of the "crystal lattice" of the polyene chain: is it metallic or not?

7.1.3 The Current in the Case of Metal Type of the "Crystal Lattice" of the Polymer Chain of the Polyene Type

Let us consider for illustration the metallic type of the "crystal lattice". Definition (7.26) is used for the current which, in terms of the mode equivalence condition (of "under-chains"), reduces to the form, that coincides with (7.27), but depends on the arbitrary impulse p, however not the constant p_0. Impulse p, in turn, is determined by the dynamical equation (14.46) [14]. However, dissipative losses, which under normal conditions are always determined by the interaction of the electron and phonon subsystem, are not taken into consideration in (14.46) [14]. On a phenomenological level, the phonon subsystem can be considered as a media, similar to liquid or gas. So, the dissipative losses should be proportional to some degree of the dimensionless velocity $\beta = \sin(p)$. Within the framework of the phenomenological approach, we can consider them as just proportional to β. Then Eq. (14.46) [14] is modified to the form:

$$\mathring{p} = \Pi - \Lambda \sin(p), \tag{7.28}$$

where Λ is a parameter of the dissipative losses ($\Lambda \ll 1$), which should grow with temperature. For the considered situation of the maximally feeble fields [15, 16] $\Pi \to 0$; $p \to 0$ and $\beta \to 0$, Eq. (7.28) is simplified to the form $\mathring{p} = \Pi - \Lambda p$. Its solution at the initial condition $p(0) = p_0$ is the function:

$$p(\tau) = p_0 e^{-\Lambda\tau} + \frac{\Pi}{\Lambda}\left(1 - e^{-\Lambda\tau}\right).$$

At $\tau \to \infty$ we obtain a stationary impulse $p_\infty \Rightarrow \frac{\Pi}{\Lambda}$, which determines a stationary current. Substitution of the impulse p_∞ instead of p_0 in (7.27) and taking into account of the condition $\Pi \to 0$ provides

$$I = \frac{2e|M_\parallel|}{\hbar N\Lambda}\left[1 - \frac{g}{32\left[1 - (\tau_\parallel \nu_\perp)^2\right]}\right]\Pi.$$

Now for the obtaining of the volt-ampere characteristics and conductivity, it is necessary to switch from the dimensionless force constant Π to the dimensional energy W. It can be done according to determinations (14.25) [14] and taking into account that $|M_\alpha| \equiv |M_\parallel|$ and $\Pi = \frac{W}{|M_\parallel|}$.

The explicit form of the energy W

$$W = 2eR_0F_x \equiv 2eR_0F$$

should be taken into account, where the simplification concerning the field strength $F \equiv F_x$ takes place. Then the current I reduces itself to the form:

$$I = \frac{4e^2R_0}{\hbar N\Lambda}\left[1 - \frac{g}{32\left[1 - (\tau_\parallel \nu_\perp)^2\right]}\right]F$$

In the simplest case, the field strength F is connected with the potential difference, which generates it U, by an obvious relationship $F = \frac{U}{R_0N}$. Then the current, at last, reduces itself to the form:

$$I = \frac{4e^2}{\hbar N^2\Lambda}\left[1 - \frac{g}{32\left[1 - (\tau_\parallel \nu_\perp)^2\right]}\right]U.$$

The factor, which stands near the potential difference U, determines the conductivity for the case of the metal type of the "crystal lattice" of the polyene chain. Its inverse value is resistance of this chain:

$$\mathfrak{R} = \frac{\hbar N^2 \Lambda}{4e^2} \left[1 - \frac{g}{32\left[1 - \left(\tau_{\|}\nu_{\perp}\right)^2\right]} \right]^{-1}$$

It is seen that the resistance grows with the growth of the square of molecule number N in the chain and with the growth of the dissipative losses, which are determined by the electron-phonon interaction.

Direct measurements of the resistance by the volt-ampere characteristics for the very low potentials and, respectively, for the currents, enable us to investigate experimentally one of four parameters N, Λ, g, and product $\tau_{\|}\nu_{\perp}$, under the condition that other three are known from other experiments or sources. For example, if the number of "lattice cells" N in polyene molecule, electron-phonon interaction, and product $\tau_{\|}\nu_{\perp}$ are known, one can answer the question concerning a possibility of realization of the specified excited soliton. If it is possible to reach non-zero values of the parameter g, soliton can be realized.

7.2 Conclusions

In the second part of this work, the general conductivity analysis of the polymer chain of the polyene type is performed. The general expression of the current, caused by the injected electron under condition of the excitation of soliton states, is found. Particular cases, which correspond to the different types of conductivity (metal, semiconductor), are analyzed as well. Experimental inspection concerning the possibility of realization of this excitation is proposed.

References

1. Bredas JL, Belionne D, Cornil J, Calhert JP, Shuai Z, Silbey R (2002) Electronic structure π-conjugated oligomers and polymers: a quantum-chemical approach to transport properties. Synth Met 125:107–116
2. Bredas JL, Street GB (1985) Polaron, bipolaron and solitons in conducting polymers. Acc Chem Res 18:309–315
3. Kachkovsky OD (2005) Soliton nature of the electronic structure of ions linear conjugated systems. Theor Exp Chem 41:139–164
4. Davydov AS (1984) Solitons in molecular systems. Scientific dumka, Kiev. 288 p
5. Mishra A (2000) Cyanine during 1990s: a review. Chem Rev 100:1973–2011
6. Meyers F, Marder SR, Perry JW (1998) Chapter 6: introducing to nonlinear optical properties organic materials. In: Interrante LV, Hampden-Smith J (eds) Chemistry advanced materials. An

overview. Wiley-VCH. Inc., New York/Chicherster/Weinheim/Brisbane/Singapore/Toronto, pp 207–268
7. Kachkovskyy AD (1997) Nature electron transitions in linear conjugated systems. Successes chemistry 66(8):715–734
8. Bach G, Daehne S (1997) Cyanine dyes and related compounds. In: Sainsbury M (ed) ROOD'S chemistry carbon compounds, 2nd suppl. to 2nd edn, Vol. IVB, Het. Comp. Elsevier science, Amsterdam, chapter 15, pp 383–481 (443)
9. Dewar M (1972) Theory of molecular orbitals in organic chemistry. Mir, Moscow. 590 p
10. Lepkowich RS, Przhonska OV, Hales JM, Hagan DJ, Van Sryland EW, Bondar MV, Slominski YL, Kachkovski AD (2004) Nature electron transitions in thiacyanines with a long polymethine chain. Chem Phys 305:259–270
11. Kachkovski AD, Tolmachev AI, Slominski YL, Kudinova MA, Derevyanko NA, Zhukova OO (2005) Electronic properties polymethine systems. Soliton symmetry breaking and spectral features dyes with a long chain. Dyes Pigments 64:207–216
12. Kachkovski AD, Przhonska OV, Ryabitzki AB (2007) Symmetry breaking in cationic and anionic polymethine dyes. J Molec Struct (THEOCHEM) 802:75–83
13. Reimers JR, Hush NS (1993) Hole, Electron and energy transfer through bridged systems. VIII. Soliton molecular switching in symmetry-broken Brooker (polymethinecyanine) cations. Chem Phys 176:407–420
14. Vasylyuk SV, Suprun AD, Yashchuk VN (2017) About possible mechanisms of nanoconductivity in polyenes polymers: the charge solitons at extremely weak external fields. In: Fesenko O, Yatsenko L (eds) Nanophysics, nanomaterials, interface studies, and applications. NANO 2016. Springer Proceedings in Physics, vol 195. Springer, https://doi.org/10.1007/978-3-319-56422-7_14
15. Suprun AD (2002) Funct Mater 9:389
16. Nogami Y, Toyama FM (1994) Phys Rev E 49:4497–4501

Chapter 8
Nanosized Oxides of Different Compositions as Adsorbents for Hazardous Substances Removal from Aqueous Solutions and Wastewaters

Małgorzata Wiśniewska, Monika Wawrzkiewicz, Anna Wołowicz, and Olena Goncharuk

8.1 Introduction

Due to industrialization and urbanization processes, large quantities of effluents containing hazardous substances are discharged into the environment. Not only inorganic contaminants such as heavy metals, e.g. Cd, Cr, Cu, Ni, As, Pb and Zn, but also organic compounds like phenol (and its salts), polyalcohols, polyacids (and other macromolecular compounds), azo dyes, dioxins, furanes as well as many others are generated by metallurgical, engineering, mining, electroplating, nuclear, chemical, textile, petroleum, plastic, cellulose, etc. industries. Many of them are known to be toxic or carcinogenic. Thus removal of such hazardous substances is of crucial importance to protect the human and the environment. Several techniques have been used to remove organic and inorganic impurities from industrial wastewaters. Recently increasing attention has been focused on adsorption techniques using metal and semimetal oxide sorbents such as aluminium oxide, iron oxide, titanium oxide, manganese oxide, zirconium oxide and silica oxide. The nanosized

M. Wiśniewska (✉)
Department of Radiochemistry and Colloid Chemistry, Faculty of Chemistry, Maria Curie-Sklodowska University, Lublin, Poland
e-mail: wisniewska@hektor.umcs.lublin.pl

M. Wawrzkiewicz · A. Wołowicz
Department of Inorganic Chemistry, Faculty of Chemistry, Maria Curie-Sklodowska University, Lublin, Poland

O. Goncharuk
O. O. Chuiko Institute of Surface Chemistry, National Academy of Sciences of Ukraine, Kyiv, Ukraine

O. Fesenko, L. Yatsenko (eds.), *Nanooptics, Nanophotonics, Nanostructures, and Their Applications*, Springer Proceedings in Physics 210,
https://doi.org/10.1007/978-3-319-91083-3_8

metal, semimetal and mixed oxides are classified as the promising sorbents because of their large surface areas and pore sizes as well as mechanical and thermal stability.

8.2 Properties and Synthesis of Individual and Mixed Nanosized Oxides

Individual and mixed nanosized oxides (NMOs) have become increasingly popular in recent decades due to their unique properties, which ensure the expansion of their application in various fields of chemistry, materials science and industry [1–4]. High specific surface area and high concentration of active surface centres of nanosized oxides provide good adsorption properties with respect to low- and high-molecular-weight adsorbents, such as polymers [5–8], metal ions, dyes, etc. [9–17]. The active surface centres in such nanooxide composites are terminal ≡M-OH groups and bridging groups ≡M-O(H)-M≡ [12, 17–20], which are hydrophilic in nature and possess different Brönsted and Lewis acidities, which ensure the catalytic activity of such nanocomposites in different catalytic processes [21–26] and determine the properties of their surface when interacting with various adsorbents in the atmosphere and aqueous medium. The ability of such groups to dissociate in an aqueous medium provides the formation of a surface charge and a double electric layer [18], which determines such surface properties as ability to adsorb metal ions [16, 17] and dyes, as well as such important properties as electrokinetic potential and colloidal stability of dispersions. In the adsorption interaction with polymers, such a surface charge can play a positive role in the case of interactions with an oppositely charged polyelectrolyte and a negative role in the case of charged electrolyte of the same sign or a nonionic polymer for which such mechanism is realized as the hydrogen bonding with undissociated surface groups. It should be noted that bridging groups of the type $\equiv M_{(1)}$-O(H)-$M_{(2)}\equiv$, which are formed on the surface of mixed oxide composites, differ significantly in properties from the bridging groups of the $\equiv M_{(1)}$-O(H)-$M_{(1)} \equiv$ type. For example, often their Brönsted acidity is often much higher, which determines noticeable differences in the properties of individual highly dispersed oxides and mixed nanocomposites.

Formation of the surface structure is significantly affected by the method of synthesis of mixed oxides: pyrogenic, sol-gel, chemical vapour deposition (CVD), etc. The pyrogenic method is one of the most common due to the convenience of its industrial application [27, 28]. A pyrogenic method of the synthesis of individual and mixed oxides consists in high-temperature hydrolysis of the corresponding metal or metalloid tetrachlorides in a hydrogen-oxygen flame at 1100–1400 °C. The variations of the initial component ratios, temperature and flow rate make it possible to obtain fumed mixed oxide of binary or triple compositions with high specific surface area and different phase ratios. The most common oxides produced by this process are individual and mixed fumed oxides of silicon, titanium and aluminium [17–20]. Also many other fumed oxides, such as GeO_2, Fe_2O_3, Cr_2O_3, MoO_3, SnO_2, V_2O_5, etc., have been produced by Degussa using the pyrogenic method [29].

Fumed nanooxides have proven to be good adsorbents of metal cations and dyes [10, 11, 16, 17], polymers [5–8], etc. Especially effective are mixed systems due to the presence of bridging groups of ≡Si-O(H)-Ti ≡ or ≡Si-O(H)-Al ≡ on their surface, which are significantly stronger Brönsted and Lewis acid sites than the terminal OH groups and bridging groups of the $\equiv M_{(1)}$-O(H)-$M_{(1)} \equiv$ type and vastly affect the properties of the nanocomposite surface [17–20]. Mixed fumed oxides are unevenly distributed in the bulk and on the surface. Thus for the Al_2O_3/SiO_2 and ternary $Al_2O_3/TiO_2/SiO_2$ systems, the phase of Al_2O_3 is maximally concentrated in the surface layer of the nanocomposite with a low Al_2O_3 content (3–10% wt.). The surface concentration of TiO_2 increases monotonously with the increasing TiO_2 content in the TiO_2/SiO_2 and ternary nanocomposites. There is mutual influence of the components on formation of the structure in synthesis process. Thus Al_2O_3 has a completely amorphous state in the Al_2O_3/SiO_2 nanocomposite in contrast to the individual fumed Al_2O_3, which includes 20% of the γ-phase. The presence of anatase and rutile in the phase structure is observed in TiO_2 and TiO_2/SiO_2, but their ratio can vary depending on the TiO_2 content in the composite. The high temperature during the pyrogenic synthesis prevents the formation and growth of large crystallites, thereby achieving a nanoscale structure. In the case of tendency to form a crystalline structure in binary oxides, for example, as in the case of TiO_2/SiO_2, each oxide forms its own phase with formation of bridge bonds of the $\equiv M_{(1)}$-O(H)-$M_{(2)} \equiv$ type at the phase boundary.

At the same time, in the formation of an amorphous structure, as in the case of alumina-silica, a more uniform distribution of oxides occurs in the surface layer of the composite with formation of a significant number of bridge bonds of the$\equiv M_{(1)}$-O(H)-$M_{(2)} \equiv$ type.

To accomplish obtaining the core-shell nanoscale structure, the CVD method [23–32] is applied, i.e. a layer of the second oxide is deposited from the vapour phase on the already synthesized nanoparticles of the carrier oxide. This method has many features associated with the structure formation of the second oxide on the core-oxide surface. Thus in most cases, it is not possible to form a continuous layer of the second oxide since it forms a crystalline or amorphous cluster structure [32]. Thus the surface structure of the CVD-synthesized oxide composite includes both the support and supported oxide surface areas, the ratio of which determines the surface properties of the mixed oxide.

The deposition of the second oxide onto the carrier oxide can also be carried out from the liquid phase using water, hexane, etc. as the solvents. Silica-supported nanocomposites with CeO, ZrO, TiO_2 and Fe_2O_3 were synthesized by this method [21, 33–37]. It was reported also in [22] as regards some other oxides, such as Mg, Mn, Ni, Cu and Zn oxides, deposited on fumed silica. Depending on the deposited oxide concentration, it is possible to obtain a layer of another oxide on the carrier oxide surface, or the carrier becomes a matrix for the growth of particles of a separate second phase at a higher content of the deposited oxide. In this case only a small fraction of the surface of the carrier oxide interacts with the synthesized oxide. This method is generally used to control the growth of the second phase particles due to the presence of a carrier oxide.

Another common method of synthesis using solvent is the sol-gel synthesis. The sol-gel process is a well-established colloidal chemistry technology; the process occurs in the liquid solution of organometallic precursors (TMOS, TEOS, Zr (IV) propoxide, Ti(IV) butoxide, etc.) and results in formation of a new phase of nanooxides by means of hydrolysis and condensation reactions. The main advantages of the sol-gel method are high purity and homogeneous nanostructure with predefined material properties at relatively low process costs. A lot of papers reported on the study of sol-gel synthesis conditions and properties of the obtained silica nanoparticles [38], zinc oxide [39], titanium dioxide [40] and mixed oxide nanocomposites [41].

Thus the variety of synthesis methods provides the possibility of obtaining highly dispersed oxide materials with diverse morphology and surface properties that are promising sorbents for such types of pollutants as ions of heavy metals, dyes, etc.

8.3 Single, Mixed and Modified Nanosized Oxides as Adsorbents

8.3.1 Metal Ions

Among the toxic pollutants introduced into the environment by humans as a result of progressing urbanization and industrialization processes, heavy metals are particularly dangerous. In the second half of the twentieth century, increased interest in heavy metals and their impact on human organisms as well as on aquatic ecosystems were observed. The main sources of heavy metals in the environment are natural and anthropogenic ones. Circulation and migration of heavy metals in the natural environment are mainly related to such processes as weathering rocks, eruption of volcanoes, evaporation of oceans, forest fires and soil-forming processes. On the other hand, anthropogenic sources of heavy metals included activity of various branches of industries, transport, municipal economy, waste landfill, chemical industry, fertilizers and wastes used for fertilization, energetics based on coal and lignite combustion, ore mining, metallurgy and exhaust fumes [42, 43]. The problem of heavy metal pollution largely concerns industrialized areas which produce a large amount of wastewaters coming from mining, hydrometallurgy, textiles, paint and pigment production, mineral fertilizers and plant protection products, rubber and plastics, wood products, etc. [44–46] (Fig. 8.1).

The heavy metals that come from these sources are dispersed into the environment in an uncontrolled way resulting in atmosphere, surface, groundwater and soil pollution [47]. Their ability to circulate in the environment and enter the food chain results in a particular risk to plant, animal and human organisms. Heavy metals are involved in the biogeochemical circulation which is part of the biological circuit (flow of elements in the trophic chain) where the first link being a plant, the next animal and the last one human (soil → plant → animal → human). Moving the

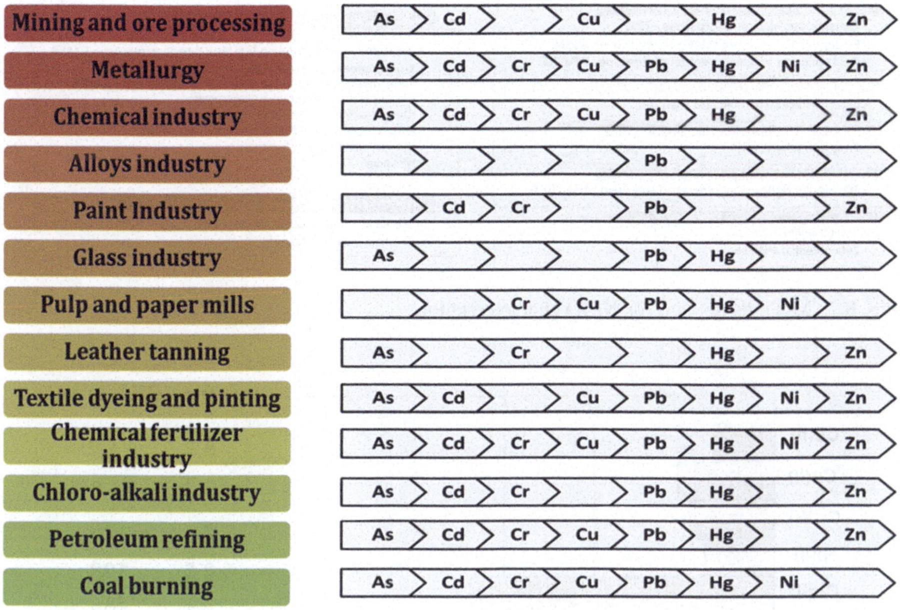

Fig. 8.1 Sources of heavy metals

metals to the next link results in an increase in concentration, resulting in a partial accumulation of these elements in a given cell. Heavy metals accumulate in the last link, the human body. Moreover, their high durability and long half-life, e.g. duration in soil >1000 years, non-biodegradability, biomagnification, chronicity, synergic effect, teratogenicity and mutagenicity result in serious health problems [42, 44, 48]. Long-term effects of heavy metals on the human body or acute toxicity can cause damage to the central nervous system, problems with the cardiovascular and gastrointestinal systems and damage to the internal organs, e.g. the liver, kidneys and lungs, and can enhance the risk of some cancers, etc. [43, 44, 49]. Taking into account danger of heavy metals, they can be divided into (a) very high, Cd, Hg, Pb, Cu, Zn and Cr; (b) high, Mo, Mn and Fe; (c) medium, Ni and Co; and (d) low degrees of potential risk, Sr and Zr. Hazards of bioaccumulation of heavy metals in the environment determine the need for their removal. Despite the existence of various methods of wastewater treatment, among others, membrane filtration, electrodialysis, photocatalysis, chemical precipitation, coagulation, flocculation, reverse osmosis or oxidation and new and effective ways of removing heavy metal ions from wastewaters are still searched for [44, 45, 50–52]. Among these techniques, sorption has come to the forefront as one of the major techniques for heavy metal removal from wastewaters due to flexibility in design and operation, possibility of generating high-quality treated effluents, high efficiency, metal selectivity, ease of operation, high regeneration and reuse of sorbent, low-cost maintenance, etc. [52, 53]. Heavy metal removal from water and wastewaters can be accomplished by sorption

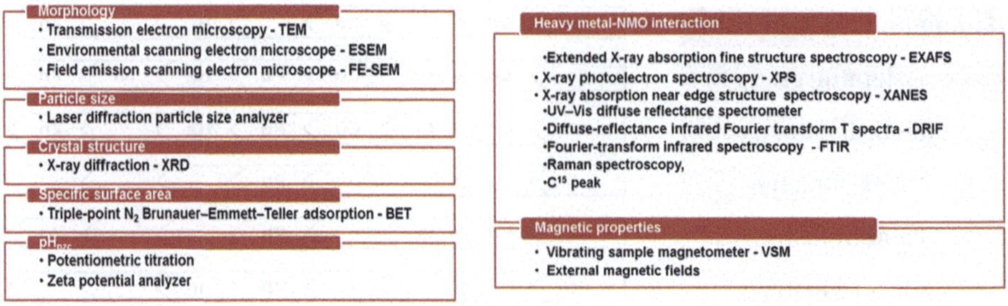

Fig. 8.2 Methods applied for NMO characterization

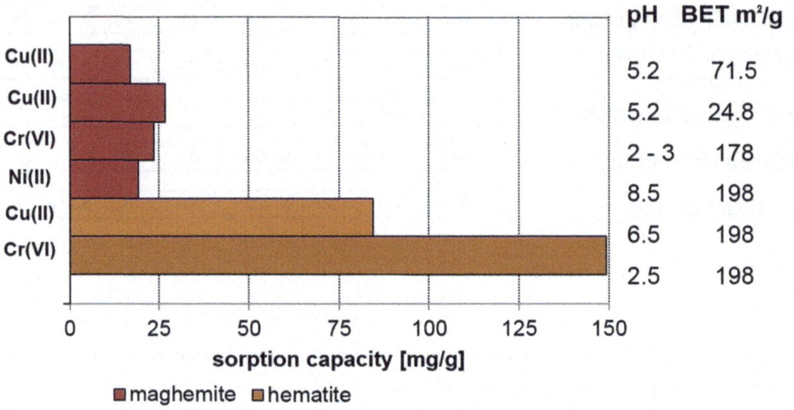

Fig. 8.3 Comparison of NFeO capacities for heavy metal ions [50, 57, 58]

processes using promising nanosized metal oxides such as aluminium, silica, ferric, manganese, titanium, magnesium, cerium oxides, etc. of which the most widely studied are iron oxides, manganese oxides, aluminium oxides and titanium oxides [50]. In order to understand the mechanism of the sorption process on oxides (interactions in the NMO-heavy metal system) and to compare the removal efficiency of these pollutants, it is extremely important to gather detailed information about the surface morphology of oxides, particle size, BET surface area, magnetic properties and crystalline structure. Such information about NMOs can be obtained using various techniques and methods presented in Fig. 8.2 [50, 54, 55].

Due to the wide use of NMOs as effective sorbents of heavy metal ions in the literature, a number of papers devoted to this subject and review ones can be found [50, 54–56]. For this reason only selected NMO applications are presented and cited below. Widespread of iron, reuse of resources, being eco-friendly and ease in the synthesis make nanosized ferric oxide (NFeOs) a common sorbent for heavy metal removal. In this group of NFeOs, goethite (α-FeOOH), hematite (α-Fe_2O_3), magnetite (Fe_2O_4), maghemite (γ-Fe_2O_3), amorphous hydrous FeOs and iron/iron oxide are frequently used [50, 57, 58] (Fig. 8.3). As it was observed, divalent metal cations usually form inner-sphere surface complexes with NMOs. In the case of hydrous ferric oxide (HFOs), the sorption processes seem to be poorly sensitive to ionic strength variation, e.g. Cu(II) and Pb(II) sorption on HFOs (0.0005–0.5 M $NaClO_4$) [59] and Pb(II) sorption on ferrihydrite (0.001–0.1 M $NaNO_3$) [60], which might suggest also inner-sphere complex formation. Sorption of Cr(VI), Cu(II) and Ni(II) on maghemite is strongly pH-dependent [57], and under optimal pH electrostatic attraction, a mechanism similar to Cr(VI) sorption onto maghemite was observed [50].

Nanosized manganese oxides (NMnOs), e.g. hydrous manganese oxide (HMO) [61–65], mixed-valence manganese oxide, nanotunnel manganese oxide (octahedral molecular sieve, OMS-1 or OMS-2) [66, 67], were applied for Cs(I) [61], Hg(II) [62], As [63], Cd [64], Pb(II), Cd(II), Zn(II) [65], radionuclides [66] and hydrometallurgical wastewater (Al, Ca, Fe, Mg, Mn, Na) [67] removal. As was pointed out, hydrous manganese oxide similar to HFO forms inner-sphere complexes with metal (M(II)), and ion exchange process takes place during the sorption processes. Moreover, the sorption process includes two steps such as rapid metal adsorption on the external surface and slow intraparticle diffusion along the micropore walls [65]. Additionally, HMO active sites are heterogeneous due to the fact that the sorption is better represented by the Freundlich isotherm model [65]. In the case of tunnelled manganese oxide, their sorption ability is strongly dependent on their structure [66].

The other popular oxides for heavy metal ion removal are aluminium, titanium and zinc ones [68]. As it was found, aluminium oxide can be chemically or physically modified which allows to introduce additional functional groups containing donor atoms, e.g. O, N, S and P, which improve removal efficiency of heavy metal ions. Modification of γ-Al_2O_3 by γ-mercaptopropyl-trimethoxysilane (γ-MPTMS) in the presence of sodium dodecyl sulphate (SDS) improves removal efficiency of Zn(II), Pb(II), Cu(II) and Fe(III) from 55, 36, 27 and 24% (Al_2O_3 1 g, SDS 50 mg, without MPBIM) to 100, 98, 97 and 96% (Al_2O_3 1 g, MPBIM 40 mg, SDS 50 mg). Surfactant-coated alumina modified by dithizone for Pb(II) sorption [69] or surfactant-coated alumina with immobilized 1,10-phenantroline for Cu(II) and Cd(II) determination [70], γ-Al_2O_3 modified by dinitrophenylhydrazine (DNPH) for Pb(II), Co(II), Cr(III), Cd(II), Ni(II) and Mn(II) sorption [71] was also applied. During the sorption process, metal sorbs on the surface through surface interactions or/and chemical-bonding interactions take place. In the case of alumina modified by γ-MPTMS, three mechanisms of sorption can be found: (a) interactions between thiol groups and metal ions, (b) hydrolyzation of metal ions and (c) electrostatic sorption. Depending on pH of the solution, mechanism (a) plays a more significant role in the case of acidic solutions, whereas for basic solutions mechanisms (b) and (c) are dominant.

As it was mentioned above, titanium and zinc oxides are also found in application in heavy metal removal due to their unique advantages, e.g. simple and cheap to prepare and convenient to tailor morphologically. Some examples of the results of heavy metal sorption on titanium and zinc oxide can be found in Table 8.1.

Table 8.1 Titanium and zinc oxide application in heavy metal removal

Metal oxide	Characteristics	Heavy metals	Performance	Additional information	References
TiO_2 Commercially available	33% titanium and 67% oxygen BET surface area 185.5 m^2/g Particle sizes 8.3 nm	Pb(II), Cd(II), Ni(II)	Pb(II) removal efficiency 100% Cd(II) removal efficiency 99.9% Ni(II) removal efficiency 99.2% 0.5 g/L TiO_2, nano- TiO_2	Langmuir isotherm/modified first order	[72]
TiO_2	Pore diameter 10–50 nm, BET surface area 208 m^2/g	Zn(II), Cd(II)	Zn(II) 15.3 mg/g Cd(II) 7.9 mg/g, pH = 9.0	–	[73]
TiO_2	Pore diameter 12 nm, BET surface area 45.4 m^2/g	Cd(II),Cu(II), Ni(II), Pb(II)	pH_{PZC} values 7.4 120.1, 50.2, 39.3, 21.7 mg/g for Cd(II), Cu(II), Ni(II), Pb(II)	Freundlich model in single- and multiple- component solutions	[74]
ZnO	Pore diameter 5–20 nm, BET surface area 147 m^2/g	Cu(II)	> 1600 mg/g	Freundlich model	[75]
ZnO/PbS	Heterostructured functional nanocomposite, nanosheets	Pb(II)	Pb(II) 6.7 mg/g	–	[76]
ZnO	50 nm (spheroid)	Cd(II), Cu (II), Ni(II), Pb(II)	360.6 mg/g (sum of four metals) in multiple- component solutions	Selectivity series Cd(II) > Pb(II) > Cu(II) – single solutions, Pb(II) > Cu(II) > Cd(II) > Ni(II) – multiple-component solutions	[77]

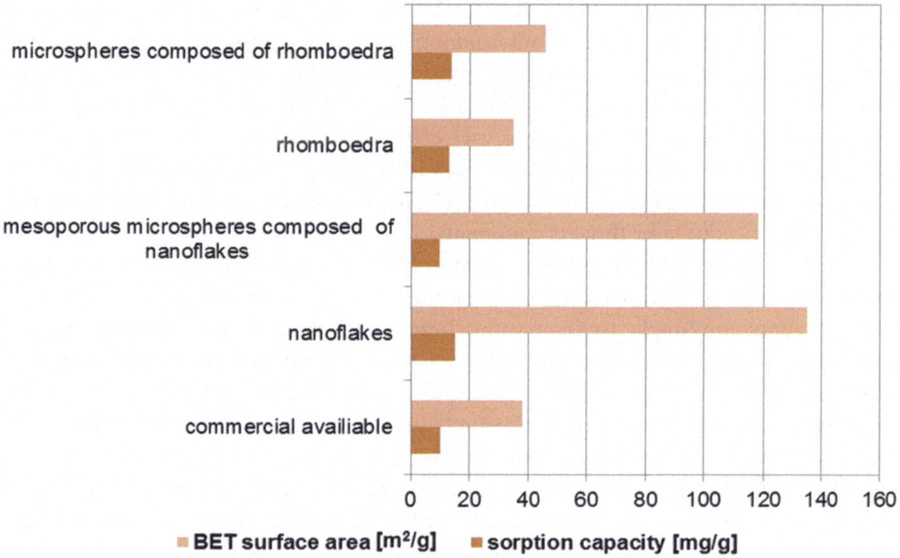

Fig. 8.4 Comparison of sorption capacities obtained for Cr(VI) on NMgOs of various morphologies

Nanosized magnesium oxides (NMgOs), e.g. nanorods, nanowires, nanotubes, nanobelts, nanocubes, etc., of various morphologies were obtained. Then their sorption behaviours towards heavy metal ions were examined, e.g. chromium removal by MCH (magnesium carbonate hydrate) of the phase structure from monocyclic ($Mg_5(CO_3)_4(OH)_2(H_2O)_4$) to hexagonal ($MgCO_3$) [78]. As it was found, both nanoflakes and flowerlike microspheres ensure very good adsorption efficiency (Fig. 8.4). After 120 min of phase contact time, the concentration of Cd(II) and Pb(II) decreased significantly from 100 mg/L to 0.007 mg/L and 0.05 mg/L, respectively.

Nanosized cerium oxide sorption behaviour depends on the size, shape, morphology and surface area [50, 79]. Cr(VI) sorption on the CeO_2 nanoparticles (synthesized by oxidation of Ce(III) nitrate under basic conditions using hexamethylenetetramine) of the mean size 12 nm and the BET surface area shows that after the sorption process, Cr(VI) was not present in the solid phase (only Cr(III) was present), whereas in the liquid phase, Cr(VI) exists. This fact indicates that on the surface of nanoparticles, the Ox-Red process can be possible and reduction of chromium takes place. The sorption process of chromium can be described by the pseudo-second-order kinetic equation and the Freundlich model [80]. Similarly to magnesium oxide, cerium oxide can be also obtained in different morphological forms. Ceria hollow nanospheres obtained by the hydrothermal method of nanocrystal size 14 nm and the BET surface area of 72 m^2/g give nearly 70 times higher sorption capacities for Cr(VI) 15.4 mg/g and Pb(II) 9.2 mg/g than the commercial bulk ceria material. The isotherm was well described by the Langmuir isotherm model. Removal of Cd(II), Pb(II) and Cr(VI) ions in single- and three-component solutions using cerium oxide (CeO_2) nanoparticles was carried out by Contreras et al. [81]. The highest percentage of removal was obtained in the case of Pb(II) adsorption capacity which was equal to 128.1 mg/g, then for Cd(II) 93.4 mg/g and finally for chromium(VI) 34.4 mg/g. pH effect (pH = 5, pH =7) on % removal efficiency was observed for Cd(II) and Cr(VI), whereas in the case of Pb(II), the pH effect was not pointed out. Additionally, sorption capacities were not affected by the type of system (single or three component).

Due to the possibility of agglomeration of NMOs, which results in loss of activity and pressure drop when used in the column systems as well as difficulty in separation, attempts are made to overcome these disadvantages by obtaining hybrid adsorbents impregnating or coating NMOs into/onto a larger size porous support forming an NMOs-supported host [50]. As the host support of NMOs for heavy metal removal, there can be applied natural supports, e.g. raw bentonite (RB), sand, montmorillonite and metallic oxide, as well as manufactured polymer supports [50]. Eren applied iron-coated (ICB, iron-coated bentonite) or magnesium-coated (MCB, magnesium-coated bentonite) RB for Pb(II) removal. The sorption capacities for Pb(II) were in the following order: RB 16.7 mg/g < ICB 22.2 mg/g < MCB 31.86 mg/g (Langmuir capacities) [50, 82]). The sorption yield of Pb(II) using bentonite increases with the pH increase (competition between H^+ and Pb(II) of exchange site can proceed), whereas with the ionic strength increase from 0.01 to 0.1 M Pb(II), the adsorption yield significantly decreases [50]. The presence of Cl^- anions also affects the Pb(II) sorption efficiency which results in different Pb (II) complexes. MCB was also applied for Cu(II) sorption. In this case sorption also depends on the kind of adsorbent surface; Cu(II) forms in the solutions as well as on pH of solutions. For Cu(II) on MCB, the sorption capacities can be presented in the following order: RB 42.41 mg/g < MCB 58.44 mg/g [83]. The other examples of supported NMOs for heavy metal removal from water, e.g. anodic alumina membrane, activated carbon, aluminium oxide, polymeric cation exchanger, etc., are reported in [50].

Magnetic sorbents based on NMOs for heavy metal removal can be also found in the literature [50, 84–86]. In the case of magnetic sorbents, modification of magnetic NFeOs can proceed by the amino, zeolite, poly(3,4-ethylenedioxythiophene) (PEDOT) groups, carbon nanotubes, humic acid, acrylic acid, alginic acid, hydrogel, SiO_2, graphene nanosheets, etc. Such modification prevents from air oxidation and particle aggregation.

8.3.2 *Dyes*

Intense development of many industries using dyes and pigments to colour their products directly contributes to pollution of the aquatic environment with these substances. Figure 8.5 shows the dyes consumption in various branches of industry in recent years [87–93].

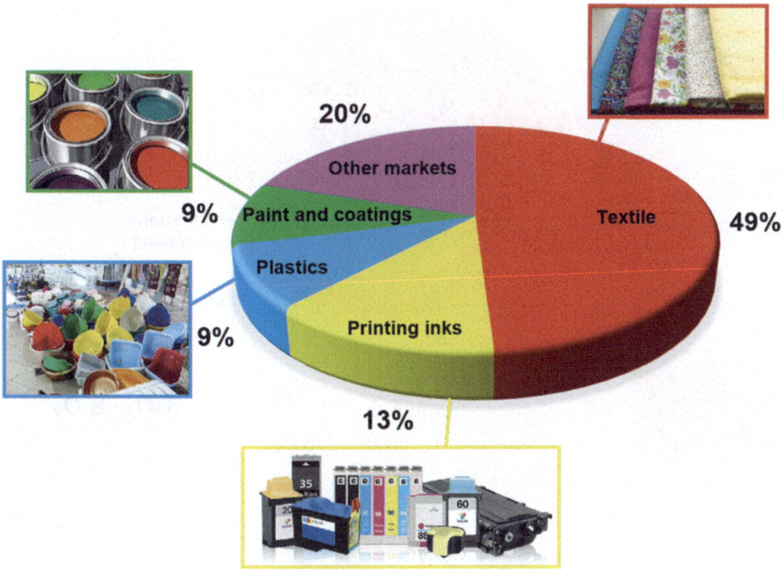

Fig. 8.5 Dyes consumption in different industries

Adsorption methods have a well-established position in the dyes wastewater treatment technology. Adsorbents composed of inorganic oxides such as SiO_2, TiO_2, Al_2O_3, Fe_2O_3, MgO or CaO are frequently used in removal processes of dyes from aqueous solutions. By altering the composition of oxide adsorbents, it is possible to remove different types of dyes, i.e. acid, reactive, direct or basic. An important feature of NMOs is the ability to modify their surface with different reagents, which allows to introduce new functional groups and obtain totally different materials of high quality and functionality. Taking into account the valuable morphological and microstructural properties of NOMs as well as mechanical strength and non-toxicity in many cases, these sorbents become increasingly popular every year in textile effluent treatment [94]. In addition, they are increasingly used as substrates for the synthesis of modern hybrid materials applied for wastewater treatment.

Decolourization of model solutions and raw wastewaters containing dyes from different industrial branches has been studied by many research teams all over the world. The selected results of studies on the use of mixed oxides for dyes removal are presented below.

The mixed nanooxides of Ti and Si (241 m^2/g) and Ti, Al and Si (433 m^2/g) are characterized by high adsorption capacities for basic dyes such as methylene blue and rhodamine 6G [95]. According to Pal et al. [95], the maximum adsorption capacity of the mixed oxide containing Ti and Si for methylene blue was found to be 162.96 mg/g. It is worth noting that the maximum adsorption capacity of the mixed Ti and Si oxide is almost 2.5 times higher than the adsorption capacity of the

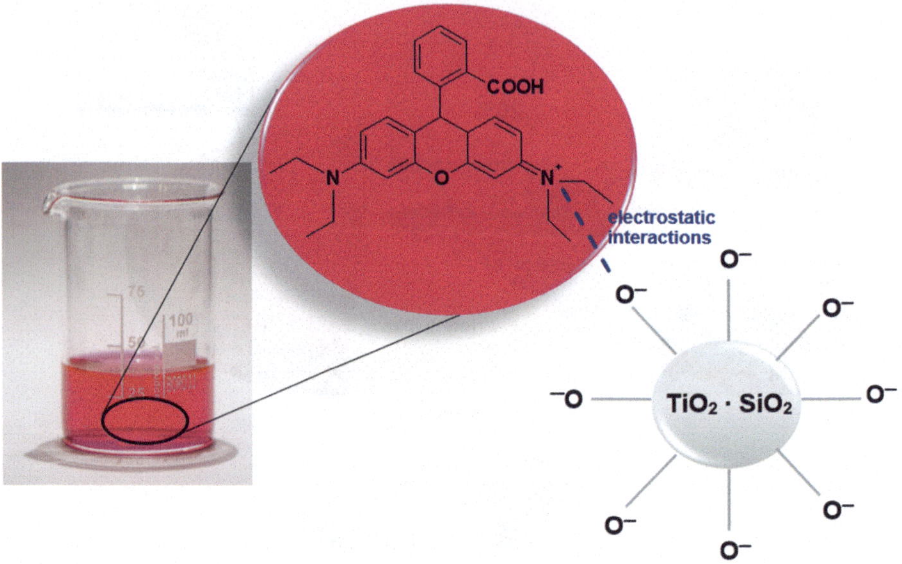

Fig. 8.6 Mechanism of interactions of rhodamine B with the surface hydroxyl groups of mixed TiO_2-SiO_2 oxides of different compositions under the experimental conditions at pH 5–7

mixed oxide containing Ti, Al and Si (65.78 mg/g). The slightly lower value of the adsorption capacity of the mixed Ti and Si oxide equal to 142.9 mg/g was found.

Rasalingam et al. [96] synthesized mesoporous titania-silica oxides of different compositions for rhodamine B adsorption. Among the prepared titania-silica oxides, the highest adsorption capacities were observed for the X-TiSi-03 (TiO_2:SiO_2 = 1:2) and X-TiSi-04 (TiO_2:SiO_2 = 1:3) sorbents which are related to large pore volumes of 0.56 and 0.58 cm^3/g, respectively. The monolayer sorption capacities calculated from the Langmuir adsorption model were found to be 41.2 mg/g for X-TiSi-03 and 44.1 mg/g for X-TiSi-04. The sorption of rhodamine B (Fig. 8.2) under pH 5–7 on these materials occurs as a result of the electrostatic interactions between the surface hydroxyl groups (–OH) of TiO_2·SiO_2 and the positively charged diethylamino groups of the dye as shown below (Fig. 8.6). The presence of oxygen bridges Ti-O-Si also favours adsorption of rhodamine B.

A more complex mechanism of interactions between the dye molecules and silica-based nanocomposite was postulated by Tanzifi et al. [97]. Adsorption of the diazo dye Amido Black B and polyaniline/SiO_2 is a result of interactions between hydroxyl, imine and amine functionalities of the oxide-based sorbent and the amine, hydroxyl and azo groups of Amido Black B. Not only the electrostatic but also van der Waals and hydrogen bonding can exist between these groups. Chemisorption of Amido Black B on the polyaniline/SiO_2 was confirmed by the kinetic studies. The applicability of the pseudo-second-order equation for description of the data was verified by the value of the determination coefficient (r^2 = 1). The surface area, total pore volume and pore diameter for the polyaniline/SiO_2 were determined as 13.36 m

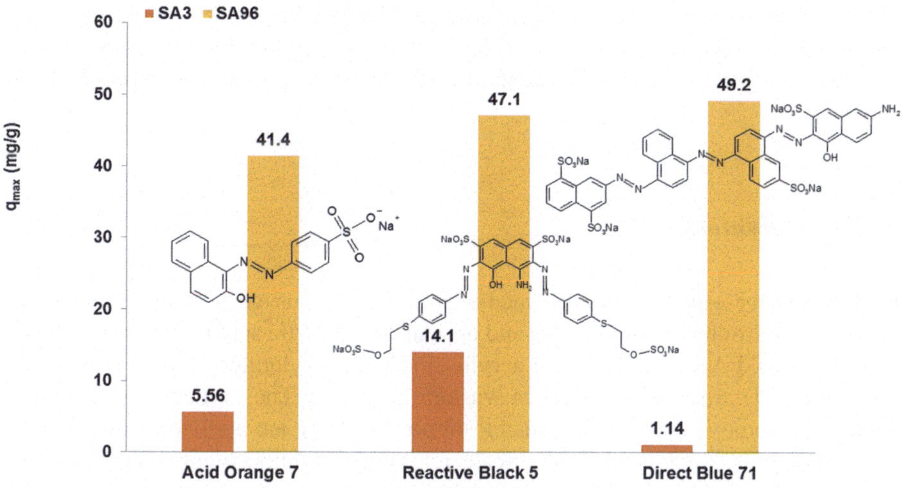

Fig. 8.7 Comparison of maximum sorption capacities of SA3 and SA96 oxides for three textile dyes

^{2}/g, 3.07 cm^3/g and 15.37 nm, respectively. Not only these parameters but also medium pH, adsorption time, temperature and mass of the sorbent influenced Amido Black B adsorption. The adsorption efficiency decreased from 98.6% to 96.8% with the increasing solution pH from 2 to 10. The equilibrium time of adsorption was equal to 60 min. Thermodynamic studies revealed that the retention of Amido Black B on the polyaniline/SiO_2 was endothermic.

SiO_2-Al_2O_3 mixed oxide modified by ferrocene was applied for methyl orange removal by Arshadi et al. [98]. The modified SiO_2-Al_2O_3 (Si/Al-Fe) was a more efficient adsorbent for methyl orange removal (89.0%) after 30 min compared with the non-modified SiO_2-Al_2O_3 (only 4.2% of dye was removed). The maximum adsorption capacity of Si/Al-Fe was equal to 381.0 mg/g. The process followed the pseudo-second-order kinetic model and was spontaneous and endothermic.

Two NMOs composed of SiO_2 and Al_2O_3 were applied for three textile dyes such as Acid Orange 7, Reactive Black 5 and Direct Blue 71 removal from aqueous solutions and wastewaters [10, 11, 99]. The percentage of silicon dioxide in the sorbents was 97% (SA3) and 4% (SA96). The mixed oxides were synthesized using the pyrogenic method which allows to obtain high concentration of Si-O-Al groups on the surface. The maximum sorption capacities (q_{max}) of both oxides for the above-mentioned textile dyes are shown in Fig. 8.7 [10, 11, 99].

It is important to emphasize the highest SA96 sorption capacity for the direct dye (49.2 mg/g). It was noticed that Direct Blue 71 adsorption on SA96 decreased in the presence of the anionic surfactant sodium dodecyl sulphate (SDS), whereas sodium chloride did not influence the retention of the dye [11]. The impact of auxiliaries such as surfactants and electrolytes on dye removal is very important because these substances are frequently present in industrial wastewaters originating from textile plants. The proposed adsorption mechanism of Direct Blue 71 dye on SA96 involved both the electrostatic interactions and hydrogen bonding. In paper [11] it was also confirmed that the mixed oxide SA96 can be applied for decolourization of raw textile wastewaters; 30% reduction of colour was observed after 96 h of contact time.

8.3.3 Polymers

Polymers soluble in the aqueous medium find a wide usage in many branches of industry, environmental protection and agriculture. Acrylic acid polymers (e.g. poly (acrylic acid), PAA, copolymer of acrylic acid with acrylamide-anionic poly(acryl-amide), PAM) are used as fillers in washing powders. They function as chelating agents, i.e. remove calcium and magnesium ions. They are applied as dispersants, foaming agents, thickeners, boiler scale inhibitors and flocculants. They are also used as auxiliaries for the paper and textile industries and wastewater treatment and as aids for the extraction of crude oil [100, 101]. The high-molecular-weight poly-acrylamides (especially of ionic character) act as effective agents preventing soil erosion and improving textural structure of arable soils. Additionally, poly(vinyl alcohol), PVA polymers and copolymers have important biomedical and pharmaceutical applications [102]. The PVA hydrogels have been studied as candidates for the tissue replacement material for soft contact lenses, artificial heart linings, artificial cartilages, catheters, skin and pancreas membranes. They are also components of drug delivery systems in oral, transdermal, buccal, intramuscular and rectal routes of administration. Moreover, PAA and PVA are excellent polymeric carriers for proteins, enzymes, drugs and other biologically active substances.

Such extensive use of polymers is associated with their presence in municipal and industrial effluents. The basic method of their removal is adsorption on different kinds of adsorbents, including mixed oxides.

The adsorption properties of four mixed oxides of the Me_xO_y-SiO_2 type (Me, metal atom: Mg, Cu, Mn or Zn) in relation to poly(vinyl alcohol) with the average molecular weight 100,000 and the content of acetate groups equal to 14% were examined [5, 6, 103]. NMOs including silica and a given metal oxide were characterized by different contents of metal oxide, i.e. 0.2 and 1 mmol/g SiO_2. Thus they were designated as 0.2 Me_xO_y-SiO_2 and 1 Me_xO_y-SiO_2, respectively.

As can be seen in Fig. 8.8 at pH 6, PVA shows similar affinity for the surfaces of all examined mixed oxides of which the highest adsorption of the polymer was obtained for 1 Mn_xO_y-SiO_2 (approx. 0.4 mg/m^2). Additionally, for all studied systems, the adsorbed amount of polyalcohol increases with the increasing content of metal oxide in the mixed oxide structure. The electrokinetic data proved that the solid surface group containing metal atoms (-Si-O-Me) [104] exhibits preferential adsorption properties of the polymer in relation to the silica hydroxyl ones (-Si-O-H). For this reason a greater number of possible connections between the polymer segments and the -Si-O-Me groups can be formed. As a result, the polymer adsorption increases with the increasing metal oxide content in the silica hybrid material.

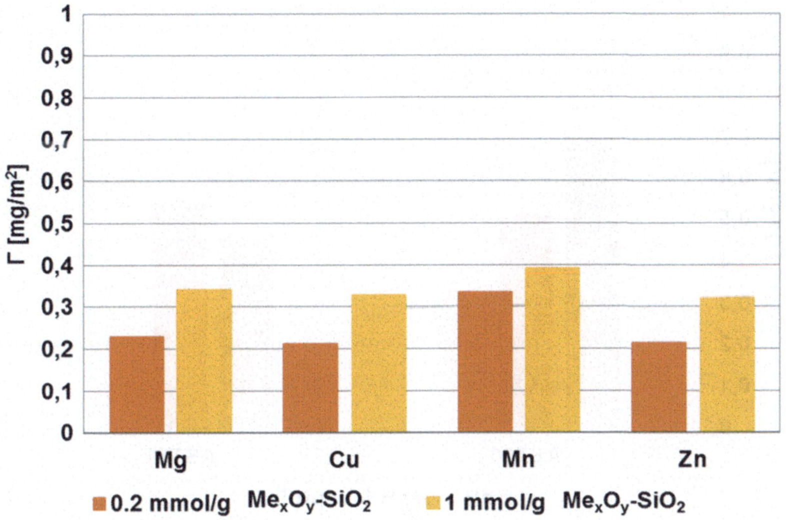

Fig. 8.8 Adsorbed amounts of poly(vinyl alcohol) PVA 100,000 on Me_xO_y-SiO_2 surface (initial polymer concentration 100 ppm, pH 6)

Under natural pH conditions (pH 6), the PVA binding with the mixed oxide surface is mainly of electrostatic nature (points of zero charge (pzc) for all Mn_xO_y-SiO_2 adsorbents are at the pH values higher than 6) [5, 6, 103]. Under such conditions the electrostatic attraction between the negatively charged macromolecules (containing ionized acetate groups) and the positive solid surface (coming from the silica $SiOH_2^+$ groups and metal -Si-O-$MeOH_2^+$ ones) occurs.

The significantly greater adsorption of poly(vinyl alcohol) reaching 0.7 mg/m^2 was obtained on the surface of ternary alumina-silica-titania (Al_2O_3-SiO_2-TiO_2, AST) mixed oxide (Fig. 8.9). Two AST oxides differed in composition – AST50 consisted of 50% TiO_2, 28% SiO_2 and 22% Al_2O_3 whereas AST71 71% TiO_2, 8% SiO_2 and 21% Al_2O_3 were applied (Fig. 8.9).

The amount of adsorbed PVA on the AST surface depends on the solution pH – the decrease of poly(vinyl alcohol) adsorption with the increasing pH for AST50-containing system and the increase of PVA adsorbed amounts with the pH rise on the AST70 surface occur. This is due to specific conformation of the adsorbed macromolecules on the surfaces of examined oxides (with different types of active sites) [105, 106]. The concentration of the hydroxyl surface groups of the mixed oxide, both neutral (-MeOH) and charged (-$MeOH_2^+$, -MeO^-), where Me means Al, Si and Ti atoms, changes with the pH increase. AST50 oxide was characterized by the pH_{pzc} and pH_{iep} (isoelectric point) values equal to 4.8 and 5.8 whereas AST70 by 4.2 and 8.6, respectively. These differences, especially considerable between the diffusion parts of electrical double-layer composition (manifested by pH_{iep} position), change significantly adsorption properties of AST mixed oxides and influence efficiency of PVA removal from aqueous solutions of varying pH values.

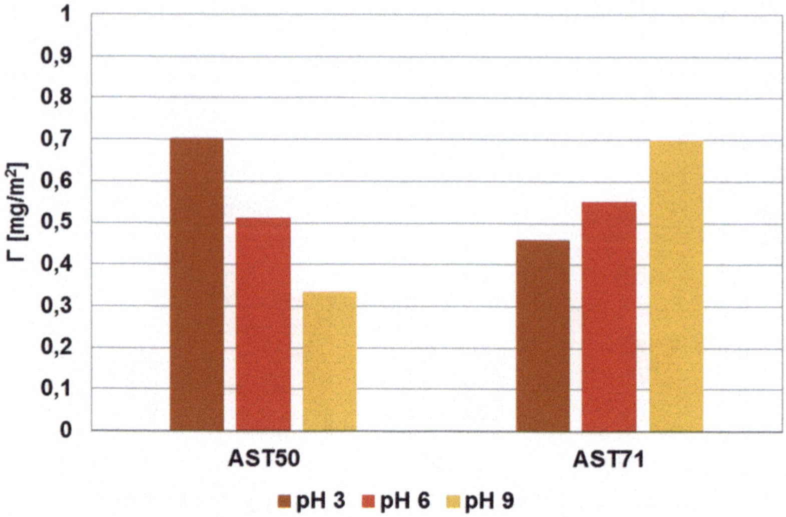

Fig. 8.9 Adsorbed amounts of poly(vinyl alcohol) PVA 100,000 on the Al_2O_3-SiO_2-TiO_2 surface (initial polymer concentration 100 ppm)

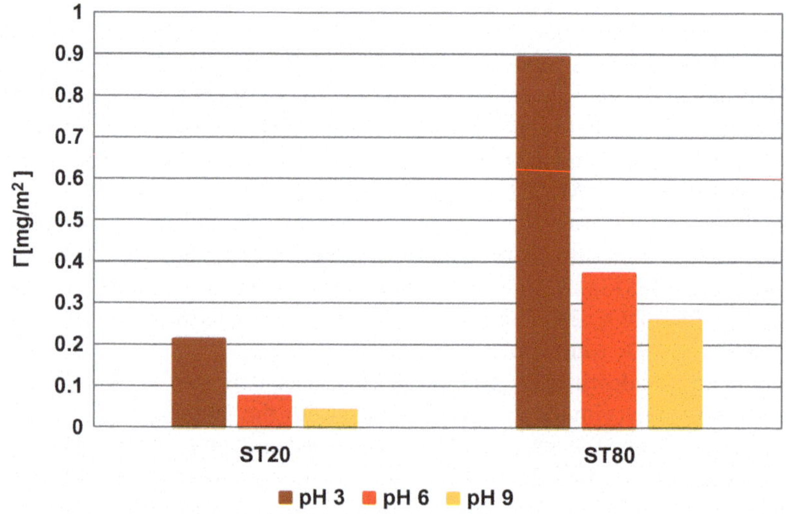

Fig. 8.10 Adsorbed amounts of anionic polyacrylamide PAM 15,500,000 on the SiO_2-TiO_2 surface (initial polymer concentration 100 ppm)

Two mixed silica-titania oxides were also used for the anionic polyacrylamide, PAM adsorption (Fig. 8.10). The polymeric substance was characterized by the average molecular weight 15,500,000 Da and the content of carboxyl groups equal to 50%. The composition of SiO_2-TiO_2, ST oxides was as follows: ST20, 80% SiO_2 and 20% TiO_2 and ST80, 20% SiO_2 and 80% TiO_2.

As can be seen in Fig. 8.9, the adsorption properties of silica-titania mixed oxide in relation to anionic polyacrylamide are determined by the solid composition. This polymer exhibits significantly greater affinity for the surface of the mixed oxide containing 80% of TiO_2. The main reason for this can be a greater tendency of polymeric functional groups (especially carboxyl and amide – to the lesser extent [107]) to interact with the surface groups combined with titania atoms. In such a case, the PAM macromolecules assume flatter conformation on the ST20 oxide surface (characterized by a small content of TiO_2). It limits considerably the accessibility of other polymeric chains to the solid surface and leads to the decrease in PAM adsorbed amounts. Due to the fact that anionic polyacrylamide contains as much as 50% of dissociable carboxyl groups, its interfacial behaviour is influenced by the solution pH. For both examined ST oxides, the greatest adsorption level of PAM is observed at pH 3, at which the favourable electrostatic interactions between the solid and the polymer occur (pH_{pzc} values were 3.8 and 4.8 for ST20 and ST80, respectively). The attractive interactions were observed between the slightly negatively charged macromolecules (small dissociation of PAM carboxyl groups) and the oppositely charged surface hydroxyl groups.

Similar tendencies were observed in the SiO_2-Al_2O_3/poly(acrylic acid) system [108], which is presented in Fig. 8.11. Each segment in the poly(acrylic acid) PAA chain contains the carboxyl functional group (characterized by the pK_a value 4.5). The composition of the SiO_2-Al_2O_3, SA adsorbents was SA3, 97% SiO_2 and 3% Al_2O_3 and SA96, 4% SiO_2 and 96% Al_2O_3. These oxides had the following pH_{pzc} values: 3.4 for SA3 and 7.6 for SA96.

It should be noted that under unfavourable electrostatic forces between the adsorbate and the adsorbent, the polymer adsorption on the mixed oxide surface

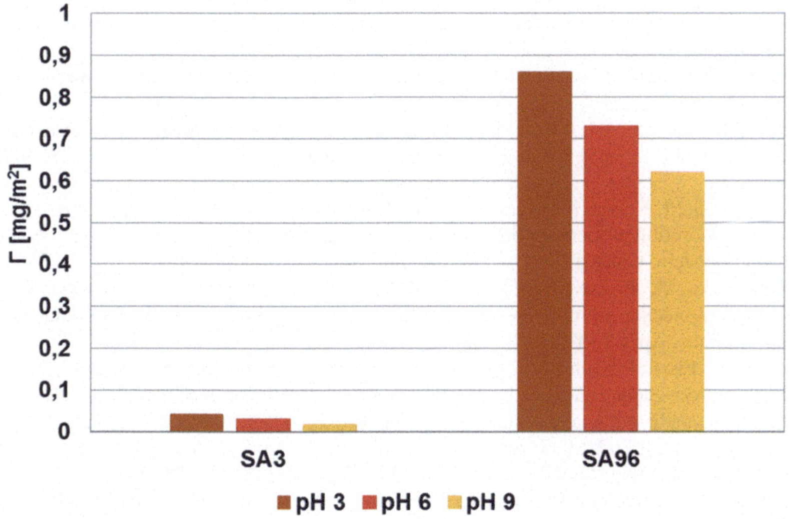

Fig. 8.11 Adsorbed amounts of poly(acrylic acid) PAA 240,000 on the SiO_2-Al_2O_3 surface (initial polymer concentration 100 ppm)

proceeds through the hydrogen bridges and chemical bonds [109]. Carboxyl groups of the polymer can act both as a donor and an acceptor of protons. In such a situation, their binding with the solid surface can take place between all types of adsorbent groups (negative, positive and neutral) and undissociated and dissociated PAA functional groups.

8.4 Conclusions

In the development of inorganic and organic substances such as metal ions, dyes and polymers, the adsorbents possessing not only high adsorption capacity but also by relatively low cost, are largely expected in both science and technology. As was shown above, different NMOs are widely used for their removal from water and wastewaters. Despite various advantages of NMOs, there still exist some technical bottlenecks which should be solved, e.g. problems with aggregation in aqueous solution, capacity loss, pressure drop in the column operation and economic and efficient and easy separation of NMOs from aqueous solutions. Such problems still remain an interesting and challenging task. Therefore fabrication of new NMOs-based composites of much better properties has been still in progress.

References

1. Hubbard AT (ed) (2002) Encyclopedia of surface and colloid science. Dekker, New York
2. Bergna HE (ed) (2005) Colloidal silica: fundamentals and applications. Taylor & Francis, Salisbury
3. Shpak AP, Gorbik PP (eds) (2010) Nanomaterials and supramolecular structures. Springer, Dordrecht
4. Blitz JP, Gun'ko VM (eds) (2006) Surface chemistry in biomedical and environmental science, NATO science series II: mathematics, physics and chemistry, vol 228. Springer, Dordrecht
5. Wiśniewska M, Nowicki P, Bogatyrov VM, Nosal-Wiercińska A, Pietrzak R (2016) Comparison of adsorption properties of Mg_xO_y–SiO_2 and Zn_xO_y–SiO_2 in the mixed oxide-poly(vinyl alcohol) system. Colloids Surf A Physicochem Eng Asp 492:12–18
6. Wiśniewska M, Bogatyrov V, Szewczuk-Karpisz K, Ostolska I, Terpiłowski K (2015) Adsorption mechanism of poly(vinyl alcohol) at the mixed oxide Cu_xO_y-SiO_2/aqueous solution interface. Appl Surf Sci 356:905–910
7. Klonos P, Pissis P, Gun'ko V, Kyritsisa A, Guzenko V, Pakhlov E, Zarko V, Janusz W, Skubiszewska-Zięba J, Leboda R (2010) Interaction of poly(ethylene glycol) with fumed silica and alumina/silica/titania. Colloids Surf A Physicochem Eng Asp 360(1–3):220–231
8. Voronin EF, Gun'ko VM, Guzenko NV, Pakhlov EM, Chuiko AA (2004) Interaction of poly (ethylene oxide) with fumed silica. J Colloid Interface Sci 279(2):326–340
9. Gun'ko VM, Zarko VI, Mironyuk IF, Goncharuk EV, Guzenko NV, Borysenko MV, Gorbik PP, Mishchuk OA, Janusz W, Leboda R, Skubiszewska-Zięba J, Grzegorczyk W, Matysek M, Chibowski S (2004) Surface electric and titration behaviour of fumed oxides. Colloids Surf A Physicochem Eng Asp 240(1–3):9–25

10. Wawrzkiewicz M, Wiśniewska M, Wołowicz A, Gun'ko VM, Zarko VI (2017) Mixed silica-alumina oxide as sorbent for dyes and metal ions removal from aqueous solutions and wastewaters. Micro Meso Mater 250:128–147
11. Wawrzkiewicz M, Wiśniewska M, Gun'ko VM, Zarko VI (2015) Adsorptive removal of acid, reactive and direct dyes from aqueous solutions and wastewaters using mixed silica-alumina oxide. Powder Technol 278:306–315
12. Gun'ko VM, Yurchenko GR, Turov VV, Goncharuk EV, Zarko VI, Zabuga AG, Matkovsky AK, Oranska OI, Leboda R, Skubiszewska-Zięba J, Janusz W, Phillips GJ, Mikhalovsky SV (2010) Adsorption of polar and nonpolar compounds onto complex nanooxides with silica, alumina, and titania. J Colloid Interface Sci 348(2):546–558
13. Di G, Zhu Z, Zhang H, Zhu J, Lu H, Zhang W, Qiu Y, Zhu L, Küppers S (2017) Simultaneous removal of several pharmaceuticals and arsenic on Zn-Fe mixed metal oxides: combination of photocatalysis and adsorption. Chem Eng J 328:141–151
14. Gao L, Li Q, Hu X, Wang X, Song H, Yan L, Xiao H (2016) One-pot synthesis of biomorphic Mg-Al mixed metal oxides with enhanced methyl orange adsorption properties. Appl Clay Sci 126:299–305
15. Lei C, Zhu X, Zhu B, Yu J, Ho W (2016) Hierarchical NiO–SiO_2 composite hollow microspheres with enhanced adsorption affinity towards Congo red in water. J Colloid Interface Sci 466:238–246
16. Skwarek E, Matysek–Nawrocka M, Janusz W, Zarko VI, Gun'ko VM (2008) Adsorption of heavy metal ions at the Al_2O_3-SiO_2/$NaClO_4$ electrolyte interface. Physicochem Probl Miner Process 42:153–164
17. Gun'ko VM, Nychiporuk YM, Zarko VI, Goncharuk EV, Mishchuk OA, Leboda R, Skubiszewska-Zieba J, Skwarek E, Janusz W, Yurchenko GR, Osovskii VD, Ptushinskii YG, Turov VV, Gorbik PP, Blitz JP, Gude K (2007) Relationships between surface compositions and properties of surfaces of mixed fumed oxides. Appl Surf Sci 253(6):3215–3230
18. Gun'ko VM, Zarko VI, Leboda R, Chibowski E (2001) Aqueous suspension of fumed oxides: particle size distribution and zeta potential. Adv Colloid Interface Sci 91:1–112
19. Gun'ko VM, Pakhlov EM, Skubiszewska-Zięba J, Blitz JP (2017) Infrared spectroscopy as a tool for textural and structural characterization of individual and complex fumed oxides. Vib Spectrosc 88:56–62
20. Gun'ko VM, Zarko VI, Leboda R, Marciniak M, Janusz W, Chibowski S (2000) Highly dispersed X/SiO_2 and C/X/SiO_2 (X=alumina, titania, alumina/titania) in the gas and liquid media. J Colloid Interface Sci 230(2):396–409
21. Sulym I, Sternik D, Oleksenko L, Lutsenko L, Borysenko M, Deryło-Marczewska A (2016) Highly dispersed silica-supported ceria–zirconia nanocomposites: preparation and characterization. Surf Interface 5:8–14
22. Gun'ko VM, Blitz JP, Bandaranayake B, Pakhlov EM, Zarko VI, Ya SI, Kulyk KS, Galaburda MV, Bogatyrev VM, Oranska OI, Borysenko MV, Leboda R, Skubiszewska-Zięba J, Janusz W (2012) Structural characteristics of mixed oxides MO_x/SiO_2 affecting photocatalytic decomposition of methylene blue. Appl Surf Sci 258:6288–6296
23. Reddy BM, Thrimurthulu G, Saikia P, Bharali P (2007) Silica supported ceria and ceria–zirconia nanocomposite oxides for selective dehydration of 4-methylpentan-2-ol. J Mol Catal A Chem 275:167–173
24. Reddy BM, Saikia P, Bharali P, Katta L, Thrimurthulu G (2009) Highly dispersed ceria and ceria–zirconia nanocomposites over silica surface for catalyticapplications. Catal Today 141:109–114
25. Navío JA, Colón G, Macías M, Sánchez-Soto PJ, Augugliaro V, Palmisano L (1996) ZrO_2–SiO_2 mixed oxides: surface aspects, photophysical properties andphotoreactivity for 4-nitrophenol oxidation in aqueous phase. J Mol Catal A Chem 109:239–248
26. Gao X, Fierro JLG, Wachs IE (1999) Structural characteristics and catalytic properties of highly dispersed ZrO_2/SiO_2 and V_2O_5/ZrO_2/SiO_2 catalysts. Langmuir 15:3169–3178

27. Basic characteristics of aerosil, Technical bulletin pigments, no. 11, Degussa AG, Hanau, 1997
28. https://www.wacker.com/cms/en/products/brands_2/hdk/hdk.jsp
29. https://www.aerosil.com/sites/lists/RE/DocumentsSI/Technical-Overview-AEROSIL-Fumed-Silica-EN.pdf
30. Pierson HO (1999) Handbook of chemical vapor deposition: principles, technology and applications. Noyes Publications, New York
31. Voronin EF, Pakhlo EM, Chuiko AA (1995) Chemisorption and hydrolysis of $TiCl_4$ on the surface of pyrogenic silica. Colloids Surf A Physicochem Eng Asp 101(2–3):123–127
32. Gun'ko VM, Zarko VI, Turov VV, Leboda R, Chibowski E, Holysz L, Pakhlov EM, Voronin EF, Dudnik VV, Gornikov YI (1998) CVD-titania on fumed silica substrate. J Colloid Interface Sci 198(1):141–156
33. Sulym I, Goncharuk O, Skwarek E, Sternik D, Borysenko MV, Derylo-Marczewska A, Janusz W, Gun'ko VM (2015) Silica-supported ceria–zirconia and titania–zirconia nanocomposites: structural characteristics and electrosurface properties. Colloids Surf A Physicochem Eng Asp 482:631–638
34. Sulym IY, Goncharuk O, Sternik D, Skwarek E, Derylo-Marczewska A, Janusz W, Gun'ko VM (2016) Silica-supported Titania–zirconia nanocomposites: structural and morphological characteristics in different media. Nanoscale Res Lett 11:111. https://doi.org/10.1186/s11671-016-1304-1
35. Sulim IY, Borysenko MV, Korduban OM, Gun'ko VM (2009) Influence of silica matrix morphology on characteristics of grafted nanozirconia. Appl Surf Sci 255:7818–7824
36. Bogatyrev VM, Gun'ko VM, Galaburda MV, Borysenko MV, Pokrovsky VA, Oranska OI, Polshin EV, Korduban OM, Leboda R, Skubiszewska-Zieba J (2009) Synthesis and characterization of Fe_2O_3/SiO_2 nanocomposites. J Colloid Interface Sci 338:376–388
37. Gun'ko VM, Bogatyrev VM, Borysenko MV, Galaburda MV, Sulim IY, Petrus LV, Korduban OM, Polshin EV, Zaulychnyy YV, Karpets MV, Foya OO, Myronyuk IF, Chelyadyn VL, UYa D, Leboda R, Skubiszewska-Zieba J, Blitz JP (2010) Morphological, structural and adsorption features of oxide composites with silica and titania matrices. Appl Surf Sci 256:5263–5269
38. Singh LP, Bhattacharyya SK, Ahalawat S, Kumar R, Mishra G, Sharma U, Singh G (2014) Sol-Gel processing of silica nanoparticles and their applications. Adv Colloid Interface Sci 214:17–37
39. Al Abdullah K, Awad S, Zaraket J, Salame C (2017) Synthesis of ZnO nanopowders by using sol-gel and studying their structural and electrical properties at different temperature. Energy Procedia 119:565–570
40. Imran M, Riaz S, Naseem S (2015) Synthesis and characterization of titania nanoparticles by sol-gel technique. Mater Today Proc 2(10):5455–5461
41. Harraz FA, Abdel-Salam OE, Mostafa AA, Mohamed RM, Hanafy M (2013) Rapid synthesis of titania–silica nanoparticles photocatalyst by a modified sol–gel method for cyanide degradation and heavy metals removal. J Alloys Compd 551:1–7
42. Tchounwou PB, Yedjou CG, Patlolla AK, Sutton DJ (2012) Heavy metals toxicity and the environment. EXS 101:133–164
43. Sharma K, Agrwal M (2005) Biological effects of heavy metals: an overview. J Environ BioI 26(2 suppl):301–313
44. Lakherwal D (2014) Adsorption of heavy metals: a review. Intern J Environ Res Develop 4 (1):41–48
45. Fu F, Wang Q (2011) Removal of heavy metal ions from wastewaters: a review. J Environ Manag 92(3):407–418
46. Agarwal SK (2009) Heavy metal pollution. APH Publishing Corporation, New Delhi
47. Lim AP, Aris AZ (2014) A review on economically adsorbents on heavy metals removal in water and wastewater. Rev Environ Sci Biotechnol 13:163–181

48. Da'na E (2017) Adsorption of heavy metals on functionalized-mesoporous silica: a review. Micro Meso Mater 247:145–157
49. Bakhiyi B, Gravel S, Ceballos D, Flynn MA, Zayed J (2018) Has the question of e-waste opened a Pandora's box? An overview of unpredictable issues and challenges. Environ Int 110:173–192
50. Hua M, Zhang S, Pan B, Zhang W, Lv L, Zhang Q (2012) Heavy metal removal from water/wastewater by nanosized metal oxides: a review. J Hazard Mater 211–212:317–331
51. Kurniawan TA, Chan GYS, Lo WH, Babel S (2006) Physico-chemical treatment techniques for wastewater laden with heavy metals. Chem Eng J 118:83–98
52. O'Connell DW, Birkinshaw V, O'Dwyer TF (2008) Heavy metal adsorbents prepared from the modification of cellulose: a review. Bioresour Technol 99:6709–6724
53. Aderhold D, Williams CJ, Edyvean RGJ (1996) The removal of heavy metal ions by seaweeds and their derivatives. Bioresour Technol 58(1):1–6
54. Fan M, Boonfueng T, Xu Y, Axe L, Tyson TA (2005) Modeling Pb sorption to microporous amorphous oxides as discrete particles and coatings. J Colloid Interface Sci 281:39–48
55. Hu J, Chen GH, Lo IMC (2005) Removal and recovery of Cr(VI) from wastewater by maghemite nanoparticles. Water Res 39:4528–4536
56. Martynyuk O, Kotolevich Y, Pestryakov A, Mota-Morales JD, Bogdanchikova N (2015) Nanostructures constituted by unusually small silica nanoparticles modified with metal oxides as support for ultra-small gold nanoparticles. Colloids Surf A Physicochem Eng Asp 487:9–16
57. Chen YH, Li FA (2010) Kinetic study on removal of copper (II) using goethite and hematite nano-photocatalysts. J Colloid Interface Sci 347:277–281
58. Hu J, Chen G, Lo IMC (2006) Selective removal of heavy metals from industrial wastewater using maghemite nanoparticle: performance and mechanisms. J Environ Eng-ASCE 132:709–715
59. Swallow KC, Hume DN, Morel FMM (1980) Sorption of copper and lead by hydrous ferric-oxide. Environ Sci Technol 14:1326–1331
60. Trivedi P, Dyer JA, Sparks DL (2003) Lead sorption onto ferrihydrite. 1. A macroscopic and spectroscopic assessment. Environ Sci Technol 37:908–914
61. Mishra SP, Vijaya (2007) Removal behavior of hydrous manganese oxide and hydrous stannic oxide for Cs (I) ions from aqueous solutions. Sep Purif Technol 54:10–17
62. Mishra SP, Dubey SS, Tiwari D (2004) Rapid and efficient removal of hg (II) by hydrous manganese and tin oxides. J Colloid Interface Sci 279:61–67
63. Takamatsu T, Kawashima M, Koyama M (1985) The role of Mn^{2+}-rich hydrous manganese oxide in the accumulation of arsenic in lake-sediments. Water Res 19:1029–1032
64. Tripathy SS, Bersillon JL, Gopal K (2006) Adsorption of Cd^{2+} on hydrous manganese dioxide from aqueous solutions. Desalination 194:11–21
65. Misono M, Ochiai EI, Saito Y, Yoneda Y (1967) A new dual parameter scale for strength of Lewis acids and bases with evaluation of their softness. J Inorg Nucl Chem 29:2685–2691
66. Dyer A, Pillinger M, Newton J, Harjula R, Moller T, Amin S (2000) Sorption behavior of radionuclides on crystalline synthetic tunnel manganese oxides. Chem Mater 12:3798–3804
67. Tsuji M, Komarneni S (1993) Selective exchange of divalent transition-metal ions in cryptomelane-type manganic acid with tunnel structure. J Mater Res 8:611–616
68. Ghaedi M, Niknam K, Shokrollahi A, Niknam E, Rajabi HR, Soylak M (2008) Flame atomic absorption spectrometric determination of trace amounts of heavy metal ions after solid phase extraction using modified sodium dodecyl sulfate coated on alumina. J Hazard Mater 155:121–127
69. Dadfarnia S, Shabani AMH, Shirie HD (2002) Determination of lead in different samples by atomic absorption spectrometry after preconcentration with dithizone immobilized on surfactant-coated alumina. Bull Kor Chem Soc 23:545–548

70. Shabani AMH, Dadfarnia S, Dehghani Z (2009) On-line solid phase extraction system using 1,10-phenanthroline immobilized on surfactant coated alumina for the flame atomic absorption spectrometric determination of copper and cadmium. Talanta 79:1066–1070
71. Afkhami A, Saber-Tehrani M, Bagheri H (2010) Simultaneous removal of heavy-metal ions in wastewater samples using nano-alumina modified with 2,4-dinitrophenylhydrazine. J Hazard Mater 181:836–844
72. Engates KE, Shipley HJ (2011) Adsorption of Pb, Cd, Cu, Zn, and Ni to titanium dioxide nanoparticles: effect of particle size, solid concentration, and exhaustion. Environ Sci Pollut Res 18:386–395
73. Liang P, Shi T, Li J (2004) Nanometer-size titanium dioxide separation/preconcentration and FAAS determination of trace Zn and Cd in water sample. Int J Environ Anal Chem 84:315–321
74. Mahdavi S, Jalali M, Afkhami A (2013) Heavy metals removal from aqueous solutions using TiO_2, MgO, and Al_2O_3 nanoparticles. Chem Eng Commun 200:448–470
75. Wang XB, Cai WP, Lin YX, Wang GZ, Liang CH (2010) Mass production of micro/nanostructured porous ZnO plates and their strong structurally enhanced and selective adsorption performance for environmental remediation. J Mater Chem 20:8582–8590
76. Ma XF, Wang YQ, Gao MJ, Xu HZ, Li GA (2010) A novel strategy to prepare ZnO/PbS heterostructured functional nanocomposite utilizing the surface adsorption property of ZnO nanosheets. Catal Today 158:459–463
77. Mahdavi S, Jalali M, Afkhami A (2012) Removal of heavy metals from aqueous solutions using Fe_3O_4, ZnO, and CuO nanoparticle. J Nanopart Res 14:846
78. Gao CL, Zhang WL, Li HB, Lang LM, Xu Z (2008) Controllable fabrication of mesoporous MgO with various morphologies and their absorption performance for toxic pollutants in water. Cryst Growth Des 8:3785–3790
79. Zhang F, Jin Q, Chan SW (2004) Ceria nanoparticles: size, size distribution, and shape. Jpn J Appl Phys 95:4319–4326
80. Bernal S, Calvino JJ, Cauqui MA, Gatica JM, Larese C, Omil JAP, Pintado JM (1999) Some recent results on metal/support interaction effects in NM/CeO_2 (NM: noble metal) catalysts. Catal Today 50:175–206
81. Contreras AR, Casals E, Puntes V, Komilis D, Sánches A, Font X (2015) Use of cerium oxide nanoparticles for the adsorption of dissolved cadmium(II), lead(II) and chromium(VI) at two different pHs in single and multi-component systems. Global NEST J 17:536–543
82. Eren E (2009) Removal of lead ions by Unye (Turkey) bentonite in iron and magnesium oxide-coated forms. J Hazard Mater 165:63–70
83. Eren E, Tabak A, Eren B (2010) Performance of magnesium oxide-coated bentonite in removal process of copper ions from aqueous solution. Desalination 257:163–169
84. Huang SH, Chen DH (2009) Rapid removal of heavy metal cations and anions from aqueous solutions by an amino-functionalized magnetic nano-adsorbent. J Hazard Mater 163:174–179
85. Oliveira LCA, Petkowicz DI, Smaniotto A, Pergher SBC (2004) Magnetic zeolites: a new adsorbent for removal of metallic contaminants from water. Water Res 38:3699–3704
86. Shin S, Jjang J (2007) Thiol containing polymer encapsulated magnetic nanoparticles as reusable and efficiently separable adsorbent for heavy metal ions. Chem Commun 41:4230–4232
87. Pereira L, Alves M (2015) Dyes – environmental impact and remediation. In: Malik A, Grohmann E (eds) Environmental protection strategies for sustainable development, strategies for sustainability. Springer, Dordrecht/Heidelberg/London/New York, pp 111–162
88. Gupta VK, Suhas (2009) Application of low-cost adsorbents for dye removal – a review. J Environ Manag 90(8):2313–2342

89. Elwakeel KZ (2009) Removal of Reactive Black 5 from aqueous solutions using magnetic chitosan resins. J Hazard Mater 167(1):383–392
90. Padmavathy S, Sandhya S, Swaminathan K, Subrahmanyam YV, Chakrabarti T, Kaul SN (2003) Aerobic decolorization of reactive azo dyes in presence of various cosubstrates. Chem Biochem Eng Q 17(2):147–151
91. Kumar Pandey A, Dubey V (2012) Biodegradation of azo dye reactive red BL by Alcaligenes sp. AA09. Int J Eng Sci 1(12):54–60
92. Dotto GL, Vieira MLG, Esquerdo VM, Pinto LAA (2013) Equilibrium and thermodynamics of azo dyes biosorption onto Spirulina platensis. Braz J Chem Eng 30(1):13–21
93. Saranraj P (2013) Bacterial biodegradation and decolourization of toxic textile azo dyes. Afr J Microbiol Res 7(30):3885–3890
94. Ciesielczyk F, Bartczak P, Zdarta J, Jesionowski T (2017) Active MgO-SiO_2 hybrid material for organic dye removal: a mechanism and interaction study of the adsorption of C.I. Acid blue 29 and C.I. Basic blue 9. J Environ Manag 204:123–135
95. Pal U, Sandoval A, Madrid SIU, Corro G, Sharma V, Mohanty P (2016) Mixed titanium, silicon, and aluminum oxide nanostructures as novel adsorbent for removal of rhodamine 6G and methylene blue as cationic dyes from aqueous solution. Chemosphere 163:142–152
96. Rasalingam S, Peng R, Koodali RT (2013) An investigation into the effect of porosities on the adsorption of rhodamine B using titania-silica mixed oxide xerogels. J Environ Manag 128:530–539
97. Tanzifi M, Yaraki MT, Kiadehi AD, Hosseini SH, Olazar M, Bharti AK, Agarwal S, Gupta VK, Kazemi A (2018) Adsorption of Amido black 10B from aqueous solution using polyaniline/SiO_2 nanocomposite: experimental investigation and artificial neural network modeling. J Colloid Interface Sci 510:246–261
98. Arshadia M, Vahid FS, Salvacion JWL, Soleymanzadeh M (2013) A practical organometallic decorated nano-size SiO_2-Al_2O_3 mixed-oxides for methyl orange removal from aqueous solution. Appl Surf Sci 280:726–736
99. Wawrzkiewicz M, Wiśniewska M, Gun'ko VM (2017) Application of silica–alumina oxides of different compositions for removal of C.I. Reactive black 5 dye from wastewaters. Ads Sci Technol 35(5–6):448–457
100. Abdelaal MY, Makki MSI, Sobahi TRA (2012) Modification and characterization of polyacrylic acid for metal ion recovery. Am J Polym Sci 2:73–78
101. Zheng H, Ma J, Ji F, Tang X, Chen W, Zhu J, Liao Y, Tan M (2013) Synthesis and application of anionic polyacrylamide in water treatment. Asian J Chem 25:7071–7074
102. Muppalaneni S, Omidian H (2013) Polyvinyl alcohol in medicine and pharmacy: a perspective. J Develop Drugs 2:3
103. Wiśniewska M, Bogatyrov V, Ostolska I, Szewczuk-Karpisz K, Terpiłowski K, Nosal-Wiercińska A (2016) Impact of poly(vinyl alcohol) adsorption on the surface characteristics of mixed oxide Mn_xO_y - SiO_2. Adsorption 22:417–423
104. Bogatyrov VM, Oranska IO, Gun'ko VM, Leboda R, Skubiszewska-Zięba J (2011) Influence of metal content on structural characteristics of inorganic nanocomposities M_xO_y/SiO_2 and $C/M_xO_y/SiO_2$. Chem Phys Tech Surf 2:135–146
105. Wiśniewska M, Ostolska I, Szewczuk-Karpisz K, Chibowski S, Terpiłowski K, Zarko VI, Gun'ko VM (2015) Investigation of the polyvinyl alcohol stabilization mechanism and adsorption properties on the surface of ternary mixed nanooxide AST (Al_2O_3-SiO_2-TiO_2). J Nanopart Res 17:12
106. Wiśniewska M, Szewczuk-Karpisz K, Ostolska I, Urban T, Terpiłowski K, Zarko VI, Gun'ko VM (2015) Effect of polyvinyl alcohol adsorption on the mixed alumina-silica-titania suspension stability. J Ind Eng Chem 23:265–272

107. Wiśniewska M, Chibowski S, Urban T (2015) Impact of polyacrylamide with different contents of carboxyl groups on the chromium(III) oxide adsorption properties in aqueous solution. J Hazard Mater 283:815–823
108. Wiśniewska M, Urban T, Grządka E, Zarko VI, Gun'ko VM (2014) Comparison of adsorption affinity of polyacrylic acid for surfaces of mixed silica-alumina. Colloid Polym Sci 292:699–705
109. Kasprzyk-Hordern B (2004) Chemistry of alumina, reactions in aqueous solution and its application in water treatment. Adv J Colloid Interface Sci 110:19–48

Chapter 9
Electrodes Based on Carbon Nanomaterials: Structure, Properties, and Application to Capacitive Deionization in Static Cells

Yurii Volfkovich, Daniil Bograchev, Alexey Mikhalin, Alexey Rychagov, Valentin Sosenkin, Vitaly Milyutin, and Daewook Park

9.1 Capacitive Deionization

Deionization processes, which are driven by electrical field, can be divided to two types. The first type is electrodeionization (EDI) that involves ion exchange and transport of charged species through the ion-exchanger bed and membranes [1, 2]. This method allows one to remove from water not only ions of alkaline metals, sulfate, and halogenides [1–3] but also transition metal cations [4, 5]. The second type is capacitive deionization (CDI); it is a promising and the most cost-effective method of water desalination [5–14]. An electric field affects adsorption of anions and cations on positive and negative electrodes, respectively, under low applied voltage (it is not sufficiently higher than 1.2 V). Highly dispersive carbon electrodes (HDCE) with a specific surface area of 500–3000 $m^2\ g^{-1}$ are used in CDI processes. The electric double layer (EDL) is charged similarly to that in electrochemical supercapacitors (ESCs), which results in deionization of the solution. Under polarity reversal or short circuit of the electrodes, ions diffuse from the interface back to the solution. This provides energy regeneration and concentrating of the solution that is removed from the cell. The deionization stage corresponds to the charging of the supercapacitor, while the concentrating stage (regeneration) is related to its discharging.

As compared with other methods of water desalination, sufficient advantages of CDI are as follows: low cost (about one third of the value of the closest competing process, reverse osmosis), high cycleability of the electrodes, and low maintenance costs. Moreover, chemical stability of the electrodes excludes permeation of foreign

Y. Volfkovich · D. Bograchev (✉) · A. Mikhalin · A. Rychagov · V. Sosenkin · V. Milyutin
A. N. Frumkin Institute of Physical Chemistry and Electrochemistry of the Russian Academy of Sciences, Moscow, Russia

D. Park
Samsung Electronics Co., Ltd, Gyeonggi-do, South Korea

O. Fesenko, L. Yatsenko (eds.), *Nanooptics, Nanophotonics, Nanostructures, and Their Applications*, Springer Proceedings in Physics 210,
https://doi.org/10.1007/978-3-319-91083-3_9

substances in the purified water. CDI is effective for treatment of water of different salinity: this method allows reducing the ion concentration down to a very low level. Finally, practically no insoluble compounds are precipitated inside the electrode pores. CDI is applied in desalination of brackish and sea water, as well as for softening of tap water, particularly for deionization of water supplied into washing machines.

Electrode materials for CDI are developed from the past years. Different types of both single-component (activated carbon, aerogels, nanotubes, graphene, etc.) [14–17] and carbon-based composite materials (carbon-carbon composite, carbon-metal oxide composite, carbon-polymer composite, and carbon-polymer-metal oxide composite) [18–21] have been suggested. Mesoporous materials (i.e., containing nanosized pores) are considered as main suitable electrodes for CDI, since these pores make significant contribution to surface area. Moreover, ion transport in these pores is faster than that in micropores. As known, porous structure of carbon materials includes both hydrophilic and hydrophobic pores [22–24]. Electrical double layer (EDL) that is responsible for adsorption capacitance of carbon exists only in hydrophilic pores. The aim of the investigation is to establish the effect of different types of nanopores on charge capacitance of carbon electrodes. This parameter determines the efficiency of CDI processes.

A number of models have been proposed for description of the CDI processes. Classical works [25–29] are used as a basis for the modeling of processes in porous CDI electrodes. In the absence of the electrolyte flux, CDI equations are similar to those in supercapacitor models [29], which are based on the theory of porous electrodes [30]. The first models of porous electrodes were proposed by de Levie more than 50 years ago [28, 29], but analytical models for discharge curves of supercapacitors have been obtained later [30]. Equation systems that describe the charging of the supercapacitor have been solved numerically [31].

More modern approaches consider electrosorption (EDL in a single pore) [32, 33], electrolyte transport along the electrodes and separator, as well as the charging of EDL of the electrodes [34–36] and surface conductivity (SC, tangential conductivity of EDL) [37–40]. Known CDI models require calculation of EDL inside pores of the electrodes. It is a very difficult problem, since it is necessary to consider the hydrophilicity-hydrophobicity of the electrodes, tortuosity, widenings-narrowings, intersections of pores, etc. The task of the investigation involves also modeling of CDI processes only based on parameters, which can be determined experimentally.

9.2 Carbon Electrodes and Their Characterization

Activated carbon textiles, such as CH900 (Curaray Co, Japan) and VISKUMAK (Neorganica LTD, RF), were used in the studies. SAIT-type electrodes (SAIT Co, South Korea) were also used. These materials were manufactured by compacting activated carbon powder together with a binder (polytetrafluorethylene, PTFE).

The method of standard contact porosimetry (MSCP) was used to study the electrode materials similarly to [41–44]. The MSCP allows one to investigate pore structure in a very wide interval of pore radius: from 1 nm to 100 μm. This technique can be applied to all materials. Besides carbon materials, pore size distributions were obtained for gas diffusion layers of electrodes for fuel cells [22, 41, 42, 45], dispersed platinum [46], polymer membranes [47], ion exchange resins [3, 5, 48], peat [42, 44], paper [42], and many other objects.

Before the measurements using water or octane, both the tested sample and standards were dried under vacuum at 170 °C.

Morphology of the electrodes was observed by means of a JSM-U3 scanning electron microscope (JEOL, Japan).

Adsorption capacity was determined as follows. A weighed sample of air-dried material (0.065 g) was inserted into the 0.2-M NaOH solution (20 cm^3); the liquid was stirred for 24 h. Then the effluent was titrated with an HCl solution using xylenol orange as an indicator. Adsorption capacity was calculated as $\frac{(c_0 - c_e)V}{M}$ (here c_0 and c_e are the concentrations of the alkaline solution before and the treated solution, respectively; V is its volume, m is the sample mass).

Adsorption capacity with respect to Cu^{2+} ions (in fact, toward $[Cu(NH_3)_4]^{2+}$) was also determined. The 0.2 M solution of copper ammoniate (pH = 9.6) was used; the effluent was analyzed by titration with EDTA using murexide [49].

9.3 Study of CDI Process in Static Cell

In general, the electric capacitance of electrodes made of activated carbon is a sum of EDL capacitance and pseudocapacitance of Faraday reactions [27]. Hence, electric capacitance was measured and EDL capacitance was calculated on the basis of this value and used further for modeling of the CDI processes.

Galvanostatic measurements of dependences of cell voltage on time were carried out under the charging-discharging conditions in order to determine the EDL capacitance. A static cell was used, in which the electrolyte was only in the pores of the electrodes and separator. The cell design (Fig. 9.1) and methods of measurements were described in [31]. This type of cell simulates both single-cell and multicell electrochemical capacitors for bipolar assemblies with serially connected individual cells. The cell contained two electrodes of the same type with similar surface areas (2.5–3 cm^2). Graphite current collectors provided good contact with the electrodes and allowed working in strongly acidic and strongly basic media. The range of applied potentials was 0–1.5 V. The current distribution layer (foil) was located between the electrode and current collectors. The collector was manufactured by compressing the thermoexfoliated graphite powder with the further impregnation by molten paraffin.

The assembly also included a separator (Gore Co.) between two electrodes. The separator film was used as a capillary for the reference electrode. This cell design

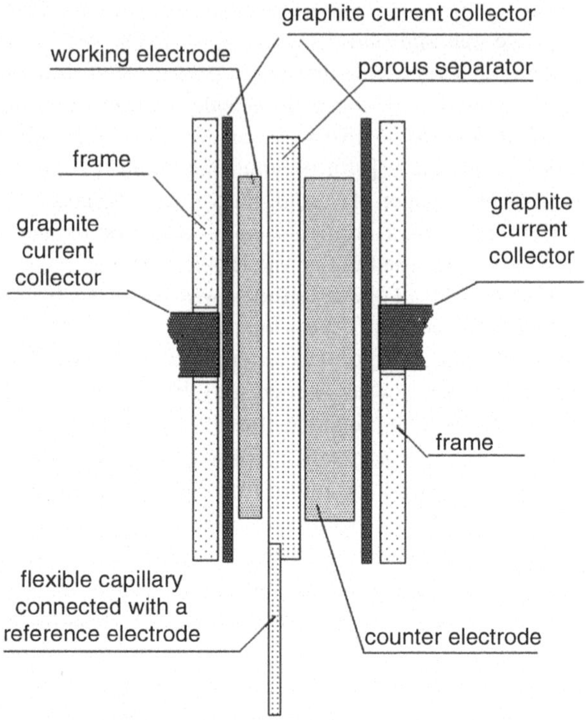

Fig. 9.1 Static electrochemical cell: scheme of electrode-separator assembly

allowed performing measurements according to both two- and three-electrode schemes. The electrodes were prewashed in a stream of deionized water and dried at 170° C under vacuum. Since the volume of electrolyte in electrode pores was extremely low (0.2–0.5 cm^3), oxygen was removed by conditioning the electrode for 10–15 min at the potential, which is close to that of hydrogen evolution. Electrochemical studies were carried out using a VoltaLab 40 potentiostat (Radiometer Analytical, France).

Galvanostatic measurements were performed in a static cell equipped by carbon electrodes. A change in the voltage during charging-discharging was controlled. Single-component $NaHCO_3$, $CaCl_2$, and $MgSO_4$ solutions of different concentrations (0.1, 0.5 and 1 g-eq dm^{-3}), as well as a mixed solution, were used as electrolytes. The composition of the mixed solution was as follows (mg-eq dm^{-3}): $NaHCO_3$ (4.4), $CaCl_2$ (11.9), $MgSO_4$ (10.7). This composition corresponded to brackish water. The capacitance values (C_c) were determined as:

$$C_{c,full} = \frac{2I_{full}\Delta t}{\Delta U} \tag{9.1}$$

where U is the cell voltage, I_{full} is current, t is the time.

Since adsorption capacity of the electrodes is proportional to capacitance of the electrode EDL ($C_{EDL,\ full}$), the change in the solution concentration (Δc) is:

$$\Delta c = \frac{C_{EDL,full}\Delta U}{2FV}, \tag{9.2}$$

where F is the Faraday constant and V is the solution volume. It was assumed in Eq. (9.2) that the adsorption efficiency of EDl charging is 100%. Expression (9.2) is also valid for electrodes of equal capacitance.

9.4 Porous Structure of Carbon Electrodes. Recognition of Micropores and Nanosized Voids

The principal MSCP data are given in Table 9.1; pore size distributions (obtained using both water and octane) are plotted vs effective pore radius (r^*) (Fig. 9.2). The r * parameter is determined as [41, 42]:

$$r^* = \frac{r}{\cos\theta} \tag{9.3}$$

where r is the true value of pore radius and θ is the wetting angle for water. According to Eq. (9.3), the θ–log$r$ curves were plotted (Fig. 9.3). Since octane wets all materials almost perfectly, $r^* \approx r$ for this liquid.

The difference between the maximal content of octane and water (maximal values of curves I and II) corresponds to hydrophobic pores. The carbon electrodes were found to include both hydrophilic and hydrophobic pores (hydrophilic and hydrophobic specific surface area). The contact angle distributions show hydrophobicity for all electrodes, especially for the SAIT material containing particles of the hydrophobic PTFE binder (see Fig. 9.3). Complex θ–log r dependences are due to inhomogeneous distribution of surface groups in pores of different size.

As follows from Fig. 9.2, the electrodes are characterized by a wide range of pore sizes: from r<1 nm to 100 μm. The porous structure of the electrodes includes both hydrophilic and hydrophobic porosity. High values of specific surface area differ noticeably (see Table 9.1). The values of hydrophilic and hydrophobic porosity are also different. The MSCP data for octane allow drawing some principal conclusions.

Table 9.1 Characteristics of porous structure of carbon electrodes

Electrode	Specific surface area, $m^2\ g^{-1}$		Ratio of hydrophilic and total surface areas	Porosity, $cm^3\ cm^{-3}$		
	Total	Hydrophilic		Total	Hydrophilic	Hydrophobic
CH900	1520	850	0.56	0.850	0.786	0.064
VISKUMAK	600	416	0.70	0.729	0.623	0.106
SAIT	940	520	0.55	0.715	0.490	0.225

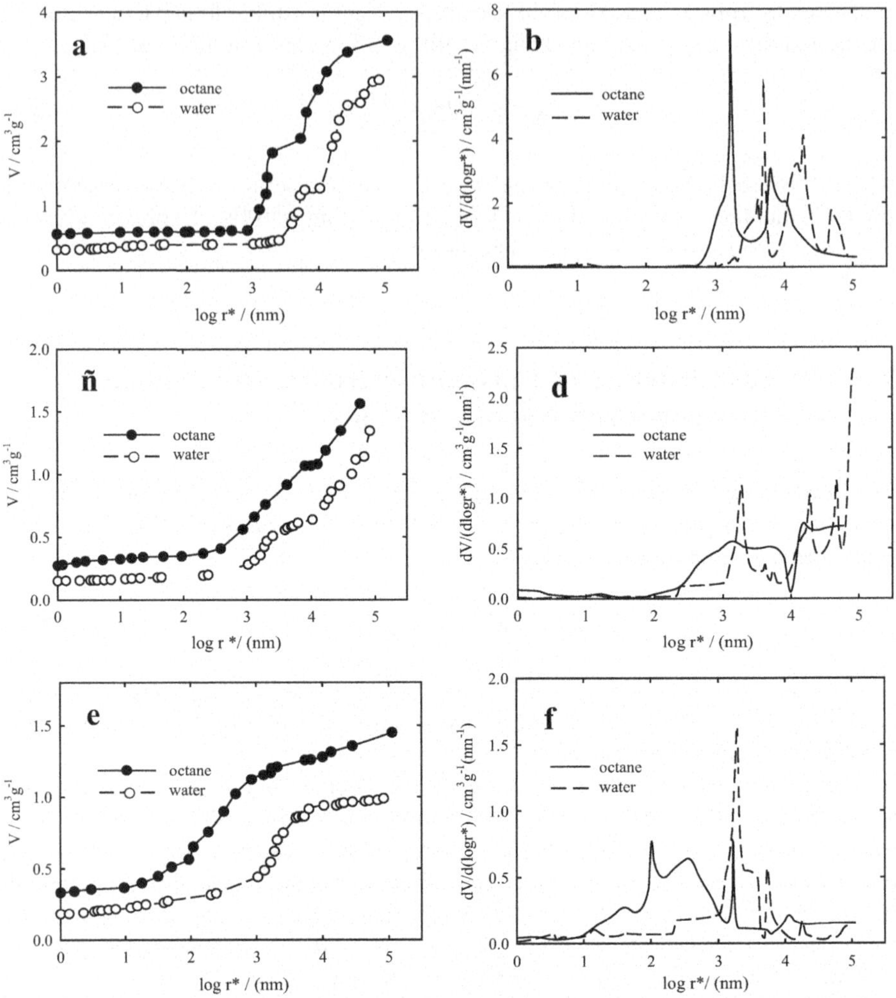

Fig. 9.2 Integral (**a**, **c**, **e**) and differential (**b**, **d**, **f**) pore size distributions for the CH900 (a, b), VISKUMAK (**c**, **d**), and SAIT (**e**, **f**) materials. The curves are plotted vs effective pore radius

Micropores, as well as large macropores, are characteristic for all materials. The volume of hydrophilic pores, which can be related to nanoobjects (r<50 nm) increases on the order: CH900 < VISKUMAK < SAIT. The total porosity (0.715–0.850) is sufficient for all electrodes.

Comparison of the data obtained in water and octane media shows higher hydrophilic porosity. The values of the total specific surface area of 600–1520 $m^2\ g^{-1}$ hydrophilic pores make sufficient contribution to it. This is very important, since EDL that determines the principal characteristics of the CDI processes exists on the

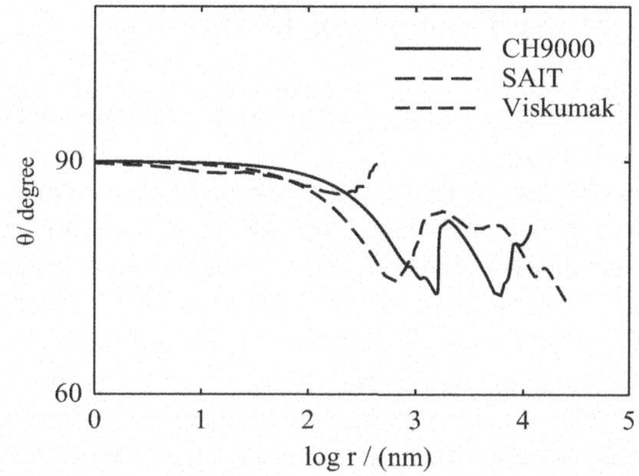

Fig. 9.3 Wetting angle distributions for the samples of carbon electrodes. The distributions were plotted vs true pore radius

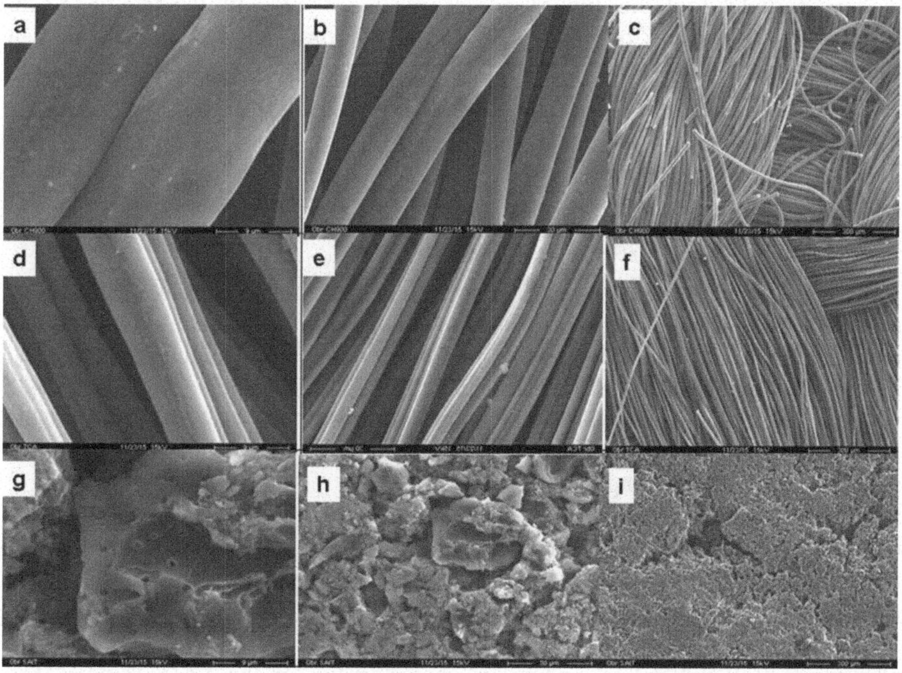

Fig. 9.4 SEM images of CH900 (**a–c**), VISKUMAK (**d–f**) textiles, and SAIT composite (**g–i**)

hydrophilic surface. The largest contribution of hydrophilic area into the total surface area has been observed for the VISKUMAK sample. Regarding the CH900 carbon material and SAIT composite, the ratios of hydrophilic and total surface areas are sufficiently lower and close to each other.

9.5 Morphology of Electrodes

Morphology of single-component materials and bicomponent SAIT composite are sufficiently different (Fig. 9.4). Large fibers of the CH900 sample are more disordered than those of the VISKUMAK, in which the fibers form ordered bundles. Voids between them correspond to the largest pores (see Fig. 9.2). Since the diameter of CH900 fibers is larger, the size of the corresponding pores is also larger as compared to the pores in the VISKUMAK sample. The specific surface area of the CH900 electrode is higher, indicating a more significant volume of micropores and small mesopores (r<5 nm) inside the fibers of this material.

The structure of the bicomponent SAIT sample is corpuscular-like. The material consists of particle agglomerates, the size of which is up to several tens of microns. These particles are evidently related to carbon bound to the polymer binder. Large massive particles (polymer binder) are also visible. Pores with the size of several microns to several tens of microns (see Fig. 9.4) correspond to spaces between agglomerates as well as to voids between them and the binder. Nanopores are mainly voids inside the particles.

9.6 Adsorption Capacity

Values of adsorption capacity are given in Table 9.2. The CH900 and VISKUMAK samples demonstrate the highest capacity with respect to Na^+. It should be noted that these values are higher in the case of fibrous single-component samples than those of Nafion-type membranes (0.8 0.9 mg-eq g^{-1} [50]). Despite a considerable difference in the hydrophilic specific surface area (see Table 9.1), the capacity values of the fibrous samples are close to each other. This means partial inaccessibility of the surface of the CH900 sample for Na^+ ions.

The difference between the values of adsorption capacity with respect to $[Cu(NH_3)_4]^{2+}$ is insignificant for all samples. Assuming ion exchange being the only mechanism, the experimental values were found to be higher than expected. The ratios of the experimental data and the calculated values are 1.3–1.4 (ACF) and 3.4 (SAIT). Thus, the larger content of surface groups that are responsible for specific adsorption of $[Cu(NH_3)_4]^{2+}$ species is characteristic for the SAIT sample. The capacity values are comparable with those for such inorganic ion exchangers as amorphous zirconium phosphate [51]. Since the structure as well as hydrophilic-

Table 9.2 Adsorption capacity of electrode materials

	Sorption capacity/mg-eq g^{-1}		
	Experimental values		Theoretical values
Material	Na^+	$[Cu(NH_3)_4]^{2+}$	
CH900	1.14	0.75	0.57
VISKUMAK	1.02	0.71	0.51
SAIT	0.37	0.62	0.18

hydrophobic and adsorption properties of the investigated materials is rather complex, it is difficult to predict their electrochemical behavior. Nevertheless, as shown in [52], adsorption capacity is proportional SC of the carbon electrodes. In its turn, SC affects conductivity of CDI cells. The maximal values of EDL capacitance are expected for the CH900 electrode due to the highest hydrophilicity and adsorption capacity.

9.7 Processes in Static Electrochemical Cell

In order to use the CDI model, it was necessary to obtain the values of specific EDL capacitance for each of the carbon electrodes and solutions of different salts of different concentrations. However, the current stage of development of electrochemistry offers no possibility for theoretical solution of this problem for microporous-mesoporous carbon electrodes. Owing to this, we have developed an experimental method to solve this task. The method involves measurements of electrochemical capacitance depending on the current. The capacitance values were determined on the basis of the measured galvanostatic curves using Eq. (9.1).

As an example, a dependence of integral capacitance of the CH900 electrode on current density is plotted in the insertion to Fig. 9.5. Decrease in current density results in growth of capacitance followed by a plateau and a rapid increase.

The increase in capacitance at 300–200 mA g^{-1} is caused by considerable ohmic losses under high current, as follows from insertion of Fig. 9.5. The decrease in current is accompanied by a decrease in ohmic losses. They become extremely low at 200–150 mA g^{-1} (plateau region) – the capacitance is determined only by the EDL. In the low current range, the contribution of pseudocapacitance of faradaic processes becomes significant. With respect to carbon electrodes, these processes are quasi-reversible redox reactions of their surface groups [27].

In the first approximation, specific capacitance of the EDL per mass unit (C_{EDL}) was estimated from the $C_{EDL,\ full}$value (see formula (9.2)), which is evaluated from the plateau data. For instance, C_{EDL}=63 (CH900, multicomponent solution) and 107 F g^{-1} (VISKUMAK, 1 N $CaCl_2$), as shown in Fig. 9.6.

The values of EDL capacitance per area unit (C^s_{EDL}) were also obtained. In this case, the C_{EDL} value was divided by the hydrophilic surface area from Table 9.1. The C_{EDL} and C^s_{EDL} values grow at an increase of the solution concentration. This is probably due the decreasing thickness of the diffuse part of EDL under these conditions. As a result, EDL can exist only in smaller pores. The values of EDL capacitance per unit of hydrophilic surface area are 9.2–26 μF cm^{-2}. With respect to CH900 and VISKUMAK textiles, the highest capacitance has been found for the 1:1 electrolyte ($NaHCO_3$), and the smallest values have been observed for the 2:2 electrolyte ($MgSO_4$). The data for the 1:2 electrolyte ($CaCl_2$) are in-between. It is possible to state the highest C_{EDL} value for the CH900 material that is characterized by the largest hydrophilic surface area (see Table 9.1) and the highest adsorption

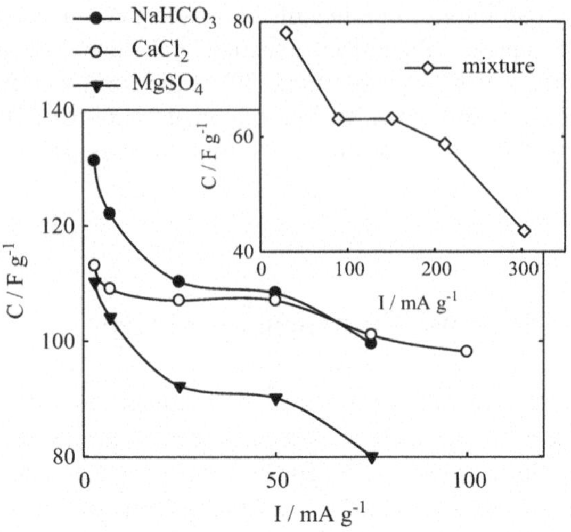

Fig. 9.5 Capacitance of the VISKUMAK electrode as a function of current density. One-component 1 N solutions were used. Masses (g) of the electrode and electrolyte were 0.076 and 0.28 ($NaHCO_3$), 0.033 and 0.19 ($CaCl_2$), and 0.058 and 0.21 ($MgSO_4$). The external electrode area was 3 ($NaHCO_3$) and 2.5 ($CaCl_2$, $MgSO_4$) cm^2. The same dependence for the CH900 electrode is plotted in the insertion. The curve was obtained for the multicomponent solution; masses of the electrode and electrolyte were 0.033 and 0.13 g, respectively

capacity (see Table 9.2). These characteristics are smallest for the SAIT composite that demonstrates the lowest values of EDL capacitance. Intermediate C_{EDL} values have been found for the VISKUMAK electrode. The highest C^s_{EDL} values have been obtained for this material. This is due to a large fraction of hydrophilic surface area that provides good connection between hydrophilic (electrochemically active) areas on the surface.

9.8 Modeling of Processes in Static Cell Involving Parameters Attributed to Hydrophilic Micropores and Nanosized Voids

Since the porous structure of AC involves a wide range of hydrophilic and hydrophobic pores (from <1 nm to 100 μm), exact calculation of C_{EDL} is practically impossible. Contrary to the known works, which involve models of electrosorption [32, 33], Donnan equilibrium [53, 54], and other approach (see above), an alternative point of view is developed. The model uses experimental SC [52] data as well as C_{EDL} values are obtained with account for hydrophilic and hydrophobic porosity of the electrodes. This approach allows one neglecting the thickness and composition of intraporous EDLs, their overlapping in weakly concentrated solutions, and absence in hydrophobic pores.

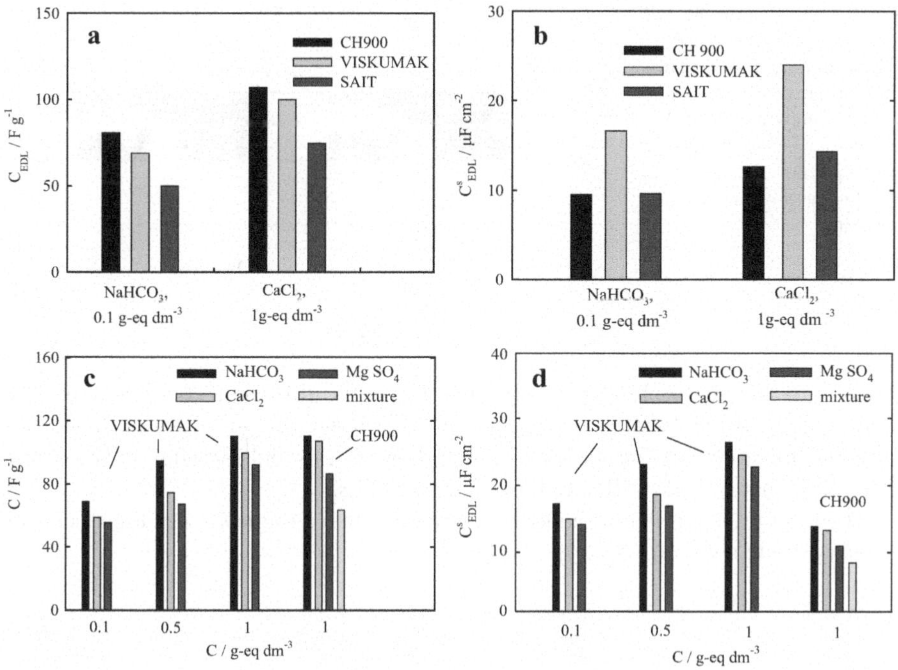

Fig. 9.6 Capacitance of EDL per mass unit (**a**, **c**) and per unit of hydrophilic surface area (**b**, **d**) for different carbon electrodes and different concentrations of electrolyte solutions

The following processes were considered: (i) charging-discharge of EDL, (ii) diffusion and migration of species, (iii) SC (the data [52] were used), (iv) specific adsorption of ions, and (v) characteristics of hydrophilic and hydrophobic pores. Both the EDL capacitance and SC are attributed to nanosized pores, which make sufficient contribution to the total specific surface area.

Application of supercapacitors to desalination of aqueous solutions is based on the properties of EDL of ideally polarized carbon electrodes. The charging of EDL causes adsorption of ions on the surface of the cathode and anode. i.e., removal of species from the solution. The main difference between an electrochemical cell for CDI and supercapacitor is usage of a diluted solution. The charging-discharging processes of EDL are characterized by reversibility and huge cycleability. Figure 9.7 illustrates the model with a simple 1D structure involving two porous carbon electrodes and a porous separator between them.

Resistance of carbon electrode is assumed to be neglected, since its conductivity is much higher than that of electrolyte. This assumption is justified due to the small concentration of hardness ions. For instance, the conductivity of the carbon electrode is about 10^5 S cm^{-1} (the conductivity of natural water containing hardness ions is about 2–3 mS cm^{-1}). Both the effective conductivity (κ) and effective diffusion coefficients D (through the electrode and separator) have to be corrected taking into consideration porous structure of the materials. The Archie's law is used [53–56]

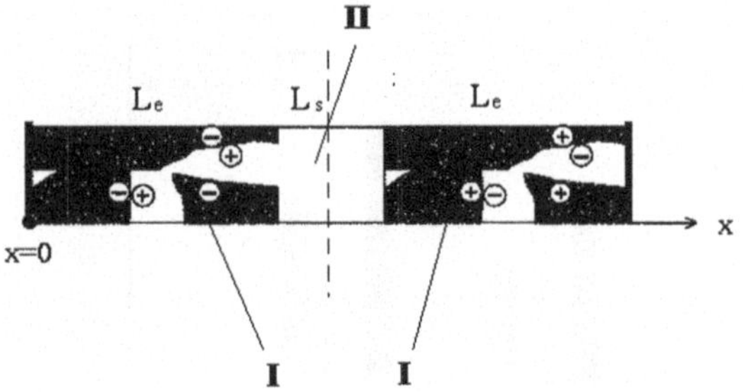

Fig. 9.7 Scheme of the cell model: (**I**) electrodes and (**II**) separator (spacer). The dashed line corresponds to the axis of symmetry of the problem. L_e is the thickness of the porous electrode. L_s is the thickness of the porous separator

$$\kappa = \kappa_1(c)\varepsilon^m; \;\; D = D_0\varepsilon^m, \tag{9.4}$$

where ε is the hydrophilic porosity, and D_0 is the diffusion coefficient of free electrolyte, m is the Archie's exponent that characterizes the influence of the porous structure on transport properties of electrolyte. $\kappa_1(c)$is bulk conductivity of free electrolyte that is a linear function of concentrations.

The mass transport in our model is described within a framework of the theory for a binary dilute solution [26, 34]:

$$\varepsilon\frac{\partial c}{\partial t} = D\frac{\partial^2 c}{\partial x^2} + \frac{C_s}{z_+F}A_s\frac{\partial \varphi}{\partial t}, \tag{9.5}$$

where φ is the electrical potential of electrolyte; C_s is the specific capacitance of electrode, which becomes zero for the separator; and A_s is the adsorption coefficient ($A_s = \left(t_+\frac{dq_+}{dq} - t_-\frac{dq_-}{dq}\right) \approx \pm\frac{1}{2}$ [34]). The charge transport in the electrolyte is described by the equation:

$$C_s\frac{\partial \varphi}{\partial t} = \frac{\partial}{\partial x}\left(\kappa_{eff}\frac{\partial \varphi}{\partial x}\right) + \frac{\partial}{\partial x}\left(\frac{\kappa_{eff}RT(t_+ - t_-)}{F}\frac{\partial \ln c}{\partial x}\right) \tag{9.6}$$

where t_+ and t_- are the transport numbers of cation and anion, respectively, and κ_{eff} is the effective conductivity. The latter term is a sum of conductivity with account for porosity and SC (κ_s) [52]. In other words, $\kappa_{eff} = \varepsilon^m\kappa_0\frac{c}{c_0} + \kappa_s$, where c_0 is the initial concentration, κ_0 is the electrolyte conductivity at the initial concentration. The method for SC measurements was described earlier [52]. The SC value is assumed to be proportional to the amount of adsorbed ions. This value is expressed as $C\sigma\frac{I_{full}t}{FV}$, where σ is the SC coefficient, which is the coefficient of proportionality between SC and adsorption capacity.

The problem is solved for a half-cell (the porous electrode and a half of the separator). The symmetry of the system with respect to boundary conditions is considered. On the left side ($x = 0$), the boundary conditions are:

$$\left.\frac{\partial c}{\partial x}\right|_{x=0} = 0; \; \left.\frac{\partial \varphi}{\partial x}\right|_{x=0} = 0. \tag{9.7}$$

In the center of separator ($x = L_e + \frac{1}{2}L_s$):

$$\left.\frac{\partial c}{\partial x}\right|_{x=L_e+\frac{1}{2}L_s} = 0; \; \kappa \left.\frac{\partial \varphi}{\partial x}\right|_{x=L_e+\frac{1}{2}L_s} = \frac{-I_{full}}{S}. \tag{9.8}$$

The initial conditions are:

$$c = c_0; \; \varphi = 0 \tag{9.9}$$

As pointed out, the initial conditions must be corrected using specific adsorption. It is assumed that only specific adsorption of species occurs before charging. This results in a decrease in the initial concentration. Furthermore, the initial concentration was also a fitting parameter; its value was found by comparing the theoretical and experimental data, as shown further.

A pdepe function of the MATLAB program package that is based on the method of lines (MOL) [57] was used to solve the system of parabolic equations. The pdepe function allows solving the initial-boundary value problems for systems of parabolic and elliptic partial differential equations in the system with a single spatial variable and the time variable.

9.9 Comparison of Theoretical and Experimental Data for Static Cell

The solution of the system (9.4)–(9.6) with boundary conditions (9.7)–(9.8) and initial condition (9.9) can be obtained using numerical methods (Tables 9.3, and 9.4). Only hydrophilic porosity (see Table 9.1) was taken into consideration. Moreover, the values of EDL capacitance per unit of hydrophilic surface area were used. The separator parameters were as follows: the thickness was 1.5×10^{-5} m, porosity was 0.46, and the Archie's exponent was 2. For example, Fig. 9.8 shows the numerical solution for the voltage and mean $NaHCO_3$ concentration in the static cell, as well as evolution of the profile concentration in the half-cell in time.

Figure 9.9 illustrates theoretical and experimental galvanostatic curves of electrode charging. The C^s_{EDL} values obtained as described in Sect. 3.4 were used for calculations. The plots are $U-t$ dependences for different currents and solutions. The

Table 9.3 Properties of 0.1 N electrolytes

Salt	$D\times10^9$, $m^2\ s^{-1}$	κ, $Ohm^{-1}\ m^{-1}$	t_+
$NaHCO_3$	1.26×10^{-9}	1.014	0.53
$CaCl_2$	1.34×10^{-9}	0.882	0.41
$MgSO_4$	8.54×10^{-9}	0.497	0.37

Table 9.4 Model parameters (current density was 28 or 100 A m^{-2})

		Measured values			Fitted values	
Electrode	Electrolyte	Electrode thickness $\times 10^4$, m	Electrode porosity	Surface capacitance $\times 10^{-5}$, F m^{-2}	Archie's exponent	SC coefficient, S m^2 mol^{-1}
CH-900	$NaHCO_3$	4.67	0.63	2.6 or 2.1	2.0	0.2
CH-900	$CaCl_2$	4.67	0.63	2.4 or 2.6	2.0	0.2
CH-900	$MgSO_4$	4.67	0.63	2.2 or 2.4	2.0	0.2
VISKUMAK	$NaHCO_3$	6.04	0.69	2.6	2.0	0.1
VISKUMAK	$CaCl_2$	6.04	0.69	2.6	2.0	0.1
VISKUMAK	$MgSO_4$	6.04	0.69	2.6 or 2.8	2.0	0.2 or 0.1
SAIT	$NaHCO_3$	4.30	0.47	2.4	2.8	0.2
SAIT	$CaCl_2$	4.30	0.47	2.6	2.0	0.1
SAIT	$MgSO_4$	4.30	0.47	2.4	2.8	0.2

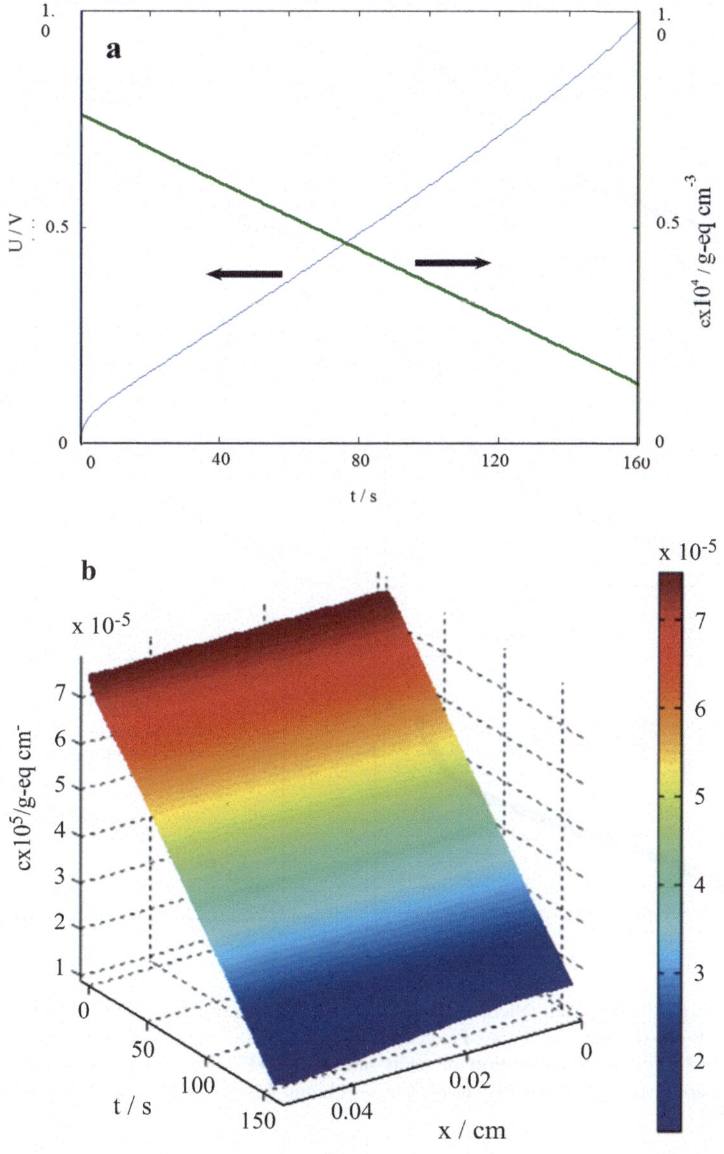

Fig. 9.8 Numerical solution of the voltage and mean concentration in the cell (**a**) and time evolution of the profile concentration in the half-cell (**b**). The CH900 electrode was used; the initial concentration of the $NaHCO_3$ solution was 0.075 mg-eq cm^{-3}

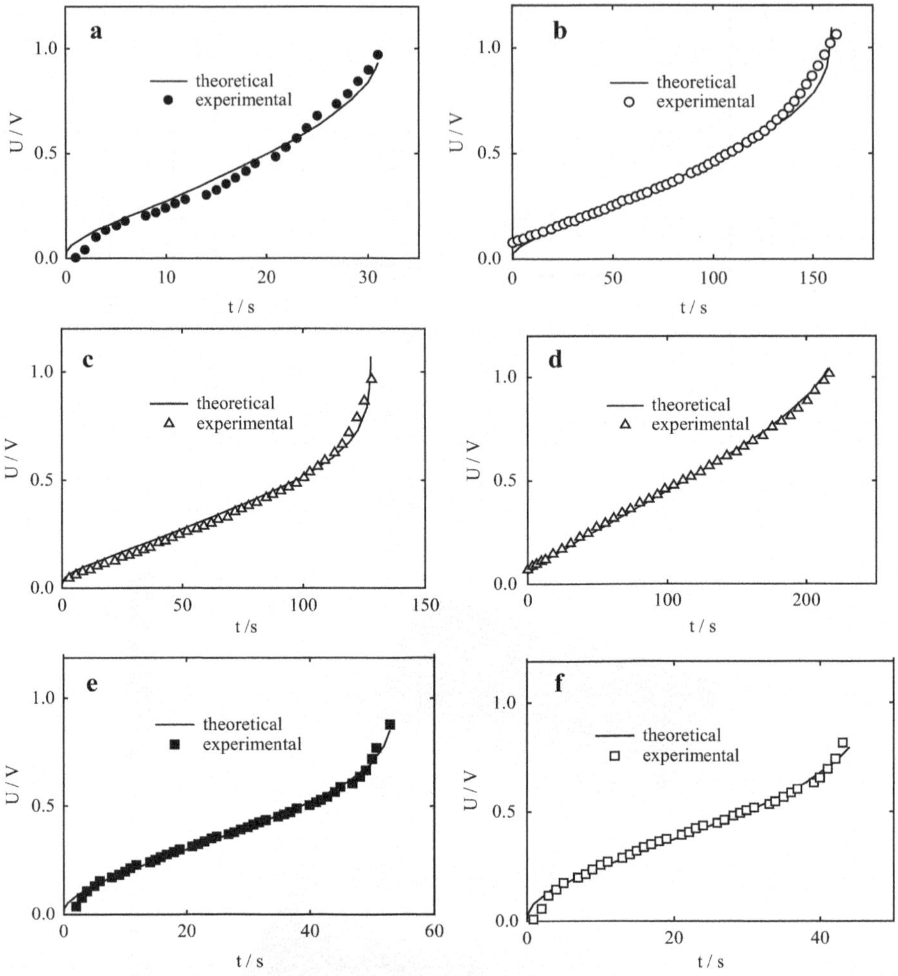

Fig. 9.9 Experimental and theoretical galvanostatic charging curves for the CH900 (**a**) VISKUMAK (**b-d**) and SAIT (**e, f**) electrodes. A solution contained a mixture of $CaCl_2$. $NaHCO_3$ and $MgSO_4$ concentration of each salt was 0.0257 M (a); 0.1 N single-component solutions were also used: $NaHCO_3$ (**b, e**), $CaCl_2$ (**c**), and $MgSO_4$ (**d, f**). $I = 2.8$ (**a–e**) and 10 (**f**) mA cm^{-2}

theoretical and experimental curves agree well with each other, which points to correctness of the model. In some cases, total experimental electrochemical adsorption capacity was slightly lower as compared with the theoretical values. This is probably due to adsorption of a certain amount of species on highly developed surface of carbon electrodes before polarization during measurements of discharge curves. Similar specific adsorption was observed earlier [46].

The charging-discharging curves simulate deionization-concentrating inside the static cell containing the CH900 electrodes (Fig. 9.10). The curves are plotted in the

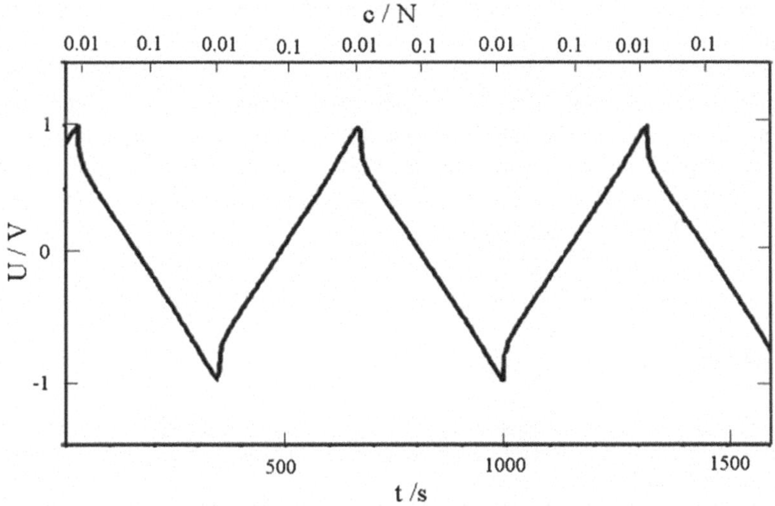

Fig. 9.10 Cyclic galvanostatic curves for the cell containing CH900 electrodes. The 0.1 N $CaCl_2$ solution was used, I=10 mA cm^{-2}

coordinates of voltage-time-concentration of the solution inside the cell. Charging is accompanied by deionization, and concentrating occurs in the course of discharge. Each semi-cycle (i.e., stages of deionization and concentrating) passes from the minimum to the maximum and further from the maximum to the minimum.

9.10 Conclusions

Among investigated AC electrodes, the CH900 sample is the most optimal material for CDI processes. This electrode is characterized by the highest value of specific surface area caused by the largest total volume of micropores and nanosized voids. Moreover, the largest total and hydrophilic porosity is attributed to the CH900 electrode. At last, the largest volume of macropores, which provide high hydrodynamical permittivity, has been found for this material. This property is very important for a dynamic cell, through which the solution passes [56]. The largest hydrophilic specific surface causes the highest values of adsorption capacity and surface conductivity [52]. This causes minimal energy consumptions for pure water production. The prospective way for optimization of AC electrodes is to develop the materials, which are characterized simultaneously by the highest nanoporosity and significant volume of macropores. Macropores are able to provide the shortest time of charging-discharging.

The simple model, which is based only on experimental data (SC and EDL capacitance), has been developed. The model allows one to avoid difficult calculations of the EDL. These calculations are complicated by features of porous structure of AC materials particularly narrowings-widenings of pores and their tortuosity.

Further investigations of CDI processes are related to development of high conductive separators that would be permeable toward cations and anions. Possible materials are polymer mosaic membranes [58]. Ceramic [59–61] or polymer membranes [62, 63] modified by nanoparticles of amphoteric inorganic ion exchanger could be considered as an alternative.

References

1. Nath K (2017) Membrane separation processes, 2nd edn. PHI Learning Private Limited, Delhi
2. Alvarado L, Chen A (2014) Electrodeionization: principles, strategies and applications. Electrochim Acta 132:583–597
3. Dzyazko YS, Ponomareva LN, Volfkovich YM et al (2013) Conducting properties of a gel ionite modified with zirconium hydrophosphate nanoparticles. Russ J Electrochem 49 (3):209–215
4. Dzyaz'ko YS, Rozhdestvenskaya LM, Pal'chik AV (2005) Recovery of nickel ions from dilute solutions by electrodialysis combined with ion exchange. Russ J Appl Chem 75:414–421
5. Dzyazko YS, Volfkovich YM, Ponomaryova LN et al (2016) Composite ion-exchangers based on flexible resin containing zirconium hydrophosphate for electromembrane separation. J Nanosci Technol 2(1):43–49
6. Oren Y (2008) Capacitive deionization (CDI) for desalination and water treatment—past, present and future (a review). Desalination 228:10–29
7. Avraham E, Noked M, Bouhadana Y et al (2009) Limitations of charge efficiency in capacitive deionization II. On the behavior of cdi cells comprising two activated carbon electrodes. J Electrochem Soc 156(10):157–162
8. Suss ME, Baumann TF, Bourcier WL et al (2012) Capacitive desalination with flow-through electrodes. Energy Environ Sci 5:9511–9519
9. Rica RA, Ziano R, Salerno D et al (2012) Thermodynamic relation between voltage-concentration dependence and salt adsorption in electrochemical cells. Phys Rev Lett 109:156103. https://doi.org/10.1103/PhysRevLett.109.156103
10. Porada S, Zhao R, van der Wal A et al (2013) Review on the science and technology of water desalination by capacitive deionization. Prog Mater Sci 58(8):1388–1442
11. Jande YAC, Kim WS (2013) Desalination using capacitive deionization at constant current. Desalination 329:29–34
12. Soffer A, Folman M (1972) The electrical double layer of high surface porous on carbon electrode. Electroanal Chem 38(1):25–43
13. Jande YAC, Kim WS (2013) Predicting the lowest effluent concentration in capacitive deionization. Sep Purif Technol 115:224–230
14. Li H, Pan L, Lu T et al (2011) A comparative study on electrosorptive behavior of carbon nanotubes and graphene for capacitive deionization. J Electroanal Chem 653(1–2):40–44
15. Gabelich CJ, Tran TD, Suffet IH (2002) Electrosorption of inorganic salts from aqueous solution using carbon aerogels. Environ Sci Technol 36(13):3010–3019
16. Li H, Lu T, Pan L et al (2009) Electrosorption behavior of graphene in NaCl solutions. J Mater Chem 19:6773–6779
17. Wang G, Dong Q, Ling Z (2012) Hierarchical activated carbon nanofiber webs with tuned structure fabricated by electrospinning for capacitive deionization. J Mater Chem 22:21819–21823

18. Gaikwad MS, Balomajumder C (2016) Polymer coated capacitive deionization electrode for desalination: a mini review. Electrochem Energy Technol 2(1). https://doi.org/10.1515/eetech-2016-0001
19. Liu Y, Nie C, Liu X et al (2015) Review on carbon-based composite materials for capacitive deionization. RSC Adv 5:15205–15225
20. Zhang D, Wen X, Shi L et al (2012) Enhanced capacitive deionization of graphene/mesoporous carbon composites. Nanoscale 4:5440–5446
21. Myint MTZ, Dutta J (2012) Fabrication of zinc oxide nanorods modified activated carbon cloth electrode for desalination of brackish water using capacitive deionization approach. Desalination 305:24–30
22. Vol'fkovich YM, Sosenkin VE, Nikol'skaya NF et al (2008) Porous structure and hydrophilic-hydrophobic properties of gas diffusion layers of the electrodes in proton-exchange membrane fuel cells. Russ J Electrochem 44(3):278–285
23. Bagotzky VS, Kazarinov VE, Vol'fkovich YM et al (1989) Macrokinetic study of thionyl chloride reduction on porous carbon electrodes. J Power Sources 26(3–4):427–433
24. Volfkovich YM, Bagotzky VS (1994) The method of standard porosimetry 2. Investigation of the formation of porous structures. J Power Sources 48(3):339–348
25. Newman J, Thomas-Alyea KE (2004) Electrochemical systems, 3rd edn. Wiley, Hoboken
26. Newman J, Tiedemann W (1975) Porous electrode theory with battery applications. AICHE J 21:25–41
27. Conway BE (1999) Electrochemical supercapacitors: scientific fundamentals and technological applications. Kluwer Academic Publishers-Plenum Publishers, New York
28. De Levie R (1963) On porous electrodes in electrolyte solutions. I. Capacitance effects. Electrochim Acta 8(10):751–780
29. De Levie R (1964) On porous electrodes in electrolyte solutions—IV. Electrochim Acta 9 (9):1231–1245
30. Srinivasan V, Weidner JW (1999) Mathematical modeling of electrochemical capacitors. J Electrochem Soc 146:1650–1658
31. Volfkovich YM, Bograchev DA, Mikhalin AA et al (2013) Supercapacitor carbon electrodes with high capacitance. J Solid State Electrochem 18(5):1351–1363
32. Yang K-L, Ying T-Y, Yiacoumi S et al (2001) Electrosorption of ions from aqueous solutions by carbon aerogel: an electrical double-layer model. Langmuir 17(6):1961–1969
33. Ying T-Y, Yang K-L, Yiacoumi S et al (2002) Electrosorption of ions from aqueous solutions by nanostructured carbon aerogel. J Colloid Interface Sci 250(1):18–27
34. Johnson AM, Newman J (1971) Desalting by means of porous carbon electrodes. J Electrochem Soc 118(3):510–517
35. Vol'fkovich YM, Mazin VM, Urisson NA (1998) Operation of double-layer capacitors based on carbon materials. Russ J Electrochem 34(8):740–746
36. Verbrugge MW, Liu P (2005) Microstructural analysis and mathematical modeling of electric double-layer supercapacitors. J Electrochem Soc 152(5):D79–D87
37. Biesheuvel PM, Bazant MZ (2010) Nonlinear dynamics of capacitive charging and desalination by porous electrodes. Phys Rev E 81(3):031502
38. Biesheuvel PM, Fu Y, Bazant MZ (2012) Electrochemistry and capacitive charging of porous electrodes in asymmetric multicomponent electrolytes. Russ J Electrochem 48(6):580–592
39. Mani A, Bazant MZ (2011) Deionization shocks in microstructures. Phys Rev E 84(6):061504. https://doi.org/10.1103/PhysRevE.84.061504
40. Biesheuvel PM, Porada S, Levi M, Bazant MZ (2014) Attractive forces in microporous carbon electrodes for capacitive deionization. J Solid State Electrochem 18(5):1365–1376
41. Volfkovich YM, Sosenkin VE (2012) Porous structure and wetting of fuel cell components as the factors determining their electrochemical characteristics. Russ Chem Rev 86(6):936–959
42. Volfkovich YM, Bagotsky VS, Filippov AN (eds) (2014) Porous materials and powders used in different fields of science and technology. Springer-Verlag, London/Heidelberg/New York/Dordrecht

43. Volfkovich YM, Sakars AV, Volinsky AA (2005) Application of the standard contact porosimetry for nanomaterials. Int J Nanotechnol 2(3):292–302
44. Rouquerol J, Baron G, Denoyel R et al (2012) Liquid intrusion and alternative methods for the characterization of macroporous materials (IUPAC technical report). Pure Appl Chem 84 (1):107–136
45. Vol'fkovich YM, Sosenkin VE, Nikol'skaya NF (2010) Hydrophilic-hydrophobic and sorption properties of the catalyst layers of electrodes in a proton-exchange membrane fuel cell: a stage-by-stage study. Russ J Electrochem 46(4):438–449
46. Gladysheva TD, Shkol'nikov EL, Vol'fkovich YM et al (1982) The porous structure of dispersed platinum. Elektrokhimiya 18(4):435–442
47. Kononenko N, Nikonenko V, Grande D et al (2017) Porous structure of ion exchange membranes investigated by various techniques. Adv Colloid Interface Sci 246:196–216
48. Dzyazko YS, Perlova OV, Perlova NA et al (2017) Composite cation-exchange resins containing zirconium hydrophosphate for purification of water from U(VI) cations. Desalin Water Treat 69:142–152
49. Jeffery JH, Bassett J, Menoham J et al (1989) Vogel's textbook on quantitative chemical analysis, 5th edn. Longman Scientific and Technical – Wiley, Harlow/New York
50. Berezina NP, Kononenko NA, Dyomina OA et al (2008) Characterization of ion-exchange membrane materials: properties vs structure. Adv Colloid Interf Sci 139(1–2):3–28
51. Dzyazko Y, Rozhdestvenska L, Palchik A, Lapicque F (2005) Ion-exchange properties and mobility of Cu^{2+} ions in zirconium hydrophosphate ion exchangers. Sep Purif Technol 45 (2):141–146
52. Vol'fkovich YM, Mikhalin AA, Rychagov AY (2013) Surface conductivity measurements for porous carbon electrodes. Russ J Electrochem 49(6):594–598
53. Galama AH, Post JW, Cohen Stuart MA et al (2013) Validity of the Boltzmann equation to describe Donnan equilibrium at the membrane-solution interface. J Memb Sci 442:131–139
54. Andersen MB, Van Soestbergen M, Mani A et al (2012) Current-induced membrane discharge. Phys Rev Lett 109(10):108301. https://doi.org/10.1103/PhysRevLett.109.108301
55. Chizmadzhev Y, Markin V, Tarasevich M et al (1971) Macrokinetics of processes in porous media. Nauka, Moscow
56. Archie GE (1952) Classification of carbonate reservoir rocks and petrophysical considerations. Am Assoc Pet Geol Bull 36(2):278–298
57. Skeel RD, Berzins M (1990) A method for the spatial discretization of parabolic equations in one space variable. SIAM J Sci Stat Comput 11(1):1–32
58. Volfkovich YM, Rychagov AY, Mikhalin AA et al (2018) Capacitive deionization of water using mosaic membrane. Desalination 426:1–10
59. Dzyaz'ko YS, Belyakov VN, Vasilyuk SL et al (2006) Anion-exchange properties of composite ceramic membranes containing hydrated zirconium dioxide. Russ J Appl Chem 79(5):769–773
60. Dzyazko YS, Volfkovich YM, Sosenkin VE et al (2014) Composite inorganic membranes containing nanoparticles of hydrated zirconium dioxide for electrodialytic separation. Nanoscale Res Lett 9:271. https://doi.org/10.1186/1556-276X-9-271
61. Martı-Calatayud MC, Garcıa-Gabaldon M, Perez-Herranz V et al (2015) Ceramic anion-exchange membranes based on microporous supports infiltrated with hydrated zirconium dioxide. RSC Adv 5:46348–46358
62. Pang R, Li X, Li J et al (2014) Preparation and characterization of ZrO_2/PES hybrid ultrafiltration membrane with uniform ZrO_2 nanoparticles. Desalination 332:60–66
63. Myronchuk VG, Dzyazko YS, Zmievskii YG et al (2016) Organic-inorganic membranes for filtration of corn distillery. Acta Periodica Technologica 47:153–165

Chapter 10
Gradient-Type Theory for Electro-Thermoelastic Non-ferromagnetic Dielectrics: Accounting for Quadrupole Polarization and Irreversibility of Local Mass Displacement

Olha Hrytsyna and Vasyl Kondrat

10.1 Introduction

This work is devoted to the development of the gradient-type continuum field theories of non-ferromagnetic thermoelastic dielectrics. In the scientific literature, such theories were constructed by extending the space of constitutive parameters with the gradients of some physical quantities, such as the strain tensor, the polarization vector, or the electric field vector [1–6]. These generalized theories allowed us to avoid singularities of solutions in boundary value problems with concentrated forces or concentrated electric charges and to describe a number of experimentally observed phenomena [7–10, etc.] that are not explained within the framework of classical theories.

Burak et al. [1, 11–13] developed the foundations of the gradient theory of dielectrics, which was based on the consideration of non-diffusive and non-convective mass flux associated with the changes in the material microstructure. However, while constructing the governing equations in the framework of this theory, only electric dipoles were taken into account. Here, we will consider the contribution of electric dipoles and quadrupoles to the polarization current as well as the irreversibility of the local mass displacement. Based on this, we construct a more general theory of dielectrics and use its relations to study the formation of the near-surface inhomogeneity of the fields.

O. Hrytsyna (✉)
Center of Mathematical Modeling of Pidstryhach Institute for Applied Problems of Mechanics and Mathematics, National Academy of Sciences of Ukraine, Lviv, Ukraine

V. Kondrat
Hetman Petro Sahaydachnyi Academy of Army Ground Forces, Lviv, Ukraine

O. Fesenko, L. Yatsenko (eds.), *Nanooptics, Nanophotonics, Nanostructures, and Their Applications*, Springer Proceedings in Physics 210,
https://doi.org/10.1007/978-3-319-91083-3_10

10.2 Equation of Entropy Balance

We consider non-ferromagnetic, electrically polarizable, heat-conducting deformable elastic solids that occupy the domain (V) of a three-dimensional Euclidian space with a smooth surface (Σ). Mechanical, thermal, and electromagnetic processes can occur within the solid. All fields that characterize these processes should obey the fundamental laws of continuum physics (first and second laws of thermodynamics, Maxwell equations, etc.).

In this section we formulate the entropy balance equation. To this end, we introduce the physical quantities related to the heat-conducting process. We take the absolute temperature T as a measure of the intensity of thermal motion. To describe the heat-conducting process, besides the field of absolute temperature, we also introduced the entropy S and the vectors of the entropy $\mathbf{J}_s$ and the heat $\mathbf{J}_q$ fluxes. These vectors can be related by the known formula: $\mathbf{J}_s = \mathbf{J}_q/T$ [14].

In an integral form, the entropy balance equation can be written as follows:

$$\frac{d}{dt}\int\limits_{(V)} S\,dV = -\oint\limits_{(\Sigma)} (\mathbf{J}_s + S\mathbf{v})\cdot\mathbf{n}\,d\Sigma + \int\limits_{(V)} \left(\sigma_s + \rho\frac{\mathfrak{R}}{T}\right)dV. \tag{10.1}$$

Here, σ_s is the entropy production per unit of volume and time, ρ is the mass density, $\mathbf{v}$ is the velocity of the center of mass, $\mathfrak{R}$ is the distributed thermal sources, t denotes the time variable, $\mathbf{n}$ is the unit vector normal to the material surface (Σ), and the dot denotes the scalar product.

In view of the divergence theorem, Eq. (10.1) can be written in the local form:

$$T\frac{\partial S}{\partial t} = -\nabla\cdot\mathbf{J}_q + \frac{1}{T}\mathbf{J}_q\cdot\nabla T - T\nabla\cdot(S\mathbf{v}) + T\sigma_s + \rho\mathfrak{R}, \tag{10.2}$$

where ∇ is the Hamilton operator.

10.3 Electrodynamics Equations

The Maxwell equations and the conservation law of electric charges in the local form are given by [15, 16]

$$\nabla\cdot\mathbf{B} = 0,\quad \nabla\cdot\mathbf{D} = \rho_e,\quad \nabla\times\mathbf{E} = -\frac{\partial\mathbf{B}}{\partial t},\quad \nabla\times\mathbf{H} = \varepsilon_0\frac{\partial\mathbf{E}}{\partial t} + \mathbf{J}_e + \mathbf{J}_{es}, \tag{10.3}$$

$$\nabla\cdot\mathbf{J}_e + \frac{\partial\rho_e}{\partial t} = 0. \tag{10.4}$$

Here, $\mathbf{E}$ and $\mathbf{H}$ are electric and magnetic fields; $\mathbf{D}$ and $\mathbf{B}$ are electric and magnetic inductions; ρ_e is the density of free electric charges; $\mathbf{J}_e$ is the density of the electric current (convection and conduction currents); $\mathbf{J}_{es}$ is the polarization current; ε_0 is electric constant; and $\times$ denotes the vector product.

We also introduce the polarization vector $\boldsymbol{\Pi}_e$, which is related to the vector $\mathbf{J}_{es}$ by the following formula [10.15]:

$$\boldsymbol{\Pi}_e(\mathbf{r}, t) = \int_0^t \mathbf{J}_{es}(\mathbf{r}, t')dt' \quad \Rightarrow \quad \mathbf{J}_{es} = \frac{\partial \boldsymbol{\Pi}_e}{\partial t}. \tag{10.5}$$

Here, $\mathbf{r}$ is the position vector. Note that the vector $\boldsymbol{\Pi}_e$ can be referred to as a vector of the local displacement of electric charges.

For non-ferromagnetic medium, the constitutive equations for the vectors of magnetic $\mathbf{B}$ and electric $\mathbf{D}$ inductions are as follows [15, 16]:

$$\mathbf{B} = \mu_0\mathbf{H}, \quad \mathbf{D} = \varepsilon_0\mathbf{E} + \boldsymbol{\Pi}_e, \tag{10.6}$$

where μ_0 is magnetic constant.

Assume that the polarization current $\mathbf{J}_{es}$ is caused by a change over time of both the dipole $\mathbf{P}$ and the quadrupole $\widehat{\mathbf{Q}}$ electric moments, namely [17]:

$$\boldsymbol{\Pi}_e = \mathbf{P} - \frac{1}{6}\nabla \cdot \hat{\mathbf{Q}}. \tag{10.7}$$

Using the formula

$$\int_{(V)} \boldsymbol{\Pi}_e dV = \int_{(V)} \rho_{e\pi}\mathbf{r}dV \tag{10.8}$$

we introduce the density of an induced charge $\rho_{e\pi}$. From Eq. (10.8) it follows that $\rho_{e\pi} = -\nabla \cdot \boldsymbol{\Pi}_e$ [15]. Based on this formula and using Eq. (10.5), one can obtain the conservation law of induced electric charges:

$$\frac{\partial \rho_{e\pi}}{\partial t} + \nabla \cdot \mathbf{J}_{es} = 0. \tag{10.9}$$

From the Maxwell equations, we get the relation:

$$\frac{\partial U_e}{\partial t} + \nabla \cdot \mathbf{S}_e + \left(\mathbf{J}_e + \frac{\partial \boldsymbol{\Pi}_e}{\partial t}\right) \cdot \mathbf{E} = 0, \tag{10.10}$$

which is known as the energy balance equation of the electromagnetic field [1, 11]. Here, $U_e = (\varepsilon_0\mathbf{E}^2 + \mu_0\mathbf{H}^2)/2$ is the energy density of the electromagnetic field and $\mathbf{S}_e = \mathbf{E} \times \mathbf{H}$ is the Poynting vector.

Note that the last term in Eq. (10.10) describes the effect of the electromagnetic field on a substance. Let us rewrite the above term in such a way that it contains the quadrupole $\widehat{\mathbf{Q}}_*$ and dipole $\mathbf{P}_*$ electric moments, the electric field vector $\mathbf{E}_*$, and the density of the electric current $\mathbf{J}_{e*}$ in the reference frame of the mass centers moving with a velocity $\mathbf{v}$ relative to the laboratory reference frame. In this co-moving frame, the vectors $\mathbf{E}$, $\mathbf{P}$, and $\mathbf{J}_e$ and the tensor $\widehat{\mathbf{Q}}$ are transformed according to the relations: $\mathbf{E} = \mathbf{E}_* - \mathbf{v} \times \mathbf{B}$, $\mathbf{P} = \mathbf{P}_*$, $\widehat{\mathbf{Q}} = \widehat{\mathbf{Q}}_*$, and $\mathbf{J}_e = \mathbf{J}_{e*} - \rho_e \mathbf{v}$. Substituting the above relations and Eq. (10.7) into (10.10), the balance equation for energy of electromagnetic field can be reduced to the following form:

$$\frac{\partial U_e}{\partial t} + \nabla \cdot \mathbf{S}_e + \left(\mathbf{J}_{e*} + \frac{\partial \mathbf{P}}{\partial t} \right) \cdot \mathbf{E}_* + \frac{1}{6} \frac{\partial \hat{\mathbf{Q}}}{\partial t}$$
$$: (\nabla \otimes \mathbf{E}_*) + \mathbf{v} \cdot \left[\rho_e \mathbf{E}_* + \left(\mathbf{J}_{e*} + \frac{\partial \mathbf{\Pi}_e}{\partial t} \right) \times \mathbf{B} \right] = 0. \tag{10.11}$$

Here, $\otimes$ is the tensor product.

10.4 Local Displacement of Mass

The mechanical, thermal, and electromagnetic processes which can occur within the solid may be accompanied by changes in the microstructure of a fixed small body particle dV. For example, such changes of a material structure can be caused by the formation of a body surface or due to the action of high-gradient fields. We characterize these changes not only by an electric flux $\mathbf{J}_{es}$ (polarization current) but also by a non-convective and non-diffusive mass flux $\mathbf{J}_{ms}$. We associate this mass flux with the process of the local mass displacement [1, 11].

Following Burak [18] we assume that the mass flux $\mathbf{J}_{ms}$ is caused by a change over time of the mass dipole moment $\mathbf{\Pi}_m$ (i.e., the vector of local mass displacement, which has a dimension of the kg $\cdot$ m/ m^3):

$$\mathbf{\Pi}_m(\mathbf{r}, t) = \int_0^t \mathbf{J}_{ms}(\mathbf{r}, t') dt' \;\Rightarrow\; \mathbf{J}_{ms} = \frac{\partial \mathbf{\Pi}_m}{\partial t}. \tag{10.12}$$

To describe the local mass displacement by the formula:

$$\int_{(V)} \mathbf{\Pi}_m dV = \int_{(V)} \rho_{m\pi} \mathbf{r} \, dV \tag{10.13}$$

we introduce the scalar quantity $\rho_{m\pi}$, which has the dimension of mass density. Note that by similar formula (10.8), in electrodynamics, the density of induced electric charge $\rho_{e\pi}$ is introduced. Hence, we call the $\rho_{m\pi}$ introduced herein the density of induced mass.

From the integral relation (10.13), it can be easily shown that [1, 11]

$$\rho_{m\pi} = -\nabla \cdot \mathbf{\Pi}_m. \tag{10.14}$$

Applying a time derivative in Eq. (10.14) and using (10.12), one can obtain the equation which has the form of the conservation law of an induced mass:

$$\frac{\partial \rho_{m\pi}}{\partial t} + \nabla \cdot \mathbf{J}_{ms} = 0. \tag{10.15}$$

10.5 Energy Balance Law

We assume that the total energy E of the system "solid-electromagnetic field" is the sum of internal energy ρu (u is the specific internal energy), kinetic energy $\rho \mathbf{v}^2/2$, and the energy U_e of the electromagnetic field: $E = \rho u + \rho \mathbf{v}^2/2 + U_e$. The total energy changes due to its convective transport through the surface $\rho(u + \mathbf{v}^2/2)$, the work $\hat{\boldsymbol{\sigma}} \cdot \mathbf{v}$ of surface tractions, the heat $\mathbf{J}_q$ and electromagnetic energy $\mathbf{S}_e$ fluxes, the work $\mu \mathbf{J}_m$ related to the mass transport relative to the center of mass of the body, the work $\mu_\pi \mathbf{J}_{ms}$ related to structure ordering (i.e., the local mass displacement), the action of the mass forces $\mathbf{F}$, and the distributed thermal sources $\mathfrak{R}$. The law of the energy balance for a dielectric body may be written as follows:

$$\frac{d}{dt}\int_{(V)} \mathcal{E}\, dV = -\oint_{(\Sigma)} \left[\rho\left(u + \frac{1}{2}\mathbf{v}^2\right)\mathbf{v} - \hat{\sigma} \cdot \mathbf{v} + \mathbf{S}_e + \mathbf{J}_q + \mu \mathbf{J}_m + \mu_\pi \mathbf{J}_{ms}\right] \cdot \mathbf{n}\, d\Sigma + \int_{(V)} (\rho \mathbf{F} \cdot \mathbf{v} + \rho \mathfrak{R})\, dV, \tag{10.16}$$

where $\hat{\boldsymbol{\sigma}}$ is the Cauchy stress tensor, μ is chemical potential, and μ_π is an energy measure of the effect of the local mass displacement on the internal energy, $\mathbf{J}_m = \rho(\mathbf{v}_* - \mathbf{v})$, where $\mathbf{v}_*$ is a velocity of the convective displacement of the fixed body particle (Fig. 10.1). Vectors $\mathbf{v}_*$ and $\mathbf{v}$ are related by the formula $\rho \mathbf{v} = \rho \mathbf{v}_* + \mathbf{J}_{ms}$. Note that from these relations and Eq. (10.12), we have the following formula:

$$\mathbf{J}_m = -\frac{\partial \mathbf{\Pi}_m}{\partial t}. \tag{10.17}$$

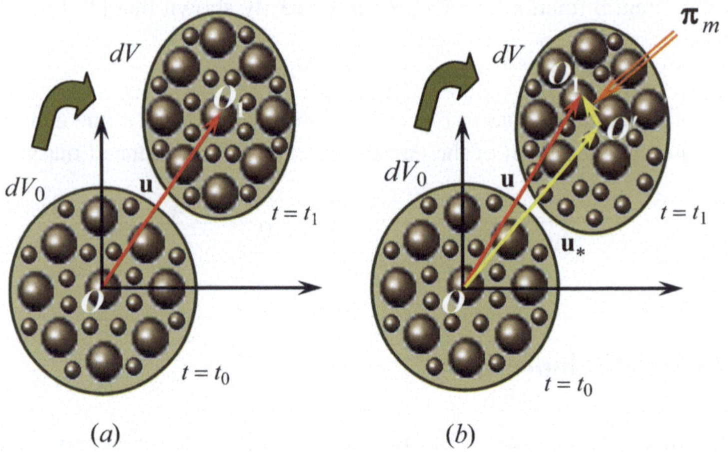

Fig. 10.1 The change of a mass center of the body particle under the action of external loads: (**a**) the classical theory (takes into consideration the convective translation only) and (**b**) theory with the local mass displacement (takes into consideration the translation, caused by the material microstructure changes, i.e., the local mass displacement, $\mathbf{u} = \mathbf{u}_* + \boldsymbol{\pi}_m$, $\boldsymbol{\pi}_m = \boldsymbol{\Pi}_m/\rho$)

Making use of the divergence theorem as well as the law of conservation of energy of electromagnetic field (10.11), and Eqs. (10.2), (10.14) and (10.17), the integral relation (10.16) can be written as follows:

$$\begin{aligned}
\frac{\partial(\rho u)}{\partial t} + \frac{1}{2}\frac{\partial}{\partial t}(\rho \mathbf{v}^2) = &-\nabla\cdot\left[\left(u + \frac{1}{2}\mathbf{v}^2\right)\rho\mathbf{v}\right] + \nabla\cdot(\hat{\sigma}\cdot\mathbf{v}) + T\frac{\partial S}{\partial t} + \mathbf{E}_*\cdot\frac{\partial \mathbf{P}}{\partial t} \\
&+ \frac{1}{6}(\nabla\otimes\mathbf{E}_*) : \frac{\partial \hat{\mathbf{Q}}}{\partial t} - \nabla\mu'_\pi\cdot\frac{\partial \boldsymbol{\Pi}_m}{\partial t} + \mu'_\pi\frac{\partial \rho_{m\pi}}{\partial t} - \frac{1}{T}\mathbf{J}_q\cdot\nabla T \\
&+ \mathbf{J}_{e*}\cdot\mathbf{E}_* - T\sigma_s + T\nabla\cdot(S\mathbf{v}) \\
&+ \mathbf{v}\cdot\left[\rho_e\mathbf{E}_* + \left(\mathbf{J}_{e*} + \frac{\partial \boldsymbol{\Pi}_e}{\partial t}\right)\times\mathbf{B} + \rho\mathbf{F}\right].
\end{aligned} \qquad (10.18)$$

Here, $\mu'_\pi = \mu_\pi - \mu$

By formulae $s = S/\rho$, $\rho_m = \rho_{m\pi}/\rho$, $\boldsymbol{\pi}_m = \boldsymbol{\Pi}_m/\rho$, $\mathbf{p} = \mathbf{P}/\rho$, and $\hat{\mathbf{q}} = \hat{\mathbf{Q}}/6\rho$, we introduce specific values of corresponding quantities. Note that vector $\boldsymbol{\pi}_m$ has a dimension of a meter while ρ_m is a dimensionless quantity. Using these formulae after some algebraic manipulations, we can write the following formulation for the conservation energy law:

$$
\begin{aligned}
\rho\frac{du}{dt} &= \hat{\sigma}_* : \frac{d\hat{e}}{dt} + \rho T\frac{ds}{dt} + \rho\mathbf{E}_* \cdot \frac{d\mathbf{p}}{dt} + \rho(\nabla \otimes \mathbf{E}_*) \cdot \frac{d\hat{\mathbf{q}}}{dt} \\
&+ \rho\mu_\pi^{\prime}\frac{d\rho_m}{dt} - \rho\nabla\mu_\pi^{\prime} \cdot \frac{d\boldsymbol{\pi}_m}{dt} - \left[\frac{\partial\rho}{\partial t} + \nabla \cdot (\rho\mathbf{v})\right] \\
&\times \left[u + \frac{1}{2}\mathbf{v}^2 - Ts - \mathbf{p} \cdot \mathbf{E}_* - \hat{\mathbf{q}} : (\nabla \otimes \mathbf{E}_*) - \rho_m\mu_\pi^{\prime} + \nabla\mu_\pi^{\prime} \cdot \boldsymbol{\pi}_m\right] \\
&+ \mathbf{J}_{e*} \cdot \mathbf{E}_* - \mathbf{J}_q \cdot \frac{\nabla T}{T} - T\sigma_s + \mathbf{v} \cdot \left(-\rho\frac{d\mathbf{v}}{dt} + \nabla \cdot \hat{\sigma}_* + \mathbf{F}_e + \rho\mathbf{F}_*\right).
\end{aligned}
\tag{10.19}
$$

Here,

$$
\hat{e} = [\nabla \otimes \mathbf{u} + (\nabla \otimes \mathbf{u})^{\mathrm{T}}]/2, \tag{10.20}
$$

$$
\hat{\sigma}_* = \hat{\sigma} - \rho[\mathbf{E}_* \cdot \mathbf{p} + \hat{\mathbf{q}} : (\nabla \otimes \mathbf{E}_*) + \rho_m\mu_\pi^{\prime} - \boldsymbol{\pi}_m \cdot \nabla\mu_\pi^{\prime}]\hat{\mathbf{I}}, \tag{10.21}
$$

$$
\mathbf{F}_* = \mathbf{F} + \rho_m\nabla\mu_\pi^{\prime} - (\nabla \otimes \nabla\mu_\pi^{\prime}) \cdot \boldsymbol{\pi}_m, \tag{10.22}
$$

$$
\begin{aligned}
\mathbf{F}_e = \rho_e\mathbf{E}_* + \left(\mathbf{J}_{e*} + \frac{\partial\boldsymbol{\Pi}_e}{\partial t}\right) \times \mathbf{B} + \rho(\nabla \otimes \mathbf{E}_*) \cdot \mathbf{p} \\
+ \rho\,\hat{\mathbf{q}} : (\nabla \otimes \nabla \otimes \mathbf{E}_*)^{\mathrm{T}(2,3)},
\end{aligned}
\tag{10.23}
$$

where $\hat{e}$ is the strain tensor defined in terms of the displacement vector $\mathbf{u}$; an upper index "T" denotes a transposed tensor; $\mathbf{F}_e$ and $\mathbf{F}_*$ are the ponderomotive and additional mass forces; $\hat{\mathbf{I}}$ is the unit tensor; $(\nabla \otimes \nabla \otimes \mathbf{E}_*)^{\mathrm{T}(2,3)}$is the third rank tensor, which is an isomer of the tensor $\nabla \otimes \nabla \otimes \mathbf{E}_*$, formed by permutation of its second and third basis vectors; and $\frac{d}{dt} = \frac{\partial}{\partial t} + \mathbf{v} \cdot \nabla$ is the material time derivative.

10.6 Balance of Mass and Momentum

Using the method presented by Green and Rivlin [19], we assume that the energy balance equation remains valid under superimposed rigid body translation. As a result, from (10.19) we get the equation of motion and the mass balance equation:

$$
\rho\frac{d\mathbf{v}}{dt} = \nabla \cdot \hat{\sigma}_* + \mathbf{F}_e + \rho\mathbf{F}_*, \tag{10.24}
$$

$$
\frac{\partial\rho}{\partial t} + \nabla \cdot (\rho\mathbf{v}) = 0. \tag{10.25}
$$

It is evident from the balance of momentum (10.24) that the electric quadrupole and mass dipole moments induce additional nonlinear body ponderomotive and mass forces, that is, $\mathbf{F}_e^{\prime} = -\rho\,(\nabla \otimes \nabla \otimes \mathbf{E}_*) : \hat{\mathbf{q}}$ and $\mathbf{F}^{\prime} = \rho_m\nabla\mu_\pi^{\prime} - \boldsymbol{\pi}_m \cdot \nabla \otimes \nabla$

μ'_π and coupled stresses $\hat{\sigma}_{*}{}' = -\rho[\,\hat{\mathbf{q}} : (\nabla \otimes \mathbf{E}_*) + \rho_m \mu'_\pi - \boldsymbol{\pi}_m \cdot \nabla \mu'_\pi]\hat{\mathbf{I}}$ within the dielectric body (see formulae (10.21) and (10.22)).

To take into account the contribution of the irreversibility of the processes of local mass displacement, we represent the vector $\nabla \mu'_\pi$ as the sum of its reversible $(\nabla \mu'_\pi)^r$ and irreversible $(\nabla \mu'_\pi)^i$ parts [13]:

$$\nabla \mu'_\pi = (\nabla \mu'_\pi)^r + (\nabla \mu'_\pi)^i. \tag{10.26}$$

In view of Eqs. (10.24), (10.25), and (10.26), we can write the conservation energy law (10.19) as follows:

$$\begin{aligned}\rho \frac{du}{dt} &= \hat{\sigma}_* : \frac{d\hat{e}}{dt} + \rho T \frac{ds}{dt} + \rho \mathbf{E}_* \cdot \frac{d\mathbf{p}}{dt} + \rho(\nabla \otimes \mathbf{E}_*) \cdot \frac{d\hat{\mathbf{q}}}{dt} \\ &+ \rho \mu'_\pi \frac{d\rho_m}{dt} - \rho(\nabla \mu'_\pi)^r \cdot \frac{d\boldsymbol{\pi}_m}{dt} + \mathbf{J}_{e*} \cdot \mathbf{E}_* - \mathbf{J}_q \cdot \frac{\nabla T}{T} \\ &- \rho(\nabla \mu'_\pi)^i \cdot \frac{d\boldsymbol{\pi}_m}{dt} - T\sigma_s.\end{aligned} \tag{10.27}$$

By means of the Legendre transformation $f = u - Ts - \mathbf{E}_* \cdot \mathbf{p} - \hat{\mathbf{q}} : (\nabla \otimes \mathbf{E}_*) + (\nabla \mu'_\pi)^r \cdot \boldsymbol{\pi}_m$, we define the generalized Helmholtz free energy. Using this new thermodynamic function, Eq. (10.27) is transformed into

$$\begin{aligned}\rho \frac{df}{dt} &= \hat{\sigma}_* : \frac{d\hat{e}}{dt} - \rho s \frac{dT}{dt} - \rho \mathbf{p} \cdot \frac{d\mathbf{E}_*}{dt} - \rho \hat{\mathbf{q}} : \frac{d(\nabla \otimes \mathbf{E}_*)}{dt} + \rho \mu'_\pi \frac{d\rho_m}{dt} \\ &+ \rho \boldsymbol{\pi}_m \cdot \frac{d(\nabla \mu'_\pi)^r}{dt} + \mathbf{J}_{e*} \cdot \mathbf{E}_* - \mathbf{J}_q \cdot \frac{\nabla T}{T} - \rho(\nabla \mu'_\pi)^i \cdot \frac{d\boldsymbol{\pi}_m}{dt} - T\sigma_s.\end{aligned} \tag{10.28}$$

While inspecting the above equation, we assume that the Helmholtz free energy is a function of $\hat{e}$, $T, \mathbf{E}_*, \nabla \otimes \mathbf{E}_*, \rho_m$ and $(\nabla \mu'_\pi)^r$. Thus, from Eq. (10.28), we get the generalized Gibbs equation

$$df = \frac{1}{\rho} \hat{\sigma}_* : d\hat{e} - s dT - \mathbf{p} \cdot d\mathbf{E}_* - \hat{\mathbf{q}} : d(\nabla \otimes \mathbf{E}_*) + \mu'_\pi d\rho_m + \boldsymbol{\pi}_m \cdot d(\nabla \mu'_\pi)^r \tag{10.29}$$

and the following relation for entropy production

$$\sigma_s = -\mathbf{J}_q \cdot \frac{\nabla T}{T^2} + \mathbf{J}_{e*} \cdot \frac{\mathbf{E}_*}{T} - \rho \frac{d\boldsymbol{\pi}_m}{dt} \cdot \frac{(\nabla \mu'_\pi)^i}{T}. \tag{10.30}$$

Note that according to (10.29), the free energy density depends not only on strain tensor $\hat{\boldsymbol{e}}$, temperature T, and electric field vector $\mathbf{E}_*$, as it follows from the classical theory, but also on the parameters, $\rho_m = -\nabla \cdot (\rho \boldsymbol{\pi}_m)/\rho$, and $(\nabla \mu'_\pi)^r$, related to the electric quadrupole and mass dipole moments. The specific electric quadrupole $\hat{\mathbf{q}}$, the potential μ'_π, and the local mass displacement vector $\boldsymbol{\pi}_m$ are the thermodynamic conjugates of the electric field gradient, the specific induced mass, and the reversible components of the gradient of a modified chemical potential.

10.7 Constitutive Equations

Since the parameters $\hat{e}, T, \mathbf{E}_*, \nabla \otimes \mathbf{E}_*, \rho_m$ and $(\nabla\mu_\pi')^r$ are independent, we obtain the following constitutive equations from the Gibbs relation (10.29):

$$\hat{\sigma}_* = \rho\frac{\partial f}{\partial \hat{e}}, \quad s = -\frac{\partial f}{\partial T}, \quad \mathbf{p} = -\frac{\partial f}{\partial \mathbf{E}_*}, \tag{10.31a}$$

$$\hat{\mathbf{q}} = -\frac{\partial f}{\partial(\nabla \otimes \mathbf{E}_*)}, \quad \mu_\pi' = \frac{\partial f}{\partial \rho_m}, \quad \boldsymbol{\pi}_m = \frac{\partial f}{\partial(\nabla\mu_\pi')^r}. \tag{10.31b}$$

We can write the Eqs. (10.31) in an explicit form. In order to obtain the linear constitutive relations, we expand f into a Taylor series about $\hat{e} = 0$, $T = T_0$, $\mathbf{E}_* = 0$, $\nabla \otimes \mathbf{E}_* = 0$, $\rho_m = 0$, $\mu_\pi' = \mu_{\pi 0}'$, and $(\nabla\mu_\pi')^r = 0$, where T_0 is a reference temperature and $\mu_{\pi 0}'$ is the modified chemical potential of an infinite medium. Denoting $\theta = T - T_0$, $I_{e1} = \hat{e} : \hat{\mathbf{I}} = e$, $I_{e2} = \hat{e} : \hat{e}$, $I_{E1} = \nabla \cdot \mathbf{E}_*$, and $I_{E2} = (\nabla \otimes \mathbf{E}_*) : (\nabla \otimes \mathbf{E}_*)$ and keeping linear and quadratic terms only, we can write the following for isotropic materials:

$$\begin{aligned} f = {} & f_0 - s_0\theta + \mu_{\pi 0}'\rho_m + \frac{1}{2\rho_0}\left(K - \frac{2}{3}G\right)I_{e1}^2 + \frac{G}{\rho_0}I_{e2} - \frac{C_V}{2T_0}\theta^2 \\ & + \frac{d_\rho}{2}\rho_m^2 - \frac{\chi_E}{2}\mathbf{E}_* \cdot \mathbf{E}_* - \frac{\chi_m}{2}(\nabla\mu_\pi')^r \cdot (\nabla\mu_\pi')^r + \frac{\chi_{q1}}{2}I_{E1}^2 - \chi_{q2}I_{E2} \\ & - \frac{K\alpha_T}{\rho_0}I_{e1}\theta - \frac{K\alpha_{E1}}{\rho_0}I_{e1}I_{E1} - \frac{K\alpha_\rho}{\rho_0}I_{e1}\rho_m - \beta_{T\rho}\rho_m\theta + \beta_{TE}I_{E1}\theta \\ & + \beta_{E\rho}I_{E1}\rho_m + \chi_{Em}\mathbf{E}_* \cdot (\nabla\mu_\pi')^r + 2G\frac{\alpha_{E2}}{\rho_0}\hat{e} : (\nabla \otimes \mathbf{E}_*). \end{aligned} \tag{10.32}$$

Here, K, G, C_V, d_ρ, α_T, α_ρ, α_{E1}, α_{E2}, χ_E, χ_m, χ_{Em}, χ_{q1}, χ_{q2}, $\beta_{T\rho}$, β_{TE} and $\beta_{E\rho}$ are material characteristics.

Formula (10.32) along with relations (10.31) leads to the following linear constitutive equations for isotropic dielectric materials:

$$\hat{\sigma} = 2G\hat{e} + 2G\alpha_{E2}\nabla \otimes \mathbf{E} + \left[\left(K - \frac{2}{3}G\right)e - K(\alpha_T\theta + \alpha_\rho\rho_m + \alpha_{E1}\nabla \cdot \mathbf{E}\right)\right]\hat{\mathbf{I}}, \tag{10.33a}$$

$$s = s_0 + \frac{C_V}{T_0}\theta + \frac{K\alpha_T}{\rho_0}e + \beta_{T\rho}\rho_m - \beta_{TE}\nabla \cdot \mathbf{E}, \tag{10.33b}$$

$$\mu'_\pi = \mu'_{\pi 0} + d_\rho \rho_m - \frac{K\alpha_\rho}{\rho_0} e - \beta_{T\rho}\theta + \beta_{E\rho} \nabla \cdot \mathbf{E}, \tag{10.33c}$$

$$\mathbf{p} = \chi_E \mathbf{E} - \chi_{Em} (\nabla \mu'_\pi)^r, \tag{10.33d}$$

$$\boldsymbol{\pi}_m = -\chi_m (\nabla \mu'_\pi)^r + \chi_{Em} \mathbf{E}, \tag{10.33e}$$

$$\hat{\mathbf{q}} = 2\chi_{q2} \nabla \otimes \mathbf{E} - 2G\alpha_{E2}\hat{\boldsymbol{e}} - \left(\chi_{q1} \nabla \cdot \mathbf{E} - \frac{K\alpha_{E1}}{\rho_0} e + \beta_{TE}\theta + \beta_{E\rho}\rho_m \right) \hat{\mathbf{I}}. \tag{10.33f}$$

Equations (10.33a–f) describe an electromechanical interaction in isotropic (centrosymmetric) materials. In the framework of the proposed theory, the body polarization is caused not only by the electric field but also by its spatial nonhomogeneity, as well as by the gradients of the strain, the temperature, and the density of induced mass. Hence, the constitutive relations (10.33a–f) for isotropic materials make it possible to describe the flexoelectric and thermopolarization effects. Note that the linear classical theories of dielectrics cannot describe these effects.

Based on Eq. (10.30) for entropy production, we can obtain kinetic relations. We represent (10.30) as follows: $\sigma_s = \frac{1}{T} \sum_{k=1}^{3} \mathbf{j}_k \cdot \mathbf{X}_k$, where $\mathbf{j}_1 = \mathbf{J}_q$, $\mathbf{j}_2 = \mathbf{J}_{e*}$, $\mathbf{j}_3 = \rho \frac{d\boldsymbol{\pi}_m}{dt}$, and $\mathbf{X}_1 = -\nabla T/T$, $\mathbf{X}_2 = \mathbf{E}_*$, and $\mathbf{X}_3 = -(\nabla \mu'_\pi)^i$ are the thermodynamic fluxes and forces. Let us assume that thermodynamic fluxes are functions of thermodynamic forces:

$$\mathbf{j}_i = \mathbf{j}_i(\mathbf{X}_1, \mathbf{X}_2, \mathbf{X}_3). \tag{10.34}$$

For linear approximation we can write $\mathbf{j}_i = \sum_{j=1}^{3} L_{ij} \mathbf{X}_j$, $(i = \overline{1,3})$, where L_{ij} and $(i, j = \overline{1,3})$ are kinetic coefficients. According to the second law of thermodynamics and Onsager theorem [14], we have $L_{ii} \geq 0$ and $L_{ij} = L_{ji}$.

Using formula (10.26) we can exclude the irreversible $(\nabla \mu'_\pi)^i$ and reversible $(\nabla \mu'_\pi)^r$ components of vector $\nabla \mu'_\pi$ from Eqs. (10.33) and (10.34). As a result, we obtain the following relations for the heat flux $\mathbf{J}_q$, electric current density $\mathbf{J}_e$, and mass and electric dipoles $\boldsymbol{\pi}_m$ and $\mathbf{p}$:

$$\mathbf{J}_q = -\lambda \nabla T + \pi_t \mathbf{J}_e + (\pi_l - \eta' \pi_t)\rho_0 \frac{\partial \boldsymbol{\pi}_m}{\partial t}, \tag{10.35a}$$

$$\mathbf{J}_e = \sigma_e \mathbf{E} - \eta \nabla T + \eta' \rho_0 \frac{\partial \boldsymbol{\pi}_m}{\partial t}, \tag{10.35b}$$

$$\frac{\partial \boldsymbol{\pi}_m}{\partial t} + \frac{1}{\tau_\pi} \boldsymbol{\pi}_m = -\frac{\chi_m}{\tau_\pi} \left[\nabla \mu'_\pi + \pi_l \frac{\nabla T}{T_0} - \left(\eta' + \frac{\chi_{Em}}{\chi_m} \right) \mathbf{E} \right], \tag{10.35c}$$

$$\mathbf{p} = \chi_E\left(1 - \frac{\chi_{Em}^2}{\chi_E \chi_m}\right)\mathbf{E} + \frac{\chi_{Em}}{\chi_m}\boldsymbol{\pi}_m. \tag{10.35d}$$

Here,

$$\sigma_e = L_{22} - \frac{L_{23}^2}{L_{33}}, \quad \lambda = \frac{1}{T_0}\left[L_{11} - \frac{L_{13}^2}{L_{33}} - \frac{(L_{12}L_{33} - L_{13}L_{23})^2}{L_{33}(L_{22}L_{33} - L_{23}^2)}\right], \quad \tau_\pi = \frac{\rho_0 \chi_m}{L_{33}},$$

$$\eta = \frac{1}{T_0}\left(L_{12} - \frac{L_{23}L_{13}}{L_{33}}\right), \quad \eta' = \frac{L_{23}}{L_{33}}, \quad \pi_t = \frac{L_{12}L_{33} - L_{13}L_{23}}{L_{22}L_{33} - L_{23}^2}, \quad \pi_l = \frac{L_{13}}{L_{33}}.$$

Note that in Eq. (10.35), we have taken into account that in a linear approximation, $\mathbf{E}_* = \mathbf{E}$, $\mathbf{J}_{e*} = \mathbf{J}_e$, and $\frac{d}{dt} = \frac{\partial}{\partial t}$.

Due to accounting for the irreversibility of the local mass displacement, we obtained the rheological constitutive relations for the specific vector of local mass displacement $\boldsymbol{\pi}_m$, the electric dipoles $\mathbf{p}$, and the fluxes $\mathbf{J}_q$ and $\mathbf{J}_e$. When we describe the local mass displacement as a reversible process, the vectors $\boldsymbol{\pi}_m$ and the electric dipoles $\mathbf{p}$ depend only on the electric vector $\mathbf{E}$ and the gradient of modified chemical potential $\nabla \mu_\pi'$ [11]. In this case, vectors $\boldsymbol{\pi}_m$ and $\mathbf{p}$ depend on the temperature gradient as well.

The field Eqs. (10.2), (10.3), (10.4), (10.15), (10.24), and (10.25) and the constitutive relations (10.6), (10.31), and (10.34) with relations (10.7), (10.20), (10.21), (10.22), (10.23), (10.26) and (10.30) form a complete set of nonlinear equations which describes the electrothermomechanical processes in elastic dielectrics with quadrupole polarization and irreversible local mass displacement. The final form of the governing equations can be obtained by substituting the constitutive and geometric relations into the field equations.

10.8 Near-Surface Inhomogeneity

We use the obtained set of equations to investigate the process of formation of a near-surface inhomogeneity of fields in an infinite traction-free isotropic layer. We assume that at time $t = 0$, the layer (region $|x| \le l$) is cut out from an infinite medium in such a way that subsequently ($t > 0$) it is in contact with a vacuum. For simplicity, we shall consider an isothermal approximation and neglect the effect of electric fields on the layer stress-strain behaviour. Then, in the linearized approximation, the fundamental field equations expressed in terms of the displacement vector $\mathbf{u}$ and the modified chemical potential $\widetilde{\mu'}_\pi$ can be written as:

$$\left(K + \frac{1}{3}G - \frac{K^2\alpha_\rho^2}{\rho_0 d_\rho}\right)\nabla(\nabla \cdot \mathbf{u}) + G\Delta\,\mathbf{u} - K\frac{\alpha_\rho}{d_\rho}\nabla\widetilde{\mu}'_\pi = \rho_0\frac{\partial^2 \mathbf{u}}{\partial t^2}, \tag{10.36a}$$

$$\Delta\tilde{\mu}'_{\pi} - \frac{1}{d_{\rho}\chi_m}\left(\tilde{\mu}'_{\pi} + \tau_{\pi}\frac{\partial\tilde{\mu}'_{\pi}}{\partial t}\right) = \frac{K\alpha_{\rho}}{d_{\rho}\chi_m\rho_0}\left[\nabla\cdot\mathbf{u} + \tau_{\pi}\frac{\partial(\nabla\cdot\mathbf{u})}{\partial t}\right]. \quad (10.36b)$$

We consider the following boundary conditions at the layer surfaces $x = \pm l$: $\hat{\boldsymbol{\sigma}} \cdot\mathbf{n} = 0$ and $\tilde{\mu}'_{\pi} = -\mu'_{\pi 0}$. The solution of the boundary value problem should satisfy the homogeneous initial conditions, $\tilde{\mu}'_{\pi} = 0, \mathbf{u} = 0$ and $\frac{\partial\mathbf{u}}{\partial t} = 0$, at $t = 0$. In this case, the resulting fields ($\hat{\boldsymbol{\sigma}}$, $\mathbf{u} = (u, 0, 0)$, and $\tilde{\mu}'_{\pi}$) are functions of the space coordinate x and the time coordinate t.

The solution of Eq. (10.36) depends on the sign of the coefficient $(d_{\rho}\chi_m)^{-1}$. From the estimation of coefficients d_{ρ} and χ_m, it follows that these quantities are positive [1]; thus, $(d_{\rho}\chi_m)^{-1} = \lambda_{\mu}^2$. Note that $l_* = \lambda_{\mu}^{-1}$ is an intrinsic length scale (a material parameter with dimension of length).

To solve the boundary value problem, we used the integral Laplace transform of a time variable. If we restrict ourselves to a quasistatic case, then for non-zero components, $\sigma_{yy} = \sigma_{zz} \equiv \sigma$ of the stress tensor can be obtained:

$$\begin{aligned}\sigma(X,\tau) = \ &\breve{\sigma}\left\{ e^{-\breve{\lambda}\, l(1-X)}\mathrm{erfc}\left[\frac{\breve{\lambda}\, l(1-X)}{2\sqrt{\tau}} - \sqrt{\tau}\right] + e^{\breve{\lambda}\, l(1-X)}\mathrm{erfc}\left[\frac{\breve{\lambda}\, l(1-X)}{2\sqrt{\tau}} + \sqrt{\tau}\right]\right.\\ &\left.+ e^{-\breve{\lambda}\, l(1+X)}\mathrm{erfc}\left[\frac{\breve{\lambda}\, l(1+X)}{2\sqrt{\tau}} - \sqrt{\tau}\right] + e^{\breve{\lambda}\, l(1+X)}\mathrm{erfc}\left[\frac{\breve{\lambda}\, l(1+X)}{2\sqrt{\tau}} + \sqrt{\tau}\right]\right\},\end{aligned} \quad (10.37)$$

where $X = x/l$ and $\tau = t/\tau_{\pi}$ are the dimensionless space coordinate and time and

$$\breve{\sigma} = \frac{KG\alpha_{\rho}\mu'_{\pi 0}}{d_{\rho}(K + 4G/3)}, \quad \breve{\lambda} = \lambda_{\mu}\sqrt{1 + \mathfrak{M}}, \quad \mathfrak{M} = \left(K + \frac{4}{3}G - \frac{K^2\alpha_{\rho}^2}{\rho_0 d_{\rho}}\right)^{-1}\frac{K^2\alpha_{\rho}^2}{\rho_0 d_{\rho}}.$$

Note that $\mathfrak{M}$ is the factor describing the coupling between the mechanical fields and the local displacement of mass [12].

The quantitative analysis of the obtained solution showed that at time $\tau = 0$, in the vicinity of the layer surfaces, there is formed an inhomogeneous distribution of the stresses σ_{yy}, σ_{zz}, and potential $\tilde{\mu}'_{\pi}$. Figure 10.2 displays the distributions of the stresses in "thin" (Fig. *a*) and "thick" (Fig. *b*) layers at different times. As one can see, the process of surface formation reaches the steady-state regime at $\tau \geq 5$.

We see that the thick layer regions that being far from the surfaces are stress-free, but at the vicinity of the surfaces, these stresses rapidly increase. On the contrary, in thin layers (films) over time, there is formed an essentially heterogeneous distribution of stresses. The middle regions of thin layers are stressed. A decrease in thickness of thin films leads to an increased stress in the film cross section.

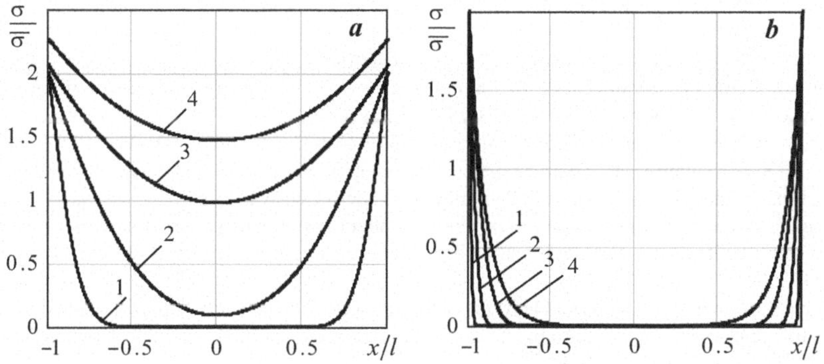

Fig. 10.2 The distribution of the stress $\sigma/\overline{\sigma}$ in thin ($l = l_*$, Fig. **a**) and thick ($l = 10l_*$, Fig. **b**) layers for different times $\tau = 10^{-2}, 10^{-1}, 5 \cdot 10^{-1}, 5$, that is, curves 1–4, respectively

10.9 Conclusions

A complete continuum nonlinear gradient-type theory of electro-thermoelastic non-ferromagnetic dielectrics was derived. The local mass displacement, the quadrupole polarization, and the irreversibility of the local mass displacement are taken into account in order for a general theory of dielectrics to be obtained. We associate the local mass displacement with the changes in the material microstructure. The electric field vector gradient, the density of induced mass, and the reversible part of the gradient of a modified chemical potential are included in the model of polarizable media as constitutive variables. It is shown that by taking into consideration the electric quadrupoles, local mass displacement, and its irreversibility, the classical continuum theory of thermoelastic dielectrics is extended to accommodate the surface, size flexoelectric, and thermopolarization effects, as well as an electromechanical interaction in centrosymmetric materials. The above effects are not accounted for in the classical theory of dielectrics. Due to an accounting of the irreversibility of the local mass displacement, the constructed theory allows us, in continuum approximation, to study the process of formation of a near-surface inhomogeneity of the stress-strain state of the solids.

References

1. Burak Ya, Kondrat V, Hrytsyna O (2011) Fundamentals of the local gradient theory of dielectrics, Lira, Uzhgorod. In Ukrainian
2. Hadjigeorgiou EP, Kalpakides VK, Massalas CV (1999) A general theory for elastic dielectrics – part I. The vectorial approach. Int J Non-Linear Mech 34:831–841
3. Kafadar CB (1971) Theory of multipoles in classical electromagnetism. Int J Engng Sci 9:831–853
4. Maugin GA (1988) Continuum mechanics of electromagnetic solids. North-Holland Publishing Company, Amsterdam

5. Mindlin RD (1972) Elasticity, piezoelectricity and crystal lattice dynamics. J Elasticity 2:217–282
6. Yang J (2006) Review of a few topics in piezoelectricity. Appl Mech Rev 59:335–345
7. Axe JD, Harada J, Shirane G (1970) Anomalous acoustic dispersion in centrosymmetric crystals with soft optic phonons. Phys Rev B 1:1227–1234
8. Mead CA (1961) Anomalous capacitance of thin dielectric structures. Phys Rev Lett 6:545–546
9. Nam CY, Jaroenapibal P, Tham D, Luzzi DE, Evoy S, Fischer JE (2006) Diameter-dependent electromechanical properties of GaN nanowires. Nano Lett 6:153–158
10. Tang C, Alici G (2011) Evaluation of length-scale effects for mechanical behaviour of micro- and nanocantilevers: I. Experimental determination of length-scale factors. J Phys D Appl Phys 44:335501
11. Burak Y, Kondrat V, Hrytsyna O (2008) An introduction of the local displacements of mass and electric charge phenomena into the model of the mechanics of polarized electromagnetic solids. J Mech Mater Struct 3(6):1037–1046
12. Kondrat VF, Hrytsyna OR (2010) Mechanoelectromagnetic interaction in isotropic dielectrics with regard to the local displacement of mass. J Math Sci 162(1):150–158
13. Kondrat V, Hrytsyna O (2012) Local gradient theory of dielectrics with polarization inertia and irreversibility of local mass displacement. J Mech Mater Struct 7(3):285–296
14. De Groot SR, Mazur P (1962) Non-equilibrium thermodynamics. North-Holland Publishing Company, Amsterdam
15. Bredov MM, Rumyantsev VV, Toptyhin IN (1985) Classic electrodynamics. Nauka, Moscow. In Russian
16. Landau LD, Lifshitz EM (1984) Electrodynamics of continuous media. Pergamon, Oxford
17. Fedorchenko AM (1988) Theoretical physics. Electrodynamics: Textbook, Vyshcha shkola, Kyiv. In Russian
18. Burak Ya (1987) Constitutive equations of locally gradient thermomechanics, *Proc. Acad. Sci. Ukrain.* SSR, 12, 19–23. In Ukrainian
19. Green AE, Rivlin RS (1964) On Cauchy's equations of motion. Z Angew Math Phys 15:290–293

Chapter 11
Graphene Quantum Dots in Various Many-Electron π-Models

Anatoliy Luzanov

11.1 Introduction

After the discovery of graphene, there has been a rising interest in its fundamental electronic properties. Up to now, graphene nanoparticles, in particular graphene quantum dots (GQDs), remain an extremely important topics in solid-state physics and material sciences [1–5]. There are many useful models and techniques for studying GQD electronic structure, and among them are the conventional tight-binding (TB) model and its Hückel version for π-electron shells. Certainly, simple Hückel and TB computations provided many important results for understanding the unique physics of graphene systems (e.g., see [2]). At the same time, accounting for many-electron effects is essential in this case of extended delocalized systems containing a lot of conjugated bonds.

The present work continues our previous study on the electronic structure of carbon-containing nanoclusters [6–11]. One of the principal aims of this paper is to show that some well-known quantum chemistry models which are less familiar to the graphene research community can be served as a suitable and feasible tool for exploring nanographenes at the semiempirical many-electron level. In particular, we apply here the so-called spin-extended Hartree-Fock (EHF) and restricted active space configuration interaction (RAS-CI) methods.

Another goal of the work is to elucidate, for the selected graphene-like molecules, a behavior of π-electrons in extreme electric fields. In this problem, the electron unpairing (see review [8]) is a particularly exciting issue. At last, the problem of

A. Luzanov (✉)
State Scientific Institution "Institute of Single Crystals", National Academy of Sciences of Ukraine, Kharkiv, Ukraine
e-mail: luzanov@xray.isc.kharkov.com

O. Fesenko, L. Yatsenko (eds.), *Nanooptics, Nanophotonics, Nanostructures, and Their Applications*, Springer Proceedings in Physics 210,
https://doi.org/10.1007/978-3-319-91083-3_11

graphene aromaticity continues to attract interest [1, 12–14], and here we will also focus on it using the previously given approach [15] .

11.2 Computational Schemes

11.2.1 Electronic Instabilities and Lowest Excitations

In this and next subsections, we shortly sketch the main electronic models that we apply to GQDs. First, we consider rather qualitative schemes. Usually, the frontier orbital energy gap (the so-called HOMO-LUMO gap) is considered as a descriptor of the molecular stability. But this electronic indicator is too crude. More correct and more informative is the Hartree-Fock (HF) stability theory [16] (see also [17], Sect 10.10). We recall that the HF stability of the given molecular system means that the HF stability matrix (matrix of second derivatives with respect to variational parameters of HF wave function) has only positive eigenvalues $\lambda^{\mathrm{stable}}$. An occurrence of zero or negative $\lambda^{\mathrm{stable}}$ values indicates on the HF instability of the system under study.

For closed-shell singlet states, the restricted HF (RHF) model is usually employed, and in this case one can define the Hartree-Fock stability of two types, singlet one and triplet one [16]. The corresponding eigenvalues of the HF stability matrix will be denoted by $\lambda^{\mathrm{stable}}_{s=0}$ and $\lambda^{\mathrm{stable}}_{s=1}$, respectively. Typically, minimum value of $\lambda^{\mathrm{stable}}_{s=1}$ is lesser than minimum $\lambda^{\mathrm{stable}}_{s=0}$, that is, the triplet RHF instability appears before the singlet HF instability does. In fact, the negative $\lambda^{\mathrm{stable}}_{s=0}$ is an indicator of a very strong electronic instability. To be more specific, consider π-electrons in the hexacene molecule $C_{26}H_{24}$ for which we find $\lambda^{\mathrm{stable}}_{s=1} = -1.16$ eV, and $\lambda^{\mathrm{stable}}_{s=0} = 2.10$ eV (as for the π-parameterization adopted here see Sect. 11.3). We see that even this not so long oligocene structure is quite unstable and exhibits sufficiently strong electron correlation.

Along with the HF stability analysis, here we invoke typical models of lowest-energy electronic transitions. The simplest is the configuration interaction (CI) singles method, or CIS method. The corresponding minimal eigenvalues, $\lambda^{\mathrm{CIS}}_{s=1}$ and $\lambda^{\mathrm{CIS}}_{s=0}$, of the CIS Hamiltonian matrix give rather crude estimates of the lowest-energy singlet-triplet and singlet-singlet transitions. Considerably more advanced is the restricted active space CI (RAS-CI) method [17, 18]. In RAS-CI one divides the full one-electron orbital space into certain active and frozen orbital subsets. To deal with this, the standard full CI (FCI) technique is applied only to electrons assigned to the chosen active orbital subset. Then, the lowest eigenvalues of the appropriate RAS-CI matrix will approximate energies of the ground and excited states. For the low-lying excitation energies, the approach provides estimates more reliable than those from CIS.

In our study we employ a nonstandard but equivalent RAS-CI formulation. It is based on the many-electron equation describing a detachment of multi-electrons (say¸ 2 k electrons), from the intermediate state with $N + 2k$ electrons where $N \equiv 2n$ is a number of electrons in an even-electron molecule under study. Thus, we start with a polyanionic $(N + 2k)-$ electron state for which we form the Slater determinant with $n + k$ occupied MOs. The aimed equation is the FCI eigenvalue problem for 2 k-electron Hamiltonian of the form

$$H^{\text{detach}} = -Q_{2k}^{\text{act}} \left[\sum_{1 \le i \le 2k} f(i) - \sum_{1 \le i < j \le 2k} g(i,j) \right] Q_{2k}^{\text{act}},$$

where f is the conventional Fock operator of the intermediate $2(n + k)$ electron singlet state (recall that $n = N/2$ is a number of electron pairs in the molecule under study); g is the two-electron Coulomb operator, and Q_{2k}^{act} is the 2 k-electron projection operator onto the chosen active subset. The lowest singlet-triplet, $\lambda_{s=1}^{\text{RAS-CI}}$, and singlet-singlet, $\lambda_{s=0}^{\text{RAS-CI}}$, excitation energies are evidently obtained from the corresponding differences of the H^{detach} eigenvalues. In practice, we carried out RAS-CI spectral calculations with $k = 6$. In terms of results, the computations are equivalent to those within the conventional RAS-CI 6 × 6 scheme. A lower part of the H^{detach} spectrum was computed by the modified Lanczos-type diagonalization algorithm described in Ref. [9], Appendix A.

11.2.2 Unrestricted and Extended Hartree-Fock Methods

Here the used HF schemes will be only briefly sketched. We begin with the unrestricted Hartree-Fock method (UHF) which is a precursor of EHF. The single-determinant UHF model is not accidentally named also as "spin-polarized Hartree-Fock method." Unlike RHF, the UHF method brings into play, if possible, spin-polarized MOs, i.e., different orbitals for different spins. The latter are usually signified by ϕ_j^{α} and by ϕ_j^{β} for spin-up and spin-down electrons, respectively. From these orbitals the fundamental projection operators, ρ^{α} and ρ^{β}, are constructed. In Dirac notation these are

$$\rho^{\alpha} = \sum_{1 \le j \le n+s} | \varphi_j^{\alpha} >< \varphi_j^{\alpha} |, \rho^{\beta} = \sum_{1 \le j \le n-s} | \varphi_j^{\beta} >< \varphi_j^{\beta} |,$$

and $n + s$ and $n - s$ are, respectively, total numbers of spin-up and spin-down electrons in the given N-electron state with a total spin s. By applying variational method, one can directly derive the conventional UHF equations in terms of ρ^{α} and ρ^{β} [19]. Although a certain correlation of opposite-spin electrons appears in the corresponding Slater determinant $|\Phi>$, the correlated state of this type does not exist in weakly correlated closed-shell systems. More exactly, in a such singlet-like case with $s = 0$, a nontrivial UHF solution is possible if the initial RHF singlet state satisfies the triplet Hartree-Fock instability condition, that is, $\lambda_{s=1}^{\text{stable}} < 0$ [20]. For strongly correlated systems (e.g., zigzag graphenes), the UHF model provides a more or less reasonable π-electron picture.

Now we proceed to the EHF approximation that was firstly formulated by Löwdin [21] for π-electron systems. There are many works on the EHF theory (see, e.g., reviews [22, 23]), and recently this theory was revived at the ab initio level [24]. In EHF the spin contamination of a trial UHF determinant |Φ> is removed by acting on |Φ> an appropriate spin-projection operator O_s. Generally, it depends on a spin z-projection value m as well. Thus, instead of |Φ>, we introduce the normalized EHF wave function as follows:

$$\left|\Phi^{\mathrm{EHF}}\right\rangle = O_s|\Phi\rangle/\sqrt{\langle\Phi|O_s|\Phi\rangle}.$$

It is important that, unlike UHF, the EHF theory imposes no restriction on electron correlation strength so that even for $s = 0$, the nontrivial EHF (i.e., $\phi_j^\alpha \neq \phi_j^\beta$) solution always exist [23]. This feature is very important for our aims because it allows us to treat weakly and strongly correlated problems on equal footing. Furthermore, there are many derivations and practical formulations of the EHF variational equations, but we prefer using matrix formulations in terms of ρ^α and ρ^β [23, 25], as they provide suitable and compact algorithms. Our matrix technique [25] is easily coded and applied to large molecules. At last, we must mention one essential drawback of EHF and related variational models (including half-projected HF and complex MO schemes). This is a lack of size consistency which precludes obtaining robust results for too large-scale systems.

11.3 π -Electron Properties of Selected Nanographenes

Now we will analyze the results of π-electron computations on small GQDs. For this study, we selected four rather typical honeycomb carbon structures which are similar in composition (Figs. 11.1 and 11.2). The first system is a nanoribbon fragment C_{130}, that is (9,6)-periacene, or shortly (9,6)-PA. Related structures are frequently considered in literature, particularly in the context of their polyradical character [8, 10]. Two other systems, GQD C_{130} and GQD C_{132}, belong to a class of recently synthesized colloidal graphene quantum dots [26] (see also Ref. [2]). For these GQDs, a simplified Hubbard-like π-electron model, with a somewhat artificial

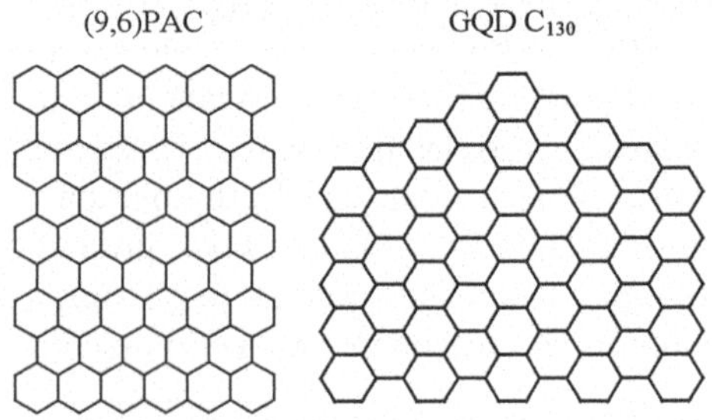

Fig. 11.1 Structure of the studied graphene molecules with strong and moderate electron correlation

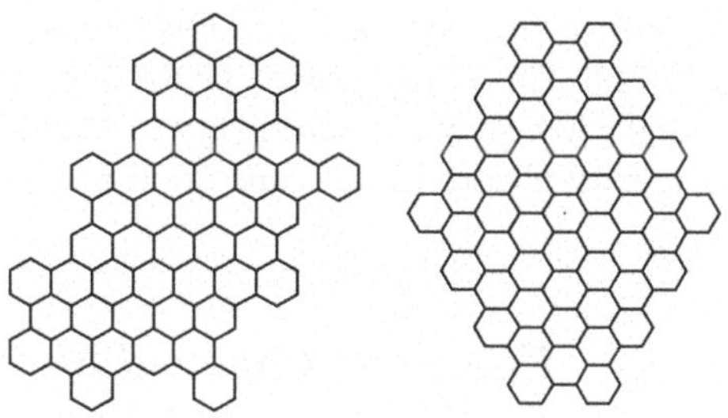

Fig. 11.2 Structure of the studied graphene molecules with weak electron correlation

Table 11.1 The π-electron RHF stability value $\lambda^{\mathrm{stable}}$, lowest CIS excitation energy λ^{CIS}, and RAS-CI excitation energy $\lambda^{\mathrm{RAS-CI}}$ (all values in eV) for singlet ($s = 0$) and triplet ($s = 1$) states in nanographenes

System	$\lambda_{s=1}^{\mathrm{stable}}$	$\lambda_{s=0}^{\mathrm{stable}}$	$\lambda_{s=1}^{\mathrm{CIS}}$	$\lambda_{s=0}^{\mathrm{CIS}}$	$\lambda_{s=1}^{\mathrm{RAS\text{-}CI}}$	$\lambda_{s=0}^{\mathrm{RAS\text{-}CI}}$
(9,6)-PA	−4.29	−0.35	−2.00	0.11	0.05	0.23
GQD C_{130}	−1.69	0.82	−0.17	1.15	0.14	0.75
GQD C_{132}	0.02	1.32	1.80	2.28	2.46	2.62
GQD C_{138}	−0.04	1.20	1.69	2.15	2.15	2.46

choice of π-parameters, was used in Ref. [2]. The last GQD C_{138} is the fully benzenoid (Clar's type) nanographene molecule, which should be particularly stable from the viewpoint of the so-called Clar aromatic sextet theory [14, 27].

Throughout the paper we apply the conventional Pariser-Parr-Pople π-electron Hamiltonian with the usual π-parameters (resonance integral $\beta_{CC} = -2.4$ eV, two-center Coulomb integrals $\gamma_{\mu\nu}$ due to Ohno, and one-center integral $\gamma_C = 11.13$ eV).

11.3.1 Hartree-Fock Instabilities and Optical Transitions

The key issue for large polycyclic hydrocarbons is their stability/instability as systems with spatially extended π-electron wave functions. The corresponding HF stability values along with CIS and RAS-CI results for excitation energies are listed in Table 11.1. The RHF stability and CIS excitations are studied here by using the conventional RHF MOs. As to our implementation of RAS-CI approach, we prefer, instead of RHF MOs, to utilize more sophisticated orbital sets such as in Ref. [28]. Only we used the EHF natural orbitals rather than the half-projected HF ones employed in [28].

From Table 11.1 we see that two first clusters, (9,6)-PA and GQD C_{130}, possess a strong triplet instability; furthermore, (9,6)-PA possesses, too, a clear singlet instability. It means that these π-structures cannot be correctly treated within RHF and related schemes. The erroneously negative value of $\lambda_{s=1}^{\mathrm{CIS}}$ for GQD C_{130} in Table 11.1 is indicative of this point. At the same time, RAS-CI data (a penultimate column in the table) show that all triplet excitation energies are positive, as it should be according to the known Ovchinnikov-Lieb rule for bipartite π-networks [29, 30]. The third system, GQD C $_{132}$, turns out to be a normal stable system with the lowest excitation energy in a visible optical region. The fully benzoid GQD C_{138} must absorb in the same optical region.

11.3.2 Effectively Unpaired Electrons

Firstly, we give some preliminary remarks. The effectively unpaired electron (EUE) theory provides a special understanding of strongly correlated spin-singlet systems in terms of loosely coupled electrons (for more detail see Ref. [8]). We only note that here we follow the Head-Gordon approach [31], but in fact the hole-particle analysis given in Ref. [6, 8] will be employed. In this approach, a special EUE number, N_{eff} (N_{U} in notation from Refs. [10, 31]), is constructed. In the hole-particle representation, N_{eff} is but a doubled occupation value of virtual particles, or explicitly,

$$N_{\text{eff}} = 2\sum_{a>n} \lambda_a,$$

with λ_a being the natural orbital occupation numbers for the particle (virtual) states of one electron. More exactly, $\{\lambda_a\}$ is understood as a set of the one-electron density matrix eigenvalues in a decreasing order. Within the EUE formalism, one can also introduce the associated EUE distribution $\{D_{\text{eff}}^A\}$ over all atoms $\{A\}$ [10, 31]. Furthermore, a suitable participation ratio index of the second order (PR2) serves to estimate an average number of the most active atoms on which unpaired electrons are localized; the PR2 precise definition is given in Ref. [8], Eq. (6.88). The above characteristics were employed in our investigation, and the main EUE results obtained at the EHF level are displayed in Fig. 11.3. In this figure the EUE

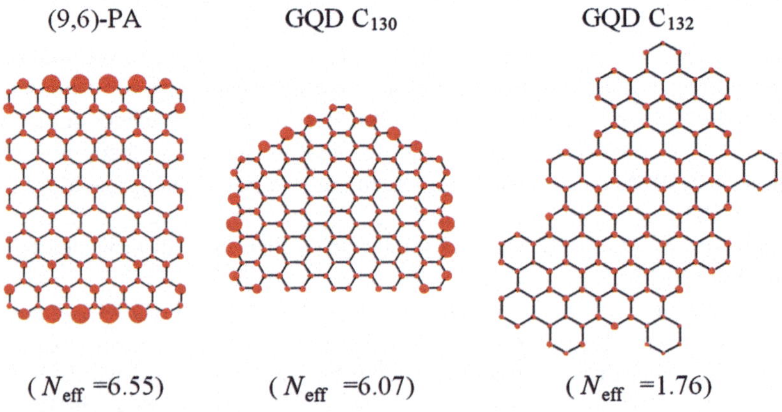

Fig. 11.3 EUE diagrams for the small graphene molecules treated by π-electron EHF method

distributions are visualized by diagrams in which atomic EUE densities are shown with red discs of radius proportional to the corresponding D^A_{eff} value. From the given diagrams, it follows that in case of the strongly correlated systems, that is, (9,6)-PA and GQD C_{130}, the number of unpaired electrons is significant ($N_{\mathrm{eff}} \approx 6$), and they are localized on the zigzag (anthracene-like) part of the total board of the graphene structures. As in the previous studies [6, 8, 10], the armchair (phenanthrene-like) part of the board is almost not populated by unpaired electrons.

Opposite to (9,6)-PA and GQD C_{130}, the third system GQD C_{132} has a small $N_{\mathrm{eff}} < 2$, and the corresponding EUE distribution is spread over the network more or less equally. These peculiarities in EUE localization are additionally reflected by the PR2 values (numbers of localization sites) which are rather small for the first two systems (PR2 = 13.8 and PR2 = 17.0, respectively) and too large for the last system (PR2 = 96.1). We do not display the EUE diagram for GQD C_{138} because in this molecule N_{eff} is too small (≈0.36), so that the atomic EUE diagram becomes inexpressive ($D^A_{\mathrm{eff}} \sim 10^{-3}$).

11.3.3 Characterizing Nanographene Aromaticity

It was long recognized that the aromaticity quantification is a too tricky matter. No wonder that there is a variety of quantum-chemical tools to understand aromaticity in a quantitative manner. Many such schemes are focused upon the so-called local aromaticity associated with a given benzenoid fragment in the whole system (e.g., see a fresh review [32] where one can find a vast literature on this issue).

In the present work, we apply a special aromaticity scale (σ_{arom} scale) proposed in Ref. [15]. It is based on comparing the atomization energies of the molecule under study and the related one produced by removing the given benzenoid ring A from the molecule. For the benzenoid cycle A, the corresponding local aromaticity index $\sigma_{\mathrm{arom}}[A]$ is computed by Eqs. (73) and (74) in Ref. [15]. In Figs. 11.4 and 11.5, the local aromaticity diagrams are built from these $\sigma_{\mathrm{arom}}[A]$ values which (in %) are inserted in benzenoid rings. The diagrams given here show, at the simple Hückel MO level, the typical features of distributing aromaticity over the graphene networks. To better understand them, it is useful to bear in mind an analogous image for the related hexacene molecule:

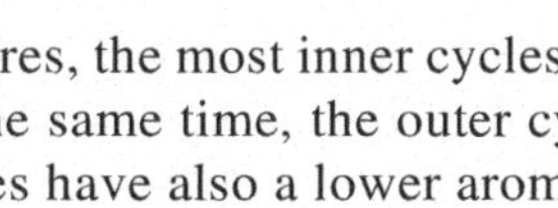

We see that in all structures, the most inner cycles are markedly less aromatic than the outer cycles. At the same time, the outer cycles belonging to a zigzag border of the nanographenes have also a lower aromaticity. We clearly observe that in all the systems, the armchair (phenanthrene-like) border regions show more aromatic, and thus more stable behavior. This fact is in conjunction to the EUE distributions given in Fig. 11.3.

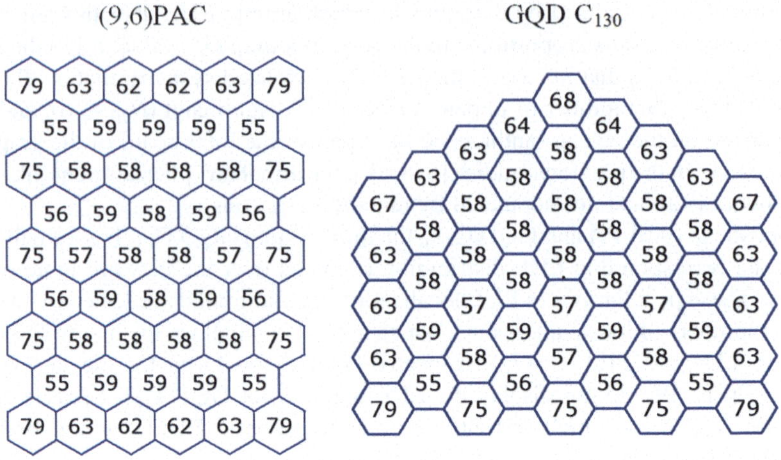

Fig. 11.4 Local aromaticity in the graphene molecules with strong and moderate electron correlation

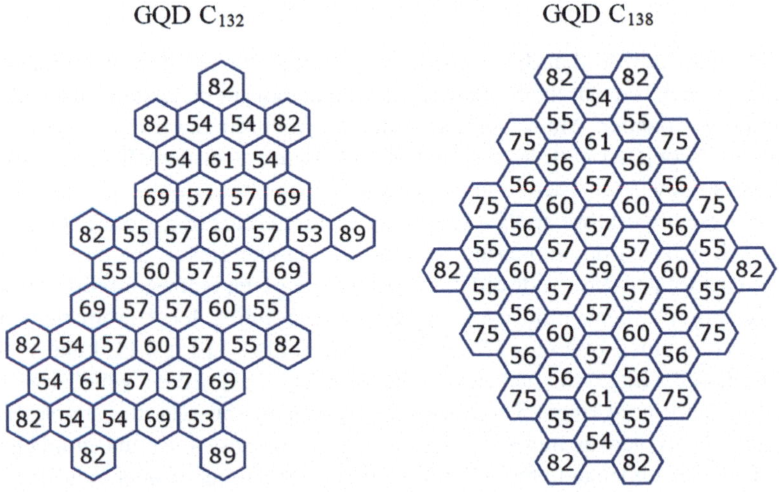

Fig. 11.5 Local aromaticity in the graphene molecules with weak electron correlation

11.3.4 Effects of Weak and Strong Electrostatic Fields

Over the last few years, there has been an upsurge of interest in the nonlinear optical properties (electric hyperpolarizabilities) of conjugated polymers and nanoclusters, including GQDs (see [33–36] and many others). In this subsection, we briefly examine the π-electron behavior of the nanographenes in presence of an electric field. The influence of electric fields, say, in *x*-direction, is usually modeled by

Table 11.2 Average π-electron polarizability and second-order hyperpolarizability (all in atomic units) for the graphene π-electron shells at the UHF level

Property	(9,6)-PA	GQD C_{130}	GQD C_{138}
α	1394	1366	1536
$\gamma \times 10^{-7}$	0.96	0.76	0.97

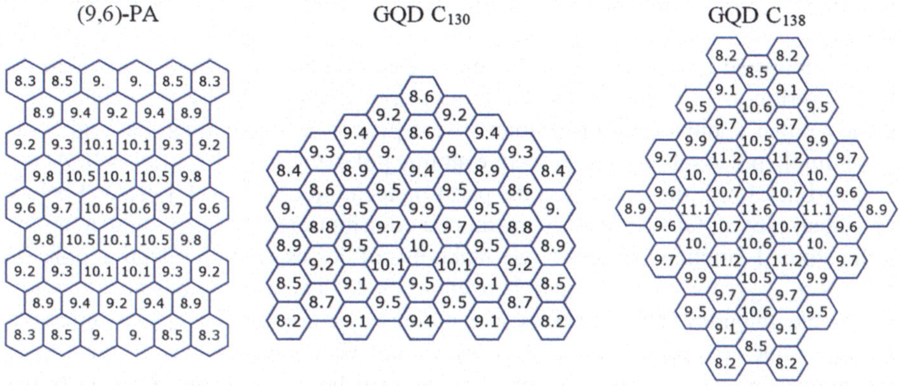

Fig. 11.6 Distribution of the average dipole π-polarizability in the small nanographenes at the UHF level

including in the full π-electron Hamiltonian the additional one-electron operator having atomic (μ,ν)-matrix elements proportional to $x_\mu\, \delta_{\mu\nu}$, with x_μ being *x*-coordinate of the μ-th atom. The proportionality constant is in fact a strength field *F*. Below in studying electric properties, we follow UHF approach rather than EHF, the latter requiring too much time to perform calculations repeatedly for many values of the field. Preliminary UHF and EHF computations on small acenes demonstrate reasonably similar field dependences of the main electronic properties. Our typical results for GQDs are presented in Table 11.2 and Fig. 11.6.

Before discussing them, recall that in weak electric fields, the key effect is well reflected through a small number of low-order electric polarizabilities. Namely, the molecular dipole moment induced by a small external electric field *F* is

$$d_{\text{ind}} \cong d + \alpha F + \beta F^2/2 + \gamma F^3/6,$$

where coefficients *d*, α, β, and γ are, respectively, the static dipole moment, (linear) polarizability, the first hyperpolarizability, and the second hyperpolarizability at the zero field. Besides, in the considered systems, which are all bipartite (alternant) ones, the dipole moment and the first hyperpolarizability are zero in standard π-electron approximations, so that $d_{\text{ind}} \cong \alpha F + \gamma F^3/6$.

The polarizabilities and hyperpolarizabilities were computed by the finite-field method (see, e.g., Ref. [37] for HF and Refs [35, 38] for FCI computations).

Table 11.2 displays the obtained α and γ data. We see that differences in the static electric properties of the molecules are not so significant. Concurrently, we must keep in mind that for the benzene (basic) molecule, the respective values are $\alpha = 25.3$ au and $\gamma = 788$ au. It means that an enormous exaltation of the static polarizabilities takes a place even in the case of small graphenes. For instance, GQD C_{138} formally contains only 23 benzene units C_6, whereas $\alpha[C_{138}]/\alpha[C_6] \cong 61$. In the case of hyperpolarizabilities, the tremendous nonadditivity effects make practically meaningless any comparison of the graphene γ values with that of the benzene subunit.

An additional interpretation of the above electric properties can be provided by their description in terms of local (atomic) contributions. Many theoretical (frequently rather sophisticated) procedures were proposed for dividing the total computed dipole polarizability into effective atomic polarizabilities $\{\alpha_\mu\}$ [39–42]. At the Hartree-Fock level, the simplest is the partition technique from Ref. [39], and below we will employ it. Only instead of the atomic contributions $\{\alpha_\mu\}$ including all atoms in the studied graphene structure, we make use of corresponding effective polarizabilities α_{Benzene} for each 6-atomic benzenoid cycle. These quantities are shown in the visible diagrams (Fig. 11.6) where numbers in hexagons are local effective α_{Benzene} values expressed in A° 3. They should be compared with the reference magnitude $\alpha^0_{\text{Benzene}} = 3.8\text{A}°\ 3$ of the isolated benzene. From Fig. 11.6 we observe that indeed the effective π-electron polarizabilities are essentially larger than $\alpha^0_{\text{Benzene}}$, and more importantly, they are markedly larger inside the graphene molecules than in peripheral hexagons. It means that in concordance with Figs. 11.4 and 11.5, in the nanographenes the aromaticity degree of inner cycles is lesser than that of border cycles. This is in line with a plausible suggestion that the aromaticity degree should be antibatic with the electronic polarizability. The assertion is supported by the so-called minimum polarizability principle [43] and its applications in many works, see, e.g., Refs. [44, 45].

To conclude the section, we shortly discuss the behavior of the same structures in strong fields. In Fig. 11.7 we show the field dependence of dipole moment d and the electron unpairing index N_{eff} (see Sect. 11.3.2). We observe in Fig. 11.7 a very large (sharp in certain field regions) increase of the dipole moment with applied field. Almost in the same field regions, N_{eff}exhibits sharp variations. Furthermore, under a very strong field, the great enhancement of radical character occurs, as it was previously reported for more simple conjugated molecules [8, 46]. The difference is that the external fields, which are causing polyradical-like structures in graphenes, are of an order of magnitude lesser as compared to those in Ref. [46]. It reflects an essentially increasing instability of nanographenes even in comparatively weak electric fields.

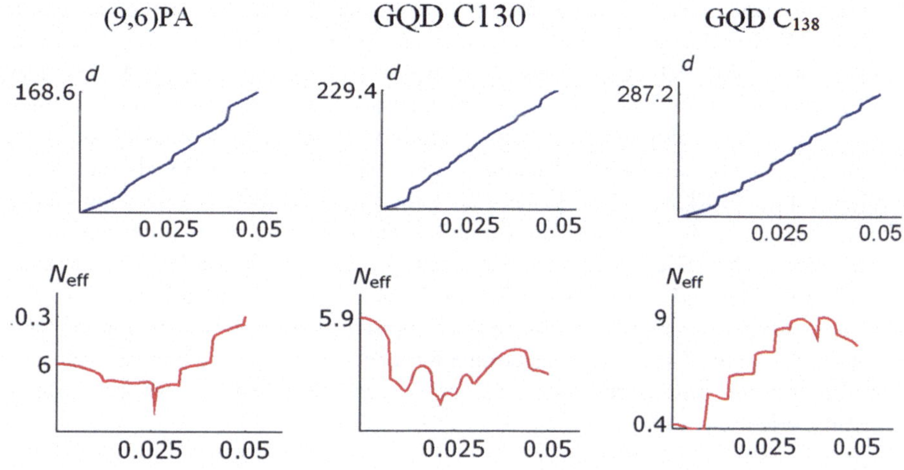

Fig. 11.7 Influence of strong electric fields on the nanographene dipole moment d (top) and effective electron unpairing N_{eff} (bottom) at the π-electron UHF level. The applied field (in atomic units) is in the long axis direction

11.4 Conclusion

We have studied the valence π-shells, i.e., the most active electron subshells, for several small graphene molecules. To this end, various approaches were employed, in each case depending on the solved problem and a feasibility of using the selected quantum chemistry method with the relatively modest hardware requirements. It was crucially important to take proper account of electron-correlation effects because many of them are found to exhibit a strong Hartree-Fock instability.

To obtain the lowest singlet and triplet excitations, we used the conventional RAS-CI technique with a special (EHF) orbital basis of natural states. The used CI technique guaranteed the non-negativity of computed triplet excitation energies $\lambda_{s\,=\,1}$ whereas the other well-known technique - CIS - did produce the wrong (negative) $\lambda_{s\,=\,1}$ in two GQDs where strong π-electron correlations take a place (Table 11.1).

Another important correlation effect is a large electron unpairing even in the ground singlet state of graphenes. The diagrams in Fig. 11.4 clearly demonstrate the unpairing effect for correlated π-systems of (9,6)-PA and GQD C_{130}. Moreover, an unusual behavior of the electron unpairing is observed in all GQDs when they are subjected to a sufficiently strong electric field. In that regard, the fully benzenoid structure GQD C_{138} turned out to be a remarkable example – a negligible initial (at the zero field) electron unpairing enormously increased even in moderate electric fields.

The last point which is worth discussing here is why we systematically use π-electron schemes whereas many authors follow seemingly more appropriate

DFT techniques. Indeed, DFT for not too huge clusters frequently gives good results (see Ref. [1] for DFT applications to graphenes). However, DFT is not an universal tool, saying nothing about a certain semiempirical character of energy functionals used in practical DFT calculations (a dependence on internal parameters of functionals, etc.). We do not also forget about a number of severe "ideological" difficulties in DFT (see review [47]), particularly those which are caused by absence of an explicit spin dependence in the general DFT theory (as McWeeny says, "unlike the classical position and momentum variables, electron spin is in a certain sense extraneous to the DFT." [48]; see also a careful analysis of Kaplan [49, 50]). In Ref. [51] one can find an interesting viewpoint on a special semiempirical use of the DFT technique as a possible tool for finding improved sets of π-electron parameters. We hope that in the future, our π-electron approaches to complex graphene networks can be further improved and extended with the help of more refined π-parameterizations [51].

References

1. Jiang DE, Chen Z (eds) (2013) Graphene chemistry: theoretical perspectives. Wiley, Puerto Rico
2. Güçlü AD, Potasz P, Korkusinski M, Hawrylak P (2014) Graphene quantum dots. Springer, Berlin/Heidelberg/New York
3. Kheirabadi N, Shafiekhani A, Fathipour M (2014) Review on graphene spintronic, new land for discovery. Superlattice Microst 74:123
4. Yin PT, Shah S, Chhowalla M, Lee K-B (2015) Design, synthesis, and characterization of graphene–nanoparticle hybrid materials for bioapplications. Chem Rev 115:2483
5. Georgakilas V, Perman JA, Tucek J, Zboril R (2015) Broad family of carbon nanoallotropes: classification, chemistry, and applications of fullerenes, carbon dots, nanotubes, graphene, nanodiamonds, and combined superstructures. Chem Rev 115:4744
6. Luzanov AV (2014) Measures of unpaired electrons for large conjugated systems. J Struct Chem 55:799
7. Luzanov AV (2014) Effectively unpaired electrons in bipartite lattices within the generalized tight-binding approximation: application to graphene nanoflakes. Funct Mater 21:414
8. Luzanov AV (2016) Effectively unpaired electrons for singlet states: from diatomics to graphene nanoclusters. In: Leszczynski J, Shukla MK (eds) Practical aspects of computational chemistry IV. Springer, Boston, p 151
9. Luzanov AV (2017) About theoretical peculiarities of lowest excitations in modified nanodiamond color centers. Funct Funct Mater 24:127
10. Luzanov AV, Plasser F, Das A, Lischka H (2017) Evaluation of the quasi correlated tight-binding (QCTB) model for describing polyradical character in polycyclic hydrocarbons. J Chem Phys 146:064106
11. Luzanov AV (2017) Localization of orbitals and electronic properties in nanodiamonds with color centers: semiempirical models. In: Fesenko O, Yatsenko L (eds) Nanophysics, nanomaterials, and applications, Springer proceedings in physics 195. Springer, Cham, p 115
12. Zdetsis AD, Economou EN (2015) A pedestrian approach to the aromaticity of graphene and nanographene: significance of Hückel's (4n+2)π electron rule. J Phys Chem C 119:16991
13. Saha B, Bhattacharyya PK (2016) Understanding reactivity, aromaticity and absorption spectra of carbon cluster mimic to graphene: a DFT study. RSC Advance 6:79768

14. Nishino N, Makino M, Aihara J-i (2016) Aromatic character of irregular-shaped nanographenes. J Phys Chem A 120:2431; Aihara J-i (2016) Graph theory of aromatic stabilization. Bull Chem Soc Jap 89:1425
15. Luzanov AV (2011) Quantum fidelity for analyzing atoms and fragments in molecule: APPLICATION to similarity, chirality, and aromaticity. Int J Quant Chem 111:2197
16. Čížek J, Paldus J (1967) Stability conditions for the solutions of the Hartree—Fock equations for atomic and molecular systems. Application to the pi-electron model of cyclic polyenes. J Chem Phys 47:3976
17. Helgaker T, Jorgensen P, Olsen J (2000) Molecular electronic-structure theory. Wiley, New York
18. Olsen J, Roos BO, Jorgensen P, HJAa J (1988) Determinant based configuration interaction algorithms for complete and restricted configuration interaction spaces. J Chem Phys 89:2185
19. McWeeny R (1992) Methods of molecular quantum mechanics. Academic Press, London
20. Koutecký J (1967) Unrestricted Hartree—Fock solutions for closed-shell molecules. J Chem Phys 46:2443
21. Löwdin P-O (1955) Quantum theory of many-particle systems. III. Extension of the Hartree-Fock scheme to include degenerate systems and correlation effects. Phys Rev 97:1505
22. Mayer I (1980) The spin-projected extended Hartree-Fock method. Adv Quant Chem 12:189
23. Mestechkin MM, Vaiman GE, Klimo V, Tino J (1983) Extended Hartree-Fock method and its application to molecules [in Russian]. Naukova Dumka, Kiev
24. Jiménez-Hoyos CA, Henderson TM, Tsuchimochi T, Scuseria GE (2012) Projected Hartree-Fock theory. J Chem Phys 136:164109
25. Luzanov AV, Ivanov VV (1991) Configuration interaction of states of the quasi-one-electron type. Theor Exp Chem 26:363
26. Yan X, Li B, Lim L-S (2013) Colloidal graphene quantum dots with well-defined structures. Acc Chem Res 46:2224
27. Clar E (1964) Polycyclic hydrocarbons. Academic Press, New York
28. Bofill JM, Pulay P (1989) The unrestricted natural orbital–complete active space (UNO–CAS) method: an inexpensive alternative to the complete active space–self-consistent-field (CAS–SCF) method. J Chem Phys 90:3637
29. Ovchinnikov AA (1978) Multiplicity of the ground state of large alternant organic molecules with conjugated bonds. Theor Chim Acta 47:297
30. Lieb EH (1989) Two theorems on the Hubbard model. Phys Rev Lett 62:1201
31. Head-Gordon M (2003) Characterizing unpaired electrons from the one-particle density matrix. Chem Phys Lett 372:508
32. Feixas F, Matito E, Poater J, Sola M (2015) Quantifying aromaticity with electron delocalisation measures. Chem Soc Rev 44:6434
33. Nagai H, Nakano M, Yoneda K, Kishi R, Takahashi H, Shimizu A, Kubo T, Kamada K, Ohta K, Botek E, Champagne B (2010) Signature of multiradical character in second hyperpolarizabilities of rectangular graphene nanoflakes. Chem Phys Lett 489:212
34. Nakano M, Champagne B (2016) Nonlinear optical properties in open-shell molecular systems. WIREs Comput Mol Sci 6:198
35. Zakharov AB, Ivanov VV, Adamowicz L (2016) Optical parameters of π-conjugated oligomer chains from the semiempirical local coupled-cluster theory. In: Leszczynski J, Shukla M (eds) Practical aspects of computational chemistry IV. Springer, Boston, p 57
36. Li H-P, Bi Z-T, Hu R-F, Han K, Li M-X, , Shen X-P, Wu Y-X (2017) Theoretical study on electronic polarizability and second hyperpolarizability of hexagonal graphene quantum dots: effects of size, substituent, and frequency. Carbon 122:756
37. Pople JA, McIver JW, Ostlund NS (1968) Self-consistent perturbation theory. I. Finite perturbation methods. J Chem Phys 49:2960
38. Pedash YF, Ivanov VV, Luzanov AV (1992) π-Electron hyperpolarizability of even-numbered polyenes in the complete configuration interaction method. Theor Exp Chem 28:19

39. Pedash VF, Luzanov AV (1981) Separation of the local contributions to the transition moments and the polarizability of a molecule in the self-consistent field method. J Struct Chem 21:439
40. Lazzeretti P, Zanasi R (1984) Resolution of molecular polarizability into atomic terms. Chem Phys Lett 109:89
41. Bader RFW (1990) Atoms in molecules – a quantum theory. Oxford University Press, Oxford
42. Otego N, Aslenoy CV, Pouchan C, Karamais P (2015) Hirshfeld-based intrinsic polarizability density representations as a tool to analyze molecular polarizability. J Comput Chem 36:1831
43. Chattaraj PK, Sengupta S (1996) Popular electronic structure principles in a dynamical context. J Phys Chem 100:16126
44. Chattaraj PK, Sarkar U, Roy DR (2007) Electronic structure principles and aromaticity. J Chem Educ 84:354
45. Santos JC, Contreras M, Merino G (2010) Structure and stability of Si 6 li 6: aromaticity vs polarizability. Chem Phys Lett 496:172
46. Luzanov AV (2013) π-system in a strong electric field. Analysis of electron unpairing. J Struct Chem 54:835
47. Kryachko ES, Ludeña EV (2014) Density functional theory: foundations reviewed. Phys Rep 544:123
48. McWeeny R (1994) Density functions and density functionals. Philos Mag B 69:727
49. Kaplan IG (2007) Problems in DFT with the total spin and degenerate states. Int J Quantum Chem 107:2595
50. Kaplan IG (2017) The Pauli exclusion principle: origin, verifications and applications. Wiley, Chichester
51. Chiappe G, Louis E, San-Fabián E, Vergés AV (2015) Can model Hamiltonians describe the electron-electron interaction in π-conjugated systems? PAH and graphene. J Phys Condens Matter 27:463001

Chapter 12
Electronic Properties of Defects in Carbon Nanotubes

H. Yu. Mykhailova, M. M. Nischenko, G. P. Prikhodko, V. V. Anikeev, and V. Yu. Koda

Defects and their influence for electrical properties were studied in the work. It is known that defects in carbon nanomaterials, in particular nanotubes, can be formed in the process of synthesis or under external influence. And they lead to a distortion of the structure of carbon nanotubes (CNT) and the differences in properties; curved, colloidal, snakelike CNTs, tubes of variable diameters, are formed only in the presence of topological defects associated with re-hybridization and unsaturated bonds [1].

The most common distortion of the framework structure is the presence of five- or seven-membered cycles in the walls of the CNT instead of the six-membered (Stone-Wales defect), a double vacancy defect (2 V), an ad-dimer defect [2, 3]. Other class of defects is manifested in the emergence of vacancies, substitutions of carbon atoms, dislocations, and others. They can be pointy and longer. Dislocations arise in the formation of structures such as roll or papier-mache and the change the number of layers of cylindrical CNT [4]. Defects are also considered carbon atoms, which are joined by a functional group. Topological defects can disappear when heated to a temperature of up to 2773 K and higher in an inert medium.

In nanostructural materials (peculiarity of an atomic structure), each element of a system or nanocomposite has a local structure of free volume, which leads to the appearance of potential barriers limiting the movement of charge carriers at the boundary.

Positrons were used for the study of such materials [4–6], which, under the action of the electric field of the ion skeletons, migrate into a free atomic volume, where they are annihilated with electrons and give information about the electronic structure. The electronic structure depends on the conditions at the boundary of the section (height and width of the potential barriers) and the sizes of the nanostructured

H. Y. Mykhailova (✉) · M. M. Nischenko · G. P. Prikhodko · V. V. Anikeev · V. Y. Koda
G. V. Kurdyumov Institute for Metal Physics, National Academy of Sciences of Ukraine, Kyiv, Ukraine

O. Fesenko, L. Yatsenko (eds.), *Nanooptics, Nanophotonics, Nanostructures, and Their Applications*, Springer Proceedings in Physics 210,
https://doi.org/10.1007/978-3-319-91083-3_12

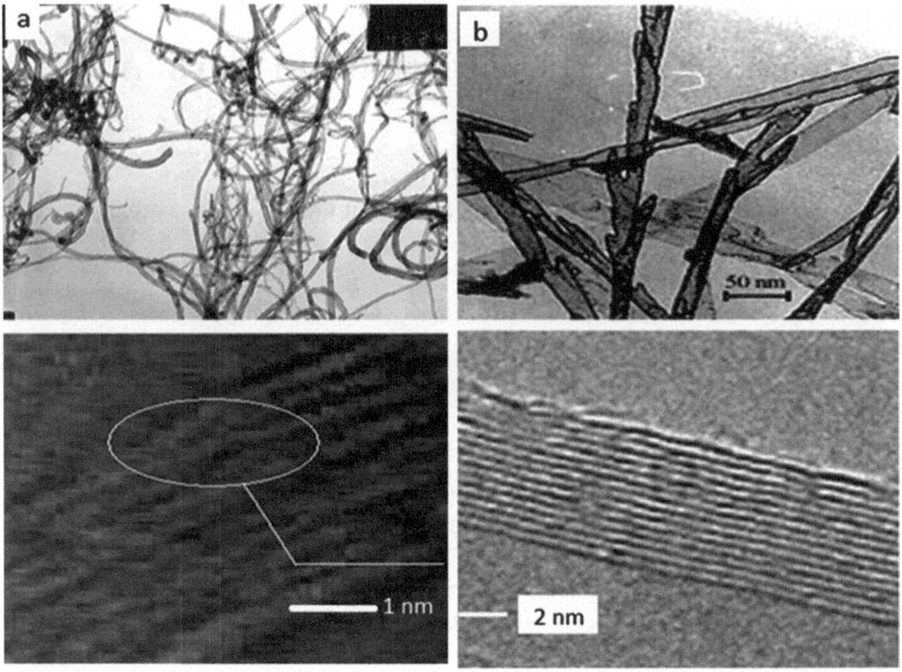

Fig. 12.1 TEM images of the CNT (**a**) with defects and boundary dislocation in CNT, (**b**) non-defect CNT and its wall

elements and determines the electronic properties of such systems and the ability to transfer electric charge [7–9]. It was shown in [4] that the free volume in nanostructured materials is characterized by the radius of localization of the wave function of the electrons, correlated with the size of interzone and intermolecular gaps and defects (vacancies, nanopores, etc.). It determines the properties of nanomaterials.

It was shown in [10] that the angular distribution of annihilation photons (ADAPh) spectra of defective tubes contains the parabolic component, in addition to the two Gaussians, and the area under a wide Gaussian is equal to 1.8 and 44% for straight and curved nanotubes, which suggests that the deviation of the form of CNT from the direct due to the presence of structural defects.

The method of the angular distribution of annihilation photons has found that in multiwalled carbon nanotubes (MWCNT), obtained under certain conditions, positrons are attracted by defects, whose transverse dimension (0.5–0.6 nm) is equal to the doubled value of the interlayer distance [4].

TEM image of nanotube shows the breakdown of the cylindrical graphene layer inside the MWCNT (Fig. 12.1a), which indicates the formation of the dislocation. Their characteristic feature is the rupture of covalent bonds on the edge of the dislocation and stretching almost twice the van der Waals bonds in its core. Such structures are quasi-uniform and extend in a circle.

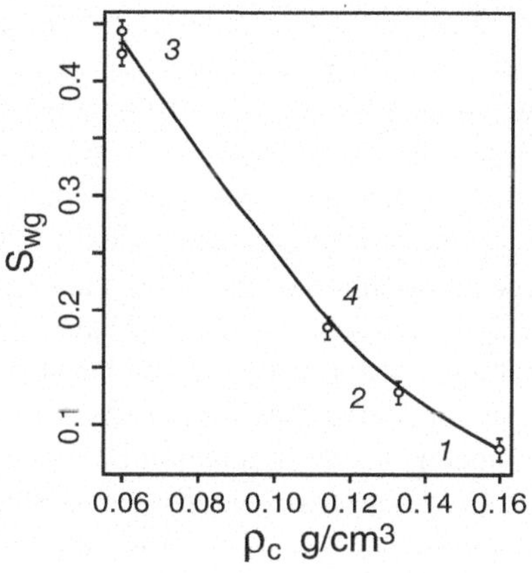

Fig. 12.2 The dependence of the area under the wide component of the ADAPh spectrum S_{wg} for CNT from the density of transition to the conductive state ρ_c. 1, 2, samples without defects; 3, sample with high concentration of dislocations; *4*, TEG

The radius of localization of the wave functions of the electrons in them is twice that of the interlayer spaces. The conditions for quantizing the energy of electrons in the nucleus of dislocations and defect-free regions are different. The interlayer distances in the nucleus of dislocations are almost twice larger than their limits, which substantially affects the positions of energy levels in the direction of dimensional quantization.

The main and excited energy levels in dislocations will be located below the positions of the corresponding levels of adjacent (defect-free) areas. This means that the electrons will be captured by the kernel of the edge dislocation. According to the electron-positron spectroscopy in nanotubes, the redistribution of electrons in the nucleus of dislocation really occurs [6]. The "capture" of free electrons in the dislocation nucleus is energy efficient. In this case, the dislocation tube can become a quasi-uniform conductor, which is locked in a circle and can form a waveguide for electrons.

Figure 12.2 shows the dependence of the area under a wide component (S_{wg}) of the ADAPh spectrum for different samples of CNT from the density of conductance state. The more defective array MWCNT has the lower density of the transition to the conductive state. Single walled carbon nanotubes - SCNT (sample 2) contains the low number of defects $S_{wg} = 7.8\%$, the transition to the conductive state is the largest $\rho_c = 0.16$ g/cm^3, and for the most defective tubes (sample 3), $S_{wg} = 42.5\%$ and $\rho_c = 0.05$ g/cm^3. This means that the most defective tubes have the most "loose" structure, in which the flow of current occurs at a lower density of the array of CNT. This is consistent with the results of work [5].

The experimental point for the TEG with a different structure compared to the nanotubes (Figs. 12.2 and 12.4) is imposed to the curve.

Properties of an array of carbon nanotubes or a material containing CNT are different from the properties of individual nanotubes. Array of CNTs doesn't conduct an electric current, in contradistinction to individual nanotubes. This is due to the fact that the connections between adjacent nanotubes are van der Waals.

It is known that in the radial (transverse) direction, nanotubes do not conduct an electric current. However, an electric current starts to pass through a CNT array when compression is under a piston, and the connection between adjacent nanotubes may even become metallic. Therefore, the measurement of the electrical properties of the CNT array was carried out under the low pressure. At the same time, the reorientation of each individual CNT, the reduction of the distances between them, the deformation of the CNT, the increase of the contact area between the nanotubes, and the reduction of the interlayer distances in the direction of the force's effect are taking place. Each nanotube has its random spatial configuration, which depends on the type and number of defects in the graphene layer and defects in the packaging of graphene layers in a nanotube (edge dislocation).

In [11], experimentally, it was shown that the bundle of oriented nanotubes under the influence of cyclic deformation retains its elastic and mechanical properties. The effect of mechanical deformation of the array of contacting nanotubes may be more complex due to the presence of contacts between adjacent nanotubes, nanotubes and metal electrodes, etc.

The features of electrical conductivity of the spatially unoriented nanotube array (sample 2) and nanotubes (sample 3) were investigated from their density σ(ρ), which varies during compression during loading and unloading in a dielectric cylinder under a piston.

These samples differ from other largest concentrations of free electrons according to EPS data [3]. The TEM image (Fig. 12.1a) shows a break in the cylindrical graphene layer inside the CNT, indicating the formation of a boundary dislocation and which is not a characteristic of defect-free CNT (Fig. 12.1b).

The statistical distribution of CNT on the outer diameter showed that it is 10–20 nm.

The dependence of the electrical conductivity of an array of bulk carbon nanotubes (sample 2) on the density of the compressed array of CNT in the direct and reverse piston movement for the first measurement cycle is given on Fig. 12.3.

In the process of lowering of the piston, the electrical conductivity is $\sim 10^{-5}$ $(\text{Ohm} \cdot \text{cm})^{-1}$ and fixed for the density $\rho_c = 0.081$ g/cm^3, and for the density $\rho = 0.11$ g/cm^3, the electrical conductivity increases up to five orders of magnitude

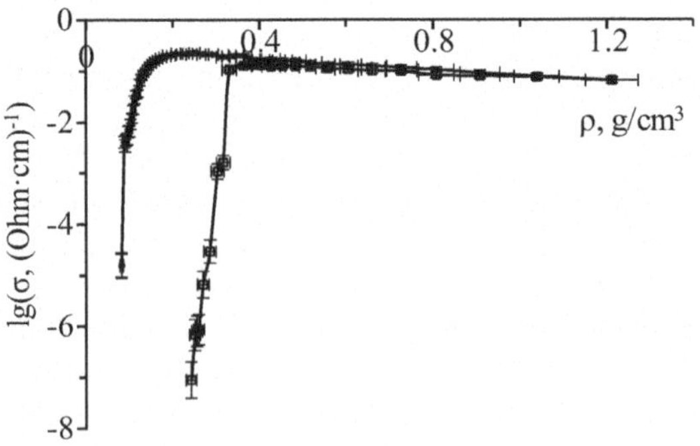

Fig. 12.3 Dependence of the logarithm of electrical conductivity lgσ of the array of multiwall carbon nanotubes without dislocations on the change in density ρ under the action of deformation and subsequent discharge. ◆, loading; □, unloading

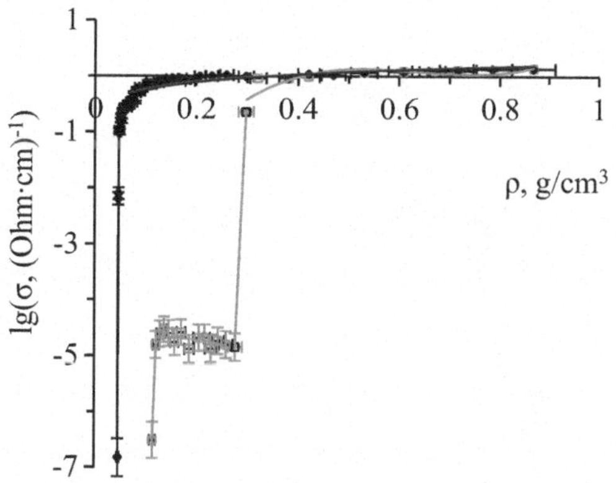

Fig. 12.4 The dependence of the logarithm of the electrical conductivity lgσ of the MWCNT array with boundary dislocations on the change in density ρ under the action of deformation and subsequent discharge. ◆, loading; □, unloading

and the value is 0.23 (Ohm cm)$^{-1}$. Further compression of the CNT to the density ρ = 1.21 g/cm^3 leads to a monotonous drop in electrical conductivity up to 0.065 (Ohm cm)$^{-1}$.

During discharge, the volume increases, and an elastic relaxation of the CNT array, which maintains contact between the nanotubes and the electrodes, is observed. From Fig. 12.4 it is evident that the electrical conductivity $\sigma(\rho)$ during unloading increases twice, to 0.13 (Ohm cm)$^{-1}$ in the range of densities ρ from 1.21 to 0.38 g/cm^3, after which the jump falls to six orders that were fixed at ρ = 0.24 g/cm^3. When further unloading, the process of elastic relaxation is completed, and there is a break in the electric circuit.

For MWCNT with dislocations, the dependence of the electrical conductivity of the bulk of CNT from their density at the forward and reverse piston stroke for the first measurement cycle is shown in Fig. 12.4. When compressing the bulk sample in the process of loading the piston, the value of the electrical conductivity, which is equal to $\sigma = 1.50 \cdot 10^{-7}$ (Ohm cm)$^{-1}$, is fixed for the density of the nanotube array ρ_c = 0.05 g/cm^3, and at density ρ = 0.11 g/cm^3, the electrical conductivity increases by 10^6 times and reaches a value equal to σ = 0.71 (Ohm cm)$^{-1}$.

Further compression of the CNT to the density ρ = 0.86 g/cm^3 leads to a gradual increase in electrical conductivity else 48%. From Fig. 12.4 it is evident that the electrical conductivity $\sigma(\rho)$ in the reverse sequence repeats the course of the curve at compression $\sigma(\rho)$ in the range of densities ρ from 0.86 to 0.30 g/cm^3, after which the electrical conductivity falls 10^4 times; then in the range ρ of 0.30 to 0.12 g/cm^3, there is an almost horizontal stair. With further increase in discharge, electrical conductivity drops rapidly due to a decrease in the total area of the contacts and a break in the electric circuit.

The density of the transition to the conductive state for CNT with dislocations is 0.05 g/cm^3 (Fig. 12.4), which is lower than for CNT without dislocations (0.08 g/cm^3): CNT with dislocations deviates to a greater extent from ideal rectilinear CNT. The deviation of their shape from the ideal is also confirmed by the presence of twisted regions and clusters on their TEM images (Fig. 12.1a).

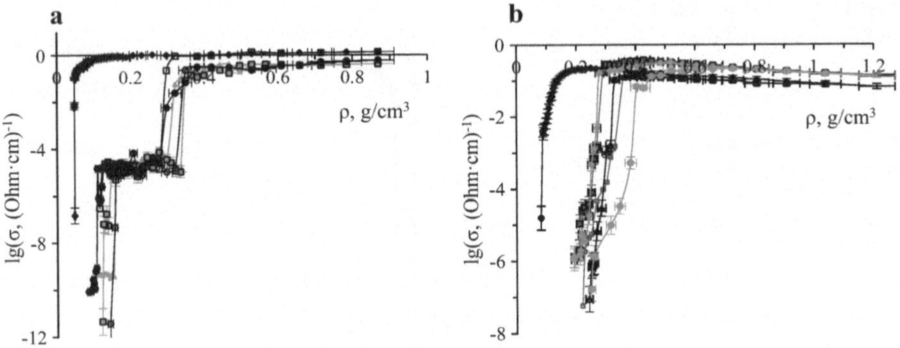

Fig. 12.5 Influence of cyclic deformations on the electrical conductivity σ(ρ) of MWCNT (**a**) with dislocations for 1, 3, and 5 measurement cycles: 1, ◆; 3, ●; 5, Δ (loading); 1, □; 3, ○; 5, ■ (unloading), (**b**) CNT without dislocations for 1, 2, and 4 measurement cycles: 1, ◆; 2, ◆; 4, ■ (loading); 1, Δ; 2, ●; 4, □ (unloading)

The array of nanotubes with boundary dislocations was exposed to loading-unloading by 12 cycles.

For the second and subsequent loading/unloading cycles for a given CNT array (Fig. 12.5a), two jumps and steps were observed both during loading and unloading, which are not observed for defect-free CNT (Fig. 12.5b).

After four cycles of deformation of the loading-unloading of the CNT without dislocations, the transition to the conducting state increases from 0.08 to 0.22 g/cm^3 (Fig. 12.5b), which is almost identical to the relaxation transition $\rho_{\mathrm{rel}} = 0.23$ g/cm^3. This testifies to the ordering of CNTs in the process of cyclic deformation.

For CNT with dislocations, from 2nd to 12th cycles of loading-unloading, the transition to the conductive state is recorded in the density range $\rho_c = 0.06$–0.10 g/cm^3. There is a tendency to increase this value with an increase of the number of cycles, which may indicate an increase in the density of the packaging of an array of CNT and their predominant orientation. For all cycles of measurement, the stair on the density curve of the electrical conductivity has an almost constant value $\sigma = (2.4 \pm 0.3)\cdot 10^{-5}$ (Ohm cm)$^{-1}$, which is stored in the density range from $\rho = 0.12$ to 0.30 g/cm^3. The return motion of the piston for the first measurement of the electric jump is observed at $\rho = 0.30$ g/cm^3, and its incidence is four orders of magnitude. For the next cycles, the density at which the jump σ is observed increases to 0.38 g/cm^3. The second jump in the reverse direction of measurements for the first measurement is observed at $\rho = 0.12$–0.28 g/cm^3 and for subsequent cycles of measurements increases to $\rho = 0.30$ g/cm^3.

The first jump of electrical conductivity is due to the transition of dielectric state to the metallic, the second one is due to the change in the mechanisms of electron transport, and the step that connects them is due to the elastic properties of twisted bundles without deformation.

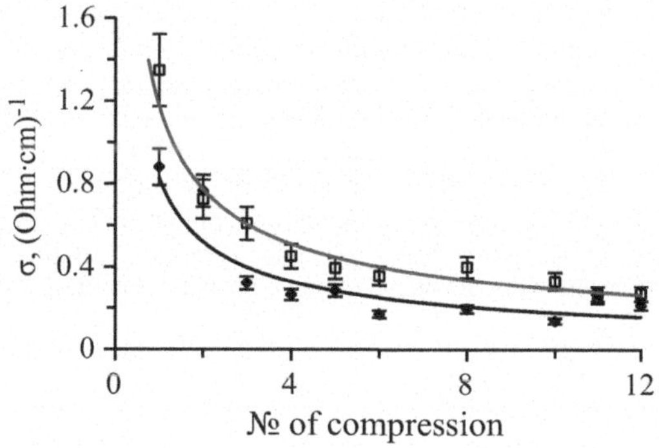

Fig. 12.6 Processes of the ordering of the CNT array under the action of cyclic loading-unloading processes. ◆, compression density $\rho = 0.50$ g/cm^3; □, $\rho = 0.86$ g/cm^3

The effect of the number of loading-unloading cycles for the electrical conductivity for compression to density $\rho = 0.86$ g/cm^3 (maximum compression) and the value of electrical conductivity for density $\rho = 0.50$ g/cm^3 is shown in Fig. 12.6.

It is evident that the electric conductivity falls to the fifth to sixth cycle, after which it remains practically unchanged, which indicates the ordering of the CNT array under the action of cyclic loading and unloading.

12.1 Conclusion

The more defective array MWCNT has the lower density of the transition to the conductive state. This means that the most defective tubes have the most "loose" structure, in which the flow of current occurs at a lower density of the array of CNT.

During cyclic deformation CNT array containing edge dislocations, nanotubes are targeted and ordering, and on the dependence of conductivity of array density, there is a two-stage jumping conductivity by several orders, combined stair. They are caused by the action of three mechanisms:

1. Increasing number of contacts and their total area
2. Elastic deformation swirled into a spiral or circle (radius of 100–1000 nm) nanotubes
3. The changing nature of electron transfer at points of tangency of the tunnel to the semiconductor and metal

References

1. Odom TW // Nature. – 1998. – V. 391. – P. 62–64
2. Charlier JC // Physical Review B. – 1996. – V. 53, №16. – P. 11108–11113
3. Nishchenko MM // Collection of scientific works: "Nanosystems, nanomaterials, nanotechnologies". – Akademperiodika, 2003, Vol. 1 (1) – P. 193–259

4. Nishchenko MM // Collection of scientific works: “Nanosystems, nanomaterials, nanotechnologies”. Akademperiodika – 2004. – Vol. 2 (3). – P. 983–991
5. Randal J // Applied physics letters. – 2005. – № 86. – P. 093104 (1–3)
6. Shantarovich VP // Physics of a solid body. – 1998. – Vol. 40 (1). – P. 164–167
7. Rakov E.G. – M .: Universitybook, 2006 – 235 pp
8. Pogrebnyak AD // Uspekhi Fizicheskikh Nauk. – 2012. – Vol. 182 (3). – P. 287–321
9. Dehtyar IYa – Kyiv: Scientific Opinion, 1971. – P. 301
10. Nishchenko MM // Perspective materials. – 2010. – Vol. 5. – P. 86–92
11. Nanomaterials, nanocoatings, nanotechnologies: Textbook - Kharkiv: KNU - 2009 - P. 209

Chapter 13
Features of Spin Transport in Magnetic Nanostructures with Nonmagnetic Metal Layers

A. M. Korostil and M. M. Krupa

13.1 Introduction

Coupling between spin currents and localized magnetic moments in magnet (M)/ normal (nonmagnetic) metal (N)-based multilayer magnetic nanostructures constitutes the basis of the mutual control between electric current and static or dynamic magnetic states. Herewith, magnetic layers include magnets with the exchange interactions both ferromagnetic (F) and antiferromagnetic (AF) types (e.g., ferrimagnetic compounds like YIG, Gd_3 Fe_5O_{12} and AFs Fe_3O_4, $NiFe_2O_4$, NiO [1, 2]), and normal metals are, usually, heavy metals with strong spin-orbit coupling (e.g., Pt, Ta, W). The mentioned interconnection in these magnetic nanostructures occurs via the interface scattering of the spin-polarized current and its *s-d* exchange interaction with static or dynamic magnetic states [3–6]. The impact of the spin current on the magnetic states is manifested through the spin-transfer torque, and the impact of the localized magnetic momentum on the spin current is manifested via the spin-dependent interface scattering accompanied by magnetoresistance effect. The spin polarization can be induced by effective bias fields of different origins including fields caused by an exchange interaction and the strong spin-orbit coupling. The entire spin-coherent region is limited in size by spin-flip relaxation processes.

In the case of static magnetic states, the mutual influence of the spin current and magnetic ordering can be manifested as the magnetoresistance effect of the dependence of the spin current on the magnetization orientation in the magnetic layer and, vice versa, the dependence of the latter on the spin current [5, 6]. Such effects can constitute the base for magnetic writing techniques in nonvolatile memory

A. M. Korostil (✉) · M. M. Krupa
Institute of Magnetism National Academy of Sciences of Ukraine and MESU, Kyiv, Ukraine

O. Fesenko, L. Yatsenko (eds.), *Nanooptics, Nanophotonics, Nanostructures, and Their Applications*, Springer Proceedings in Physics 210,
https://doi.org/10.1007/978-3-319-91083-3_13

technologies such as MRAM [7] and racetrack memories [8]. They also include the giant magnetoresistance (GMR) effect in metallic magnetic multilayers, which was commercially utilized in high-end magnetic recording media [9]. Obtaining the mentioned multilayer magnetic nanostructures with properties of electric-controlled magnetic switching and the magnetic-controlled spin current involves the description of features of the spin transport in magnetic heterogeneous nanostructures allowing for the compatibility conditions at the interfaces [4, 10]. This is usually solved within the Landauer-Büttiker formalism [11] and, more rigorously, using the nonequilibrium Keldysh-Green functions [12, 13].

In the case of the dynamic magnetic states, their interconnection with the control spin current is affected by the magnetic precession-induced spin pumping and the spin accumulation in the normal-metal layer at the interface [3, 10]. The action of the spin currents on the magnetic dynamics via the spin-transfer torque and the reciprocal process of spin pumping result in the effect of controlled magnetic auto-oscillations [14]. The magnetic dynamic damping is related to the spin-pumping effect at the M|N interface that can be compensated by the spin-transfer torque from the spin current of the converted input current. This spin transfer is governed by the reflection and transmission matrices of the system, analogous to the scattering theory of transport and interlayer exchange coupling. Due to interfacial processes, M|N coupling becomes important in the limit of ultrathin ($\leq$ 10 nm) magnetic films and can lead to a sizable enhancement of the damping constant.

The abovementioned coupling effects at interfaces can occur in the magnetic nanostructures with both ferromagnetic (F) and antiferromagnetic (AF) exchange interactions, which are realized in ferro-, ferri-, and antiferromagnetic materials. Normal metal layers are medium for the spin currents, which can be converted from the control charge current by the spin-orbit interaction, especially the spin Hall and the spin-orbit Rashba [15, 16] effects.

The paper is organized as follows. In Sect. 1, the spin-dependent transport in the F/N-based magnetic nanostructure is studied for the static magnetization. In the modified Stoner model with potential barrier dependent on the physical parameters including the magnetization directions, the chemical potentials of the layers, and the contact conductances, the parametrically dependent scattering of spin-polarized current is investigated. The mentioned parameters are determined by the spin-polarized kinetic equations in the framework of the Keldysh-Green function approach. It is considered both in single and composite F/N-based magnetic nanostructures. In Sect. 2, features of the interconnection between magnetization dynamics and the spin currents are studied in the F/N-based nanostructures. The process of the magnetization precession-induced pumping spin current in the nonmagnetic layers is considered as the result of the parametric time dependence of the interfacial scattering with the precession as the parameter. It is shown that the spin pumping slows down the precession corresponding to an enhanced Gilbert damping constant in the Landau-Lifshitz-Gilbert model. The spin current related to the spin pumping, which flows back into the ferromagnetic layers and driven by the accumulated spins in the normal-metal layers, is also discussed.

13.2 Spin Transport in the Case of Static Magnetic Field

13.2.1 Features of Spin-Dependent Electric Current in the F/N Bilayers

Characteristic features of the spin-dependent transport and the interfacial scattering in multilayer magnetic nanostructures with nonmagnetic metal (N) interlayers are manifested in the model bilayer nanostructure F/N depicted in Fig. 13.1.

These features are related to the conditions under which long-range spin effects are observable in normal metals. Spins injected into a normal-metal layer relax due to unavoidable spin-flip processes

Naturally, the dwell time on the layer must be shorter than the spin-flip relaxation time in order to observe nonlocality in the electron transport. For a simple F/N double heterostructure (F/N/F) with antiparallel magnetizations, the condition can be described as follows [17]. The spin current into the normal-metal layer is roughly proportional to the particle current, $e(ds/dt)_{tr}\sim I = V/R$, where s is the number of excess spins on the normal-metal layer, V is the voltage difference between the two reservoirs coupled to the normal-metal layer, and R is the F|N contact resistance. When the layer is smaller than the spin-diffusion length, the spin-relaxation rate is $e(ds/dt)_{rel} = - s/\tau_{sf}$, where τ_{sf} denotes the spin-relaxation time on the layer. (Otherwise, this simple approach breaks down since the spatial dependence of the spin distribution in the normal metal should be taken into account [18].) The number of spins on the normal-metal layer is equivalent to a nonequilibrium chemical potential difference $\Delta\mu = s\delta$ in terms of the energy level spacing δ(the inverse density of states) (more generally the relation between $\Delta\mu$ and s is determined by the spin susceptibility). The spin accumulation on the normal-metal layer significantly affects the transport properties when the nonequilibrium chemical potential difference is of the same order of magnitude or larger than the applied source-drain voltage, $\Delta\mu > eV$ or $\delta\tau_{sf}/\hbar > R/R_K$, where $R_K = e^2/h$ is the quantum resistance. Thus, spin accumulation is only relevant for sufficiently small normal-metal layers and/or sufficiently long spin-accumulation times and/or good contact conductances.

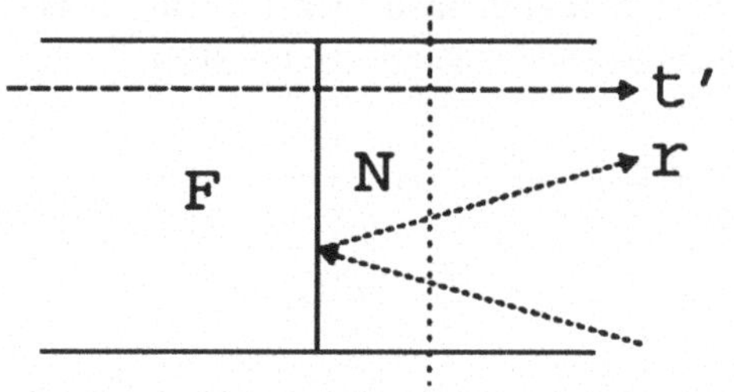

Fig. 13.1 A contact between a ferromagnetic (F) and a normal (N) metal layers. At the normal-metal side, the current is denoted as the dotted line. The transmission coefficient from the ferromagnet to the normal metal is t', and the reflection matrix from the normal metal to the normal metal is r

The spin-dependent current in the model structure F/N (Fig. 13.1) is expressed via the 2×2 distribution matrix $f(\varepsilon)$ in spin space at a given energy ε in the layer. The external reservoirs are assumed to be in local equilibrium so that the distribution matrix is diagonal in spin space and attains its local equilibrium value $f = 1f(\varepsilon, \mu_\alpha)$, where 1 is the unit matrix, $f(\varepsilon, \mu_\alpha)$ is the Fermi-Dirac distribution function, and $f(\varepsilon, \mu_\alpha)$ is the local chemical potential in reservoir α. The direction of the magnetization of the ferromagnetic layers is denoted by the unit vector $\boldsymbol{m}_\alpha$.

The 2×2 nonequilibrium distribution matrices in the layers in the stationary state are uniquely determined by current conservation

$$\sum_\alpha I_{\alpha\beta} = \left(\frac{\partial f_\beta}{\partial t}\right)_{rel}, \tag{13.1}$$

where $I_{\alpha\beta}$ denotes the 2×2 current in spin space from layer (or reservoir) α to layer (or reservoir) β and the term on the right-hand side describes spin relaxation in the normal layer. The right-hand side of Eq. (13.1) can be set to zero when the spin current in the layer is conserved, i.e., when an electron spends much less time on the layer than the spin-flip relaxation time τ_{sf}. If the size of the layer in the transport direction is smaller than the spin-flip diffusion length $l_{sf} = \sqrt{D\tau_{sf}}$, where D is the diffusion coefficient, then the spin relaxation in the layer can be introduced as $(\partial f^N/\partial t)_{rel} = (1\mathrm{Tr}(f^N)/2 - f^N)/\tau_{sf}$.

13.2.2 Passing the Electric Current Through the F|N Contact

The dependence of the spin current through the F|N interface on the distribution functions in the bilayer magnetic nanostructure F/N is described on the base of the method of the nonequilibrium Green function theory and Keldysh formalism. The Green functions are determined by the Kadanoff-Baym equations, representing the quantum kinetic equations for one-particle propagators. The electron subsystem of the bilayer F/N nanostructure subjected to an external field φ in the modified Stoner model is described by the Hamiltonian, which in the representation of the field operators of annihilation (ψ) and creation ($\psi^\dagger$) is given by

$$H(t) = \int dx \psi^\dagger(x) h(x,t) \psi(x) + \frac{1}{2} \times \int dx_1 dx_2 \psi^\dagger(x_1) \psi^\dagger(x_2) w(x_1, x_2) \psi(x_2) \psi(x_1), \tag{13.2}$$

where

$$h(r,t) = -\frac{1}{2}\nabla^2 + V^p(r) + V^s(r) + \varphi(r,t) \tag{13.2}$$

Here $x = (r, s)$ is the collective space-spin coordinate, the one-particle potential (13.2) contains spin-independent ($V^p(r)$) and spin-dependent ($V^s(r)$) parts of the barrier potentials in the interface F|N, and $w(r_1, r_2)$ is the operator of the two-particle interaction. The spin-dependent potential is $V^s(r) = (\sigma \cdot \boldsymbol{m})W^s(r)$, where σ is the Pauli matrix and $\boldsymbol{m}$ is the unit vector of the magnetization existing only in the ferromagnetic layer (F) and vanishing in the normal-metal layer (N).

The physical properties of the system are described by the one-particle nonequilibrium Green function, i.e., the expectation value of the time contour-ordered product of creation and annihilation field operators $\psi^\dagger(i)$ and $\psi(i)$ ($i = (1, 2, \ldots)$), respectively, with the collective coordinate, $i = (x_i, z_i)$, where $x_i = (r, s)$ is position-spin component and z_i is the time component. The latter takes real and imaginary values in intervals $[t_0, t]$ and $[t_0, t_0 - i\beta]$ (β is inverse temperature), respectively. The imaginary time interval corresponds to the equilibrium state. The conjugation of the mentioned time intervals forms the so-called Keldysh contour C consisting of forward and backward real-time branches and the thermal imaginary track. This nonequilibrium Green function can be presented as [12, 13]

$$G(1,2) = -i\frac{\mathrm{Tr}\left[T_C e^{-i\int_C H(z)\mathrm{d}z\psi(1)}\psi(1)\psi^\dagger(2)\right]}{\mathrm{Tr}\left[T_C e^{-i\int_C H_0(z)\mathrm{d}z\psi(1)}\right]} == -i\left\langle \psi_H(1)\psi_H^\dagger(2)\right\rangle. \quad (13.3)$$

where T_C is contour time-ordering operator on the Keldysh contour C with $H_0(z) = H(z)|_{z\in[t_0, t_0 - i\beta]}$ and the subscript H in the right-hand side denotes the Heisenberg representation. Due to (13.3), the Green function can be presented as

$$G(1,2) = \theta(z_1, z_2)G^>(1,2) + \theta(z_2, z_1)G^<(1,2), \quad (13.4)$$

where $\theta(z_1, z_2)$ is the contour step function equal to 1 or 0 versus z_1 or z_2 and is later on the contour C, and the greater ($G^>$) and lesser ($G^<$) Green functions are determined by the relations

$$G^>(1,2) = -i\left\langle \psi_H(1)\psi_H^\dagger(2)\right\rangle, \quad G^<(1,2) = i\left\langle \psi_H^\dagger(2)\psi_H(1)\right\rangle, \quad (13.5)$$

where an extra sign is introduced due to the interchange of the fermionic operators by the contour-ordering operator T_C. The Green function provides a direct access to observable physical quantities of the system. For example, the equal-time limit gives directly the particle spin density at the space-time point 1

$$\langle n(1)\rangle = \left\langle \psi_H^\dagger(1)\psi_H(1)\right\rangle = -G^<(1, 1^+) \quad (13.6)$$

(a superscript "+" means infinitesimal). The spin current density is determined as

$$I(1) = -\left\{\partial_t G^>(t, t') + \partial_{t'} G^<(t, t')\right\}_{t'=t} \quad (13.7)$$

The Green functions are described by the equation

$$\{i\partial_{z_1} - h(1)\}G(1,2) = \delta(1,2) + \int d3w(1,3)G(1,3;2,3^+) \tag{13.8}$$

following the Schrödinger equation for wave functions. Here, the two-particle Green function

$$G(1,3;2,3^+) = \left\langle \psi_H^\dagger(2)\psi_H(1)\psi_H(3)\psi_H^\dagger(3^+)\psi_H^\dagger(2)\right\rangle \tag{13.9}$$

expresses via the variation derivative with respect to the variation in the infinitesimal external potential ν of the one-particle Green function by the relation [12]

$$G(1,3;2,3^+) = i\langle n(3)\rangle G(1,2) + \frac{\delta G(1,2)}{\delta \nu(3)} \tag{13.10}$$

Consequently, Eq. (13.8) can be represented as self-contained variation derivative equation with the matrix representation

$$\left(L^0 + L^1\right)G = 1, \tag{13.11}$$

where the matrix L^0 is determined by matrix elements

$$L^0(1,2) = \left(i\partial_z - h^*(1) + \int d3w(1,3)\langle n(3)\rangle\right)\delta_{1,2} \tag{13.12}$$

not containing the variation derivative. The matrix L^1 is determined by matrix elements

$$L^1(1,2) = i\delta_{1,2}\int d3\,w(1,3)\frac{\delta}{\delta\nu(3)} \tag{13.13}$$

which are proportional to the variation derivative.

The expression for the functional derivative

$$\frac{\delta}{\delta\nu(3)}G = \frac{\delta}{\delta\nu(3)}G\left[G^{-1}\right]G \tag{13.14}$$

(the bracket separates the expression experiencing the variation differentiation) allows to represent the operation of the differential matrix L^1 on G via the self-energy matrix Σ not containing functional derivatives:

$$L^1G = \Sigma G, \quad \Sigma = L^1G\left[G^{-1}\right] \tag{13.15}$$

Finally, the matrix Eq. (13.11) reduces to the system

$$\left(L^0 + \Sigma\right)G = 1 \tag{13.16}$$

$$\Sigma = -L^1 G\left[L^0 + \Sigma\right] \tag{13.17}$$

The first-order approximation with respect to the interaction w determines the Hartree-Fock self-energy matrix $\Sigma^{\mathrm{HF}} = -L^1 G[L^0]$ with matrix elements

$$\Sigma^{\mathrm{HF}}(1,2) = -\delta(1,2)\int d3w(1,3)\langle n(3)\rangle + iw(1,2)G(1,2) \tag{13.18}$$

Here the first term describes the classical Hartree potential at 1 produced by the charge density throughout the space, and the second term is the space-nonlocal exchange potential originating from the Pauli exclusion principle and antisymmetry of the wave functions. Due to (13.16) the self-energy matrix in second-order approximation is determined by the equation

$$\Sigma = -L^1 G\left[L^0\right] + L^1 G\left[L^1 G\left[L^0\right]\right] \tag{13.19}$$

with matrix elements of the form

$$\Sigma^2 = \Sigma^{\mathrm{HF}}(1,2) + G(1,2)\int d3d4w(1,3)w(2,4)G(4,3)G(3,4^+) \\ - \int d3d4G(1,3)w(1,4)G(3,4)G(4,2)w(3,2) \tag{13.20}$$

Here the second and third terms describe the correlation and scattering effects, respectively.

In the stationary situation, the Green function is determined via its energy Fourier transform G_E which without the subscript will be considered below. Entering in the given bilayer model *F/N*, the Cartesian coordinate system with the axis x along the interface plane normal and the axes y and z in the interface plane, the Green function can be represented by decomposition into quasi-one-dimensional modes as

$$G_{ss'}(1,1') = \sum_{nm,\,\alpha\beta} G^{\alpha\beta}_{nsms'}\chi^n_s(\rho,x)\chi^{m*}_{s'}\left(\rho',x'\right)e^{i\alpha k^n_s x - i\beta k^m_{s'} x'}, \tag{13.21}$$

where the indices $\alpha, \beta = (+,-)$; the signs "+" and "−" denote right-going (+) and left-going (−) modes, respectively; $\chi^n_s(\rho,x)$ is the transverse wave function; and k^n_s denotes the longitudinal wave vector for an electron in transverse mode n with spin s. Then, from the definition of the current through the Green function, at the spatial independence of the transverse wave function $\chi^n_s(\rho,x)$, it can be obtained the expression

$$I_{ss'}(x) = ie\sum_{n\alpha\beta}\left(\alpha v^n_s - \beta v^m_{s'}\right)G^{\alpha\beta}_{nsms'}(x,x)\int d\rho\chi^n_s(\rho,x)\chi^m_{s'}(\rho,x), \tag{13.22}$$

describing the spin current, where $v^n_s = \hbar k^n_s/m$ is the longitudinal velocity for an electron in transverse mode n with spin s. In a normal metal, the transverse states and the longitudinal momentum are spin-independent, and the spin current simplifies to

$$I_{ss'}(x) = 2ie\sum_{n\alpha} \alpha v_s^n G_{nsms'}^{\alpha\alpha}(x,x), \tag{13.23}$$

which is used in the calculation of the spin current on the normal side of the contact.

Using the representation

$$G_{nsms'}^{\alpha\beta}(x,x') = -i\left(\frac{g_{nsms'}^{\alpha\beta}(x,x')}{\sqrt{v_s^n v_{s'}^m}} + 1\delta_{ss'}\frac{\alpha\delta_{\alpha,\beta}\mathrm{sign}(x\text{-}x')}{v_s^n}\right), \tag{13.24}$$

where the latter term does not contribute to the current on the normal-metal side, whence it follows the expression

$$I_{ss'}(x) = 2e\sum_{n\alpha} \alpha v_s^n g_{nsms'}^{\alpha\alpha}(x,x) \tag{13.25}$$

for the spin current on the normal-metal side.

The complete description of the spin current through the F|N interface involves taking into account the connection between waves propagating to the right (left) on the right-hand side of the contact $\psi_R^+(\psi_R^-)$ and waves propagating to the right (left) on the left-hand side of the contact $\psi_L^+(\psi_L^-)$. This is described by the transfer matrix M obeying the relation

$$\begin{pmatrix} \psi_R^+ \\ \psi_R^- \end{pmatrix} = M\begin{pmatrix} \psi_L^+ \\ \psi_L^- \end{pmatrix}, \tag{13.26}$$

which in terms of the transmission (t) and reflection (r) coefficients takes the form

$$M = \begin{pmatrix} t - r'(t')^{-1}r & r'(t')^{-1} \\ -(t')^{-1}r & (t')^{-1} \end{pmatrix}. \tag{13.27}$$

Here the transmission and reflection coefficients enter in definition of the scattering matrix

$$S = \begin{pmatrix} r & t' \\ t & r' \end{pmatrix}, \tag{13.28}$$

where $r_{nm}^{s\sigma}$ is the reflection matrix for incoming states from the left in mode m and spin σ to mode n with spin s and $t_{nm}^{s\sigma}$ is the transmission matrix for incoming states from the left transmitted to outgoing states to the right. In addition, r' is the reflection matrix for incoming states from the right reflected to the right, and t' is the transmission matrix for incoming states from the right transmitted to the left. The Green function to the left ($x = x_2$) of the interface is expressed via the Green function to the right ($x = x_1$) of the interface by the relation

$$g^{\sigma\sigma'}_{nsms'}(x = x_2, x') = \sum_{ls'', \sigma''} M^{\sigma\sigma''}_{nsls''} g^{\sigma''\sigma'}_{ls''ms'}(x = x_1, x'), \tag{13.29}$$

which after redefining, $g_{2(1)} = g(x = x_{2(1)}, x' = x_{2(1)})$, takes the matrix form $g_2 = M g_1 M^{\dagger}$.

In the approximation of isotropic quasi-classical Green functions in nanolayers 1 and 2, $(G_1)^{\alpha\beta}_{nsms'} = \delta_{n,m}\delta^{\alpha\beta}(G_1)^{\alpha\beta}_{ss'}$. In the representation of the retarded (G^R), advanced (G^A), and Keldysh (G^K) Green function, the total Green function has the form $G = \begin{pmatrix} G^R & G^K \\ 0 & G^A \end{pmatrix}$. Here $G_R = -G_A = -1$, $G_{K,\ 1(2)} = h_{1(2)}1$, where the two-dimensional matrix h is related to the nonequilibrium distribution functions $h_{1\ (2)} = 2(f(\varepsilon)_{1(2)} - 1)$. Herewith

$$g^{-+}_{R,1} = 2r, \quad g^{+-}_{R,2} = 2r' \tag{13.30}$$

and

$$g^{--}_{K,1} = t' h_2 t'^{\dagger} + h_1 r^{\dagger}, \quad g^{--}_{K,2} = t h_1 t^{\dagger} + r' h_2 r'^{\dagger} \tag{13.31}$$

Inserting the expression (13.31) into (13.25) results in the expression

$$I = \frac{e}{h}\left\{\sum_{mm}\left[t'^{nm} f^F \left(t'^{mn}\right)^{+} - \left(M f^F - \tau^{nm} f^N (\tau^{mn})^{+}\right)\right]\right\}, \tag{13.32}$$

which describes the current through interface on its normal-metal side. Here $r^{nm}_{ss'}$ is the reflection coefficient for electrons from transverse mode m with spin s' incoming from the normal-metal side reflected to transverse mode n with spin s on the normal-metal side, and $t'^{nm}_{ss'}$ is the transmission coefficient for electrons from transverse mode m with spin s' incoming from the ferromagnet transmitted to transverse mode n with spin s on the normal-metal side.

13.2.3 Spin Parametric Dependence of Spin Current

The relation (13.32) between the current and the distribution functions has a simple form after transforming the spin-quantization axis. Disregarding spin-flip processes in the contacts, the reflection matrix for an incoming electron from the normal metal transforms is

$$r^{nm} = \sum_s u^s r^{nm}_s, \quad u^{\uparrow(\downarrow)} = (1 \pm \sigma \cdot \boldsymbol{m})/2, \tag{13.33}$$

where $r_{\uparrow(\downarrow)}^{nm}$ are the spin-dependent reflection coefficients in the basis where the spin-quantization axis is parallel to the magnetization $\boldsymbol{m}$ in the ferromagnet and σ is a vector of Pauli matrices. Similarly for the transmission matrix $t_{\uparrow(\downarrow)}^{nm}$

$$t'^{nm}\left(t'^{mn}\right)^{+}=\sum_{s}u^{s}\left|t'^{nm}_{s}\right|^{2} \tag{13.34}$$

From the unitarity of the scattering matrix, it follows that the general form of the relation (13.32) reads

$$eI=\sum_{s=\uparrow,\downarrow}G^{s}u^{s}\left(f^{F}-f^{N}\right)u^{s}-G^{\uparrow\downarrow}u^{\uparrow}f^{N}u^{\downarrow}-\left(G^{\uparrow\downarrow}\right)^{*}u^{\downarrow}f^{N}u^{\uparrow}, \tag{13.35}$$

where it is introduced the spin-dependent conductance parameters

$$G^{\uparrow(\downarrow)}=\frac{e^{2}}{h}\left[M-\sum_{nm}\left|r_{\uparrow(\downarrow)}^{nm}\right|^{2}\right]=\frac{e^{2}}{h}\sum_{nm}\left|t_{\uparrow(\downarrow)}^{nm}\right|^{2} \tag{13.36}$$

and the mixing conductance

$$G^{\uparrow\downarrow}=\frac{e^{2}}{h}\left[M-\sum_{nm}r_{\uparrow}^{nm}\left(r_{\downarrow}^{nm}\right)^{*}\right] \tag{13.37}$$

The precession of spins leads to an effective relaxation of spins non-collinear to the local magnetization in ferromagnets, and consequently the distribution function is limited to the form $f^{F}=1f_{0}^{F}+\sigma\cdot\boldsymbol{m}f_{s}^{F}$. Such a restriction does not appear in the normal-metal layer and f^{N} can be any Hermitian 2×2 matrix.

The relation between the current through a contact and the distributions in the ferromagnetic layer and the normal-metal layer is determined by four parameters, the two real spin-dependent conductances ($G^{\uparrow}$, $G^{\downarrow}$) and the real and imaginary parts of the mixing conductance $G^{\uparrow\downarrow}$. These contact-specific parameters can be obtained by microscopic theory or from experiments. The spin conductances $G^{\uparrow}$ and $G^{\downarrow}$ describe spin transport for a long time [19]. The mixing conductance is relevant for transport between non-collinear ferromagnets. Note that although the mixing conductance is a complex number, the 2×2 current in spin space is Hermitian, and consequently the current and the spin current in any direction given by (13.35) are real numbers. Due to the definitions of the spin-dependent conductances (13.36) and the "mixing" conductance (13.38)

$$2\mathrm{Re}G^{\uparrow\downarrow}=G^{\uparrow}+G^{\downarrow}+\frac{e^{2}}{h}\sum_{nm}\left|r_{\uparrow}^{nm}-r_{\downarrow}^{nm}\right|^{2} \tag{13.38}$$

and consequently, the conductances should satisfy the relation 2 Re $G^{\uparrow\downarrow}\geq G^{\uparrow}+G^{\downarrow}$.

In terms of a scalar particle and a vector spin contribution $\boldsymbol{I} = (1\boldsymbol{I}_0 \pm \sigma \cdot \boldsymbol{I}_s)/2, \boldsymbol{f}^{N(F)} = 1\boldsymbol{f}_0^{N(F)} \pm \sigma \cdot s_s(\boldsymbol{m})\boldsymbol{f}_s^{N(F)}$, and the particle current is described by the expression

$$I_0 = \left(G^{\uparrow} + G^{\downarrow}\right)\left(f_0^F + f_0^N\right) + \left(G^{\uparrow} - G^{\downarrow}\right)\left(f_s^F - \boldsymbol{m} \cdot s f_s^N\right) \tag{13.39}$$

The familiar expressions for collinear transport are recovered when $\boldsymbol{m} \cdot \boldsymbol{s} = \pm 1$. The spin current is described by the expression

$$\begin{aligned} \boldsymbol{I}_s = \boldsymbol{m}\left[\left(G^{\uparrow} - G^{\downarrow}\right)\left(\left(f_0^F - f_0^N\right) + f_s^F\right) + \left(2\mathrm{Re}G^{\uparrow\downarrow} - G^{\uparrow} - G^{\downarrow}\right)\boldsymbol{m} \cdot s f_s^N\right] \\ -2s\mathrm{Re}G^{\uparrow\downarrow} f_s^N + (\boldsymbol{m} \times s) 2\mathrm{Im}G^{\uparrow\downarrow} f_s^N. \end{aligned} \tag{13.40}$$

The first two terms point in the direction of the magnetization of the ferromagnet $\boldsymbol{m}$, the third term is in the direction of the nonequilibrium spin distribution $\boldsymbol{s}$, and the last term is perpendicular to both $\boldsymbol{s}$ and $\boldsymbol{m}$. The last contribution solely depends on the imaginary part of the mixing conductance. This term can be interpreted by considering how the direction of the spin on the normal metal layer $\boldsymbol{s}$ would change in timekeeping, with all other parameters constant. The cross product creates a precession of $\boldsymbol{s}$ around the magnetization direction $\boldsymbol{m}$ of the ferromagnet similar to a classical torque while keeping the magnitude of the spin-accumulation constant. In contrast, the first three terms represent diffusion-like processes, which decrease the magnitude of the spin accumulation. Due to the abovementioned relation 2 Re $G^{\uparrow\downarrow} \geq G^{\uparrow} + G^{\downarrow}$, the nonequilibrium spin distribution f_s^N propagates easier into a configuration parallel to $\boldsymbol{s}$ than parallel to $\boldsymbol{m}$, since these processes are governed by positive diffusion-like constants 2 Re $G^{\uparrow\downarrow}$ and 2 Re $G^{\uparrow\downarrow} - G^{\uparrow} - G^{\downarrow}$, respectively.

13.2.4 Dependence of Spin Current on Contact Type

The four conductance parameters $G^{\uparrow}$, $G^{\downarrow}$, Re$G^{\uparrow\downarrow}$, and Im$G^{\uparrow\downarrow}$ depend on the microscopic details of the contact between ferromagnetic and normal-metal layers, which is characterized by elementary model contacts of a diffusive, a ballistic, and a tunnel types.

In the case a diffusive contact between a normal metal (N) and a ferromagnet (F) with conductances G_D^N and G_D^{Fs}, respectively, the spin-dependent resistance of the whole contact is the sum $\left(G_D^{Fs}\right)^{-1} + \left(G_D^N\right)^{-1}$, consequently the whole conductance $G_D^s = G_D^{Fs} G_D^N / \left(G_D^{Fs} + G_D^N\right)$, where $s = (\uparrow, \downarrow)$. These spin-dependent conductances fully describe collinear transport (in the absence of spin-flip scattering).

For non-collinear magnetizations, the mixing conductance, which is also needed, can be derived from the scattering matrix. The latter follows from the diffusion equation, describing the scattering properties of the contact by a spatially dependent distribution matrix. The current density (i) on the normal side of the contact ($x < 0$) is $i(x < 0) = \sigma^N \partial_x f$, where σ^N is the conductivity of the N layer. Consequently, the total current in the N layer with the length L^N is $I(x < 0) = G_D^N(L^N\partial_x)f$, where f is the spatially dependent distribution matrix on the normal side in the contact. In the normal-metal part, the boundary condition is $f(x = -L^N) = f^N$. In a ferromagnet, spin-up and spin-down states are incoherent, and hence spins non-collinear to the magnetization direction relax and only spins collinear with the magnetization will propagate sufficiently far away from the F|N interface. It is assumed that the ferromagnet is sufficiently strong and that the contact is longer than the ferromagnetic decoherence length $\xi = \sqrt{D/h_{ex}}$, where D is the diffusion constant and h_{ex} is the exchange splitting. The decoherence length is typically very short in ferromagnets, $\xi = 2$ nm in Ni wires.

The distribution function on the ferromagnetic side is represented by a two-component distribution function $f(x > 0) = u^{\uparrow}f^{\uparrow} + u^{\downarrow}f^{\downarrow}$, where $u^{\uparrow}$ and $u^{\downarrow}$ are the mentioned spin-projection matrices. Here a spin accumulation collinear to the magnetization direction in the ferromagnet is taken into account. The boundary condition determined by the distribution function in the ferromagnetic part is thus

$$f^{\uparrow}\left(x = L^F\right) = f^{F\uparrow},\quad f^{\downarrow}\left(x = L^F\right) = f^{F\downarrow}, \tag{13.41}$$

where L^F is the length of the F layer.

In assumption that the resistance of the diffusive region of the contacts is much larger than the contact resistance between the N and F layers, the total current in the ferromagnet is

$$I(x > 0) = G_D^{F\uparrow}u^{\uparrow}\partial_x f^{\uparrow} + G_D^{F\downarrow}u^{\downarrow}\partial_x f^{\downarrow} \tag{13.42}$$

The distribution function is continuous across the F|N interface, $f(0^+) = f(0^-)$. Current conservation on the left ($x < 0$) and on the right ($x > 0$) of the normal metal-ferromagnet interface dictates the equation, $\partial_x I = 0$, which together with the boundary conditions $f(x = -L^N) = f^N$, $f(0^+) = f(0^-)$, and (13.41) uniquely determines the distribution functions and hence the conductance in the diffusive contact. Then the current on the normal side of the contact becomes

$$\begin{aligned} eI = {} & G_D^{\uparrow}u^{\uparrow}\left(f^F - f^N\right)u^{\downarrow} + G_D^{\downarrow}u^{\downarrow}\left(f^F - f^N\right)u^{\downarrow} \\ & + G_D^N\left[u^{\uparrow}\left(f^F - f^N\right)u^{\downarrow} + u^{\downarrow}\left(f^F - f^N\right)u^{\downarrow}\right] \end{aligned} \tag{13.43}$$

The current in a diffusive contact thus takes the generic form (13.35) with $G^{\uparrow} = G_D^{\uparrow}$, $G^{\downarrow} = G_D^{\downarrow}$, and $G^{\uparrow\downarrow} = G_D^N$. The mixing conductance is thus real and only depends on the normal conductance. The latter results can be understood as a consequence of the effective spin relaxation of spins non-collinear to the local magnetization direction. Those spins cannot propagate in the ferromagnet, and consequently the effective conductance can only depend on the conductance in the normal metal.

In the case of the ballistic contact, the reflection and transmission coefficients appearing in (13.36) and (13.37) are diagonal in the space of the transverse channels since the transverse momentum is conserved. In a simplified model [20], the transmission channels are either closed $t = 0$ or open $t = 1$. The conductances (13.36) and (13.37) can then be found by simply counting the number of propagating modes. Then the spin-dependent conductance $G_B^{\uparrow(\downarrow)} = (e^2/h)N^{\uparrow(\downarrow)}$, where $N^{\uparrow}(N^{\downarrow})$ is the number of spin-up (spin-down) propagating channels. The mixing conductance is determined by $G_B^{\uparrow\downarrow} = \max\left(G_B^{\uparrow}, G_B^{\downarrow}\right)$ and is real. In a quantum mechanical calculation, the channels just above the potential step are only partially transmitting, and the channels below a potential step can have a finite transmission probability due to tunneling. Furthermore, the band structure of ferromagnetic metals is usually complicated and interband scattering exists even at ideal interfaces. The phase of the scattered wave will be relevant giving a nonvanishing imaginary part of the mixing conductance.

In the case of a tunneling contact, the transmission coefficients are exponentially small, and the reflection coefficients have a magnitude close to one. The spin-dependent conductance is

$$G_T^s = \frac{e^2}{h} \sum_{nm} \left| t_s^{nm} \right|^2. \tag{13.44}$$

For simple models of tunnel barriers, $r_s^{nm} = \delta^{nm} \exp i\phi^n = \delta r_s^{nm}$, where the phase shift ϕ^n is spin-independent. The expansion (38) in the small correction δr_s^{nm} leads to the expression $\mathrm{Re} G_T^{\uparrow\downarrow} = \left(G_T^{\uparrow} + G_T^{\downarrow}\right)/2$, where $G_T^{\uparrow}$ and $G_T^{\downarrow}$ are the spin-dependent tunneling conductances (13.44). Since the transmission coefficients in a tunnel contact are all exponentially small, the imaginary part of $G_T^{\uparrow\downarrow}$ is of the same order of magnitude as $G_T^{\uparrow}$ and $G_T^{\downarrow}$, but it is not universal and depends on the details of the contact.

13.3 Spin Transport in the Case of Dynamic Magnetic Field

13.3.1 Features of Coupling Spin Currents with Magnetic Dynamics

The interconnection between spin currents and the magnetic dynamics in F/N-based magnetic multilayer nanostructures underlies the current-controlled magnetic dynamics and utilization of the latter as new functionality in spintronic devices [21]. One is related to the *s-d* exchange interaction with localized spins and the spin-dependent scattering of spin-polarized electrons near the F|N interface. The impact of the spin current on localized spins occurs via a finite torque on the magnetic order parameter, and, vice versa, a moving magnetic order vector loses torque by emitting a spin current. The magnetic precession acts as a spin pump which transfers angular momentum from the magnetic into normal metal.

The technological potential of the mentioned magnetic nanostructures is related to utilizing transition metals (for instance, Co, Ni, Fe) that operate at ambient temperatures. Examples are current-induced tunable microwave generators (spin-torque oscillators) [22, 23] and nonvolatile magnetic electronic architectures that can be randomly read, written, or programmed by current pulses in a scalable manner [24]. The interaction between currents and magnetization can also cause undesirable effects such as enhanced magnetic noise in read heads made from magnetic multilayers [25].

In the framework of the Landau-Lifshitz-Gilbert model, the impact of the spin current on the magnetic dynamics, caused by the spin transfer, reduces to change of fundamental parameters such as the gyromagnetic ratio and Gilbert damping parameters. This spin transfer is governed by the reflection and transmission matrices of the system, analogous to the scattering theory of transport and interlayer exchange coupling. In the case when the normal-metal layers adjacent to the ferromagnetic layers are perfect spin sinks, the spin accumulation in the normal metal vanishes [26]. In the opposite case, the spin accumulation accompanies the spin diffusion, which gives essential contribution to the total spin current and its interconnection with magnetic dynamics.

Spin pumping by a precessing ferromagnet is, in some sense, the reverse process of current-induced magnetization dynamics. When the pumped spin angular momentum is not quickly dissipated to the normal-metal atomic lattice, a spin accumulation builds up and creates reaction torques due to transverse-spin backflow into ferromagnets. The interplay between magnetization dynamics and the nonequilibrium spin-polarized transport in heterostructures determining magnetic properties will be considered for the case of F/N-based nanostructures below.

13.3.2 Precession-Induced Spin Pumping Through F/N Interfaces

Characteristic properties of the precession-induced spin pumping are manifested in the model *N/F/N* magnetic junction schematic of which is displayed in Fig. 13.2. The ferromagnetic layer F is a spin-dependent scatterer that governs electron transport between (left (L) and right (R)) normal-metal reservoirs. The 2×2 operator $\boldsymbol{I}_l$ for the charge and spin current in lth lead ($l = L, R$) can be expressed in terms of operators $a_{\alpha m,\ l}(E)$ and $b_{\alpha m,\ l}(E)$ that annihilate a spin-α electron with energy E leaving (entering) the lth lead through the mth channel:

$$I_l^{\alpha\beta}(t) = \frac{e}{h}\sum_m \int dEdE' e^{i(E-E')t/\hbar}\left[a^{\dagger}_{\beta m,l}(E)a_{\alpha m,l}\left(E'\right) - b^{\dagger}_{\beta m,l}(E)b_{\alpha m,l}\left(E'\right)\right]. \tag{13.45}$$

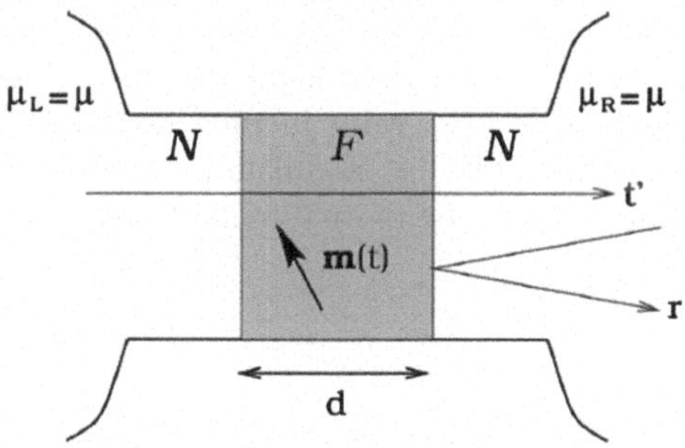

Fig. 13.2 Ferromagnetic layer (F) sandwiched between two normal-metal layers (N). The reflection and transmission amplitudes r and t' govern the spin current pumped into the right lead

When the scattering matrix $s^{\alpha\beta}_{mn,ll'}(t)$ of the ferromagnetic layer varies slowly on the time scales of electronic relaxation in the system, an adiabatic approximation may be used. The annihilation operators for particles entering the N layers are then related to the operators of the outgoing states by the instantaneous value of the scattering matrix $b_{am,l}(E) = s^{\alpha\beta}_{mn,ll'}(t)a_{\beta n,l'}(E)$. In terms of $a_{\alpha m,\ l}$ only, we can evaluate the expectation value $\left\langle I_l^{\alpha\beta}(t)\right\rangle$ of the current operator using $\left\langle a^{\dagger}_{\alpha m,l}(E)a_{\beta n,l'}\left(E'\right)\right\rangle = f_l(E)\delta_{\alpha,\beta}\delta_{m,n}\delta_{l,l'}\delta\left(E - E'\right)$, where $f_l(E)$ is the isotropic distribution function in the lth reservoir. When the scattering matrix depends on a single time-dependent parameter $X(t)$, then the Fourier transform of the current expectation value $I_l(\omega) = \int dte^{i\omega t}I_l(t)$ can be written as

$$I_l(\omega) = g_{X,l}(\omega)X(\omega) \tag{13.46}$$

In terms of a frequency ω- and X-dependent parameter

$$g_{X,l}(\omega) = -\frac{e\omega}{4\pi}\sum_{l'}\int dE\left(-\frac{\partial f_{l'}(E)}{\partial E}\right)\sum_{mn}\left(-\frac{\partial s_{mn,ll'}(E)}{\partial X}s^{\dagger}_{mn,ll'}(E) - \text{H.c.}\right). \tag{13.47}$$

Equation (13.46) is the first-order (in frequency) correction to the dc Landauer-Büttiker formula [11]. At equilibrium $f_R(E) = f_L(E)$, (13.46) is the lowest-order nonvanishing contribution to the current. Furthermore, at sufficiently low temperatures, we can approximate $\partial f_l(E)/\partial E$ by a δ-function centered at Fermi energy. The expectation value of the 2×2 particle number operator $Q_l(\omega)$(defined by $I_l(t) = dQ_l(t)/dt$) in time or by $I_l(\omega) = -i\omega Q_l(\omega)$ in frequency domain) for the lth reservoir is then given by

$$Q_l(\omega) = \left(\frac{e}{4\pi i}\sum_{mnl'}\frac{\partial s_{mn,ll'}(E)}{\partial X}s^{\dagger}_{mn,ll'}(E) + \text{H.c.}\right)X(\omega), \tag{13.48}$$

where the scattering matrices are evaluated at the Fermi energy. Because the prefactor on the right-hand side of (13.46) does not depend on frequency ω, the equation is also valid in time domain. The change in particle number $\delta Q_l(t)$ is proportional to the modulation $\delta X(t)$ of parameter X, and the 2×2 matrix current (directed into the normal-metal leads) reads

$$I_l(t) = e\frac{dn_l}{dX}\frac{dX(\mathrm{t})}{dt}, \tag{13.49}$$

where the "matrix emissivity" into lead l is

$$\frac{\partial n_l}{\partial X} = \frac{1}{4\pi i}\sum_{mnl'}\frac{\partial s_{mn,ll'}(E)}{\partial X}s^{\dagger}_{mn,ll'}(E) + \mathrm{H.c.} \tag{13.50}$$

If the spin-flip scattering in the ferromagnetic layer is disregarded, the scattering matrix $\boldsymbol{s}$ can be written in terms of the spin-up and spin-down scattering coefficients $\boldsymbol{s}^{\uparrow(\downarrow)}$ using the projection matrices $\boldsymbol{u}^{\uparrow} = (1 + \boldsymbol{\sigma} \cdot \boldsymbol{m})/2$ and $\boldsymbol{u}^{\downarrow} = (1 - \boldsymbol{\sigma} \cdot \boldsymbol{m})/2$:

$$s_{mn,ll'} = s^{\uparrow}_{mn,ll'}u^{\uparrow} + s^{\downarrow}_{mn,ll'}u^{\downarrow} \tag{13.51}$$

The spin current pumped by the magnetization precession is obtained by identifying $X(t) = \varphi(t)$, where φ is the azimuthal angle of the magnetization direction in the plane perpendicular to the precession axis. For simplicity, we assume that the magnetization rotates around the y-axis: $\boldsymbol{m} = (\sin\varphi, 0, \cos\varphi)$. Using (13.51), it is then easy to calculate the emissivity (13.50) for this process:

$$\frac{\partial n_l}{\partial \varphi} = -\frac{1}{4\pi}\left[A_r\sigma_y + A_i\left(\sigma_x \cos\varphi - \sigma_y \sin\varphi\right)\right], \tag{13.52}$$

where $A_r(A_i) =$ Re (Im)$[g^{\uparrow\downarrow} - t^{\uparrow\downarrow}]$. Expanding the 2×2 current into isotropic and traceless components,

$$I = \frac{1}{2}I_c - \frac{e}{\hbar}\sigma \cdot \boldsymbol{I}_s. \tag{13.53}$$

Here the charge current I_c and spin current $\boldsymbol{I}_s$ are identified. Due to Eqs. (13.49), (13.52), and (13.53), the charge current vanishes, $I_c = 0$, and the spin current

$$\boldsymbol{I}_s = (A_i\cos\varphi, A_r - A_i\sin\varphi)\frac{\hbar}{4\pi}\frac{d\varphi}{dt} \tag{13.54}$$

can be rewritten as

$$\boldsymbol{I}_s^{\mathrm{pump}} = \frac{\hbar}{4\pi}\left(A_r\boldsymbol{m} \times \frac{d\boldsymbol{m}}{dt} - A_i\frac{d\boldsymbol{m}}{dt}\right) \tag{13.55}$$

This current into a given N layer depends on the complex-valued parameter $A = A_r + iA_i$ (the "spin-pumping conductance") and the time-dependent order parameter of the ferromagnet $\boldsymbol{m}(t)$. In addition, $A = g^{\uparrow\downarrow} - t^{\uparrow\downarrow}$ depends on the scattering matrix of the ferromagnetic layer since

$$g^{\sigma\sigma'} = \sum_{mn} \left[\delta_{mn} - r^{\sigma}_{mn} \left(r^{\sigma'}_{mn} \right)^{*} \right] \tag{13.56}$$

is the dimensionless dc conductance matrix [10, 26] and $t^{\uparrow\downarrow} = \sum_{mn} t'^{\uparrow}_{mn} \left(t'^{\downarrow}_{mn} \right)^{*}$. Here $r^{\uparrow}_{mn}$ ($r^{\downarrow}_{mn}$) is a reflection coefficient for spin-up ~spin-down electrons on the normal-metal side, and $t'^{\uparrow(\downarrow)}_{mn}$ is a transmission coefficient for spin-up/spin-down electrons across the ferromagnetic film from the opposite reservoir into the normal-metal layer, where m and n label the transverse modes at the Fermi energy in the normal-metal films. The magnetization can take arbitrary directions; in particular, $\boldsymbol{m}(t)$ may be far away from its equilibrium value. In such a case, the scattering matrix itself can depend on the orientation of the magnetization, and one has to use $A(\boldsymbol{m})$ in Eq. (13.55).

When the ferromagnetic film is thicker than its transverse spin coherence length $d > \pi/(k^{\uparrow} - k^{\downarrow})$, where $k_F^{\uparrow(\downarrow)}$ are the spin-dependent Fermi wave vectors, $t^{\uparrow\downarrow}$ vanishes [26], the spin pumping through a given F|N interface is governed entirely by the interfacial mixing conductance $A = g^{\uparrow\downarrow} = g_r^{\uparrow\downarrow} + ig_i^{\uparrow\downarrow}$, and we can consider only one of the two interfaces, as it is presented in Fig. 13.3.

The spin current (13.55) leads to a damping of the ferromagnetic precession, resulting in a faster alignment of the magnetization with the (effective) applied magnetic field $\boldsymbol{H}_{eff}$. The pumped spins are entirely absorbed by the attached ideal reservoirs. Thereto the enhancement rate of damping is accompanied by an energy flow out of the ferromagnet, until a steady state is established in the accompanied F/N system. For simplicity, assume a magnetization which at time t starts rotating

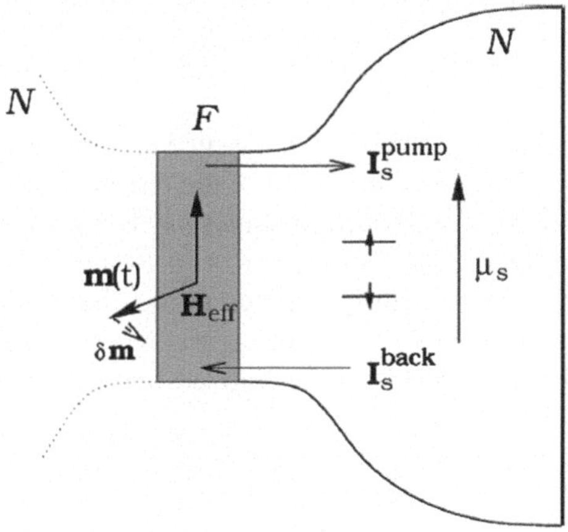

Fig. 13.3 Ferromagnetic nanostructure with the unit magnetization vector $\boldsymbol{m}(t)$ precessing around the external magnetic field $\boldsymbol{H}_{eff}$ and inducing the spin pumping $\boldsymbol{I}_s^{pump}$ through the interface F/N. The spin accumulation $\boldsymbol{\mu}_s$ results in the back spin current $\boldsymbol{I}_s^{back}$

around the vector of the magnetic field, $\boldsymbol{m}(t) \perp \boldsymbol{H}_{eff}$. In a short interval of time δt, it slowly (i.e., adiabatically) changes to $\boldsymbol{m}(t + \delta t) = \boldsymbol{m}(t) + \delta\boldsymbol{m}$. In the presence of a large but finite nonmagnetic reservoir without any spin-flip scattering attached to one side of the ferromagnet, this process can be expected to induce a nonvanishing spin accumulation $\boldsymbol{\mu}_s = \int d\varepsilon \mathrm{Tr}[\sigma f(\varepsilon)]$, where σ is the Pauli matrix vector and $f(\varepsilon)$ is the 2×2 matrix distribution function at a given energy ε. For a slow enough variation of $\boldsymbol{m}(t)$, this nonequilibrium spin imbalance must flow back into the ferromagnet, canceling any spin current generated by the magnetization rotation, since, due to the adiabatic assumption, the system is always in a steady state.

For the spins accumulated in the reservoir along the magnetic field, $\boldsymbol{\mu}_s \| \boldsymbol{H}_{eff}$ flow of N_s spins into the normal metal transfers the energy $\Delta E_N = N_s\mu_s/2$ and angular momentum $\Delta L_N = N_s\hbar/2$ (directed along $\boldsymbol{H}_{eff}$). By the conservation laws, $\Delta E_F = -\Delta E_N$ and $\Delta L_F = -\Delta L_N$, for the corresponding values in the ferromagnet. Using the magnetic energy, $\Delta E_F = -\gamma \Delta L_F H_{eff}$, where γ is the gyromagnetic ratio of the ferromagnet, it can be found that $N_s\mu_s/2 = \gamma N_s \hbar H_{eff}/2$. Then $\mu_s = \hbar\gamma H_{eff} = \hbar\omega$, where $\omega = \gamma H_{eff}$ is the Larmor frequency of precession in the effective field. The spin-up and spin-down chemical potentials in the normal metal split by $\mu_s = \hbar\omega$, the energy corresponding to the frequency of the perturbation.

The above mentioned the backflow of spin current $\boldsymbol{I}_s^{\mathrm{back}}$, which equals to the pumping current $\boldsymbol{I}_s^{\mathrm{pump}}$ described by the expression

$$\boldsymbol{I}_s^{\mathrm{back}} == \frac{1}{2\pi}\left(g_r^{\uparrow\downarrow}\boldsymbol{\mu}_s + g_i^{\uparrow\downarrow}\boldsymbol{m} \times \boldsymbol{\mu}_s\right) = \frac{\hbar}{4\pi}\left(g_r^{\uparrow\downarrow}\boldsymbol{m} \times \frac{d\boldsymbol{m}}{dt} - g_i^{\uparrow\downarrow}\frac{d\boldsymbol{m}}{dt}\right) \tag{13.57}$$

Here, it is used that $\mu_s = \hbar\omega$ and $\boldsymbol{\mu}_s \perp \boldsymbol{m}$, since by the conservation of angular momentum, the spin transfer is proportional to the change in the direction $\delta\boldsymbol{m} \perp \boldsymbol{m}$. Thus, for the case of a single and finite reservoir, Eq. (13.55) is recovered. It is easy to repeat the proof for an arbitrary initial alignment of $\boldsymbol{m}(t)$ with $\boldsymbol{H}_{eff}$.

The expressions for the adiabatic spin pumping are not the whole story, since spin-flip scattering is essential. In this case, the spin build-up occurs in the normal metal at dynamic equilibrium. Then, the contribution to $\boldsymbol{I}_s$ due to the spin-accumulation-driven current $\boldsymbol{I}_s^{\mathrm{back}}$ back into the ferromagnet:

$$\boldsymbol{I}_s = \boldsymbol{I}_s^{\mathrm{pump}} - \boldsymbol{I}_s^{\mathrm{back}}, \tag{13.58}$$

which vanishes in the absence of spin-flip scattering.

The spin current out of the ferromagnet carries angular momentum perpendicular to the magnetization direction. By conservation of angular momentum, the spins ejected by $\boldsymbol{I}_s$ correspond to a torque $\boldsymbol{T} = -\boldsymbol{I}_s$ on the ferromagnet. If possible, interfacial spin-flip processes are disregarded, and the torque t is entirely transferred to the coherent magnetization precession. The dynamics of the ferromagnet can then be described by a generalized Landau-Lifshitz-Gilbert (LLG) equation [5]

$$\frac{d\boldsymbol{m}}{dt} = -\gamma \boldsymbol{m} \times \boldsymbol{H}_{eff} + \alpha_0 \boldsymbol{m} \frac{d\boldsymbol{m}}{dt} + \frac{\gamma}{M_s V} \boldsymbol{I}_s, \tag{13.59}$$

where α_0 is the dimensionless bulk Gilbert damping constant, M_s is the saturation magnetization of the ferromagnet, and V is its volume. The intrinsic bulk constant α_0 is smaller than the total Gilbert damping $\alpha = \alpha_0 + \alpha'$. The additional damping α' caused by the spin pumping is observable in, for example, FMR spectra here.

13.3.3 Spin-Accumulation-Driven Backflow in the F/N Bilayer

The precession of the magnetization does not cause any charge current in the system. The spin accumulation or nonequilibrium chemical potential imbalance $\boldsymbol{\mu}_s$ in the normal metal is a vector, which depends on the distance from the interface x x, $0 < x < L$, where L is the thickness of the normal-metal film (see Fig. 13.4). When the ferromagnetic magnetization steadily rotates around the z axis, $\boldsymbol{m} \times \partial_t \boldsymbol{m}$ and the normal-metal spin accumulation $\boldsymbol{\mu}_s(x)$ are oriented along z. There is no spin imbalance in the ferromagnet, because $\boldsymbol{\mu}_s(x)$ is perpendicular to the magnetization direction $\boldsymbol{m}$. The time-dependent $\boldsymbol{m}_s$ is also perpendicular to $\boldsymbol{m}$ even in the case of a precessing ferromagnet with time-dependent instantaneous rotation axis, as long as the precession frequency ω is smaller than the spin-flip rate τ_{sf}^{-1} in the normal metal.

The spin accumulation diffuses into the normal metal as

$$i\omega \boldsymbol{\mu}_s = D\partial_x^2 \boldsymbol{\mu}_s - \tau_{sf}^{-1} \boldsymbol{\mu}_s, \tag{13.60}$$

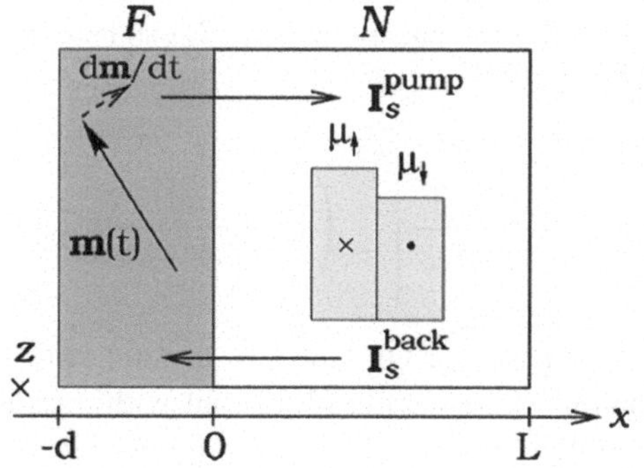

Fig. 13.4 Schematic view of the *F-N* bilayer. Precession of the magnetization direction $\boldsymbol{m}(t)$ of the ferromagnet F pumps spins into the adjacent normal-metal layer N by inducing a spin current $\boldsymbol{I}_s$ pump. This leads to a buildup of the normal-metal spin accumulation $\boldsymbol{\mu}_s$ which either relaxes by spin-flip scattering or flows back into the ferromagnet as $\boldsymbol{I}_s^{back}$

where D is the diffusion coefficient. The boundary conditions are determined by the continuity of the spin current from the ferromagnet into the normal metal at $x = 0$ and the vanishing of the spin current at the outer boundary $x = L$:

$$x = 0: \ \ \partial_x \boldsymbol{\mu}_s = -2(\hbar NSD)^{-1} \boldsymbol{I}_s, \ \ x = L: \ \ \partial_x \boldsymbol{\mu}_s = 0 \tag{13.61}$$

where N is the (one-spin) density of states in the layer and S is the area of the interface. The solution to (13.60) with the boundary conditions (13.61) is

$$\boldsymbol{\mu}_s(x) = \frac{\cosh k(x-L)}{\sinh kL} \frac{2\boldsymbol{I}_s}{\hbar NSDk} \tag{13.62}$$

Using relation $D = v_F^2 \tau_{el}/3$ between the diffusion coefficient D, the Fermi velocity v_F, and the elastic scattering time τ_{el}, we find for the spin-diffusion length $\lambda_{sd} = v_F\sqrt{\tau_{el}\tau_{sf}/3}$. An effective energy level spacing of the states participating in the spin-flip scattering events in a thick layer can be defined by $\delta_{sd} = (NS\lambda_{sd})^{-1}$. The spin-accumulation-driven spin current $\boldsymbol{I}_s^{\text{back}}$ through the interface reads

$$\begin{aligned} \boldsymbol{I}_s^{\text{back}} &= \frac{1}{8\pi}\Big[2g_r^{\uparrow\downarrow}\boldsymbol{\mu}_s(0) + 2g_i^{\uparrow\downarrow}\boldsymbol{m}\times\boldsymbol{\mu}_s(0) \\ &\quad + \left(g^{\uparrow\uparrow} + g^{\downarrow\downarrow} - 2g_r^{\uparrow\downarrow}\right)\left(\boldsymbol{m}\cdot\boldsymbol{\mu}_s(0)\right)\Big]. \end{aligned} \tag{13.63}$$

Substitution of (13.62) into (13.63) gives total spin current

$$\boldsymbol{I}_s = \boldsymbol{I}_s^{\text{pump}} - \frac{\beta}{2}\Big[2g_r^{\uparrow\downarrow}\boldsymbol{I}_s + 2g_i^{\uparrow\downarrow}\boldsymbol{m}\times\boldsymbol{I}_s + \left(g^{\uparrow\uparrow} + g^{\downarrow\downarrow} - 2g_r^{\uparrow\downarrow}\right)(\boldsymbol{m}\cdot\boldsymbol{I}_s)\boldsymbol{m}\Big], \tag{13.64}$$

where the spin current returning into the ferromagnet is governed by the backflow factor $\beta = (\tau_{sf}\delta_{\text{sd}})/(\hbar \tanh(L/\lambda_{sd}))$. When the normal metals are shorter than the spin-diffusion length ($L \ll \lambda_{sd}$), $\beta \to \tau_{sf}\delta/h$, where $\delta = (NSL)^{-1}$. Basically, β is therefore the ratio between the energy level spacing of the normal-metal layer with thickness $L_{sf} = \min(L, \lambda_{sd})$ and the spin-flip rate.

By inverting Eq. (13.64), the total spin current $\boldsymbol{I}_s$ can be expressed in terms of the pumped spin current $\boldsymbol{I}_s^{\text{pump}}$ ([26])

$$\boldsymbol{I}_s = \left[1 + 2g_r^{\uparrow\downarrow} + \frac{\left(\beta g_i^{\uparrow\downarrow}\right)^2}{1+\beta g_r^{\uparrow\downarrow}}\right]^{-1}\left(1 - \frac{\beta g_i^{\uparrow\downarrow}}{1+\beta g_r^{\uparrow\downarrow}}\boldsymbol{m}\times\right)\boldsymbol{I}_s^{\text{pump}} \tag{13.65}$$

Then, substituting (13.55) into (13.65) results in total spin current $\boldsymbol{I}_s$, which is described by the equation of the form (13.55) but with a redefined spin-pumping conductance $A' = A'_r + A'_i$

$$\boldsymbol{I}_s = \frac{\hbar}{4\pi}\left(A'_r\boldsymbol{m}\times\frac{d\boldsymbol{m}}{dt} - A'_i\frac{d\boldsymbol{m}}{dt}\right) \tag{13.66}$$

Here A' is function of the mixing conductance $g^{\uparrow\downarrow}$ and the backflow factor β,$A' = A'(g^{\uparrow\downarrow}, \beta)$. For realistic $F|N$ interfaces $g_i^{\uparrow\downarrow} \ll g_r^{\uparrow\downarrow}$, so that $g^{\uparrow\downarrow} \approx g_r^{\uparrow\downarrow}$ and, consequently, A_i' vanishes. Substituting (13.66) into (13.59) renormalizes its Gilbert dumping constant, α_0, so that $\alpha_0 \to \alpha = \alpha_0 + \alpha'$, where

$$\alpha' = \frac{g_L g^{\uparrow\downarrow}}{4\pi\mu}\left[1 + g^{\uparrow\downarrow}\frac{\left(\tau_{sf}\delta_{sd}/h\beta g_i^{\uparrow\downarrow}\right)^2}{\tanh(L/\lambda_{sd})}\right]^{-1} \tag{13.67}$$

is the additional damping constant due to the interfacial F|N coupling. Here, g_L is the g factor and μ is the total layer magnetic moment in units of μ_B. When $L \to \infty$, (13.67) reduces $\alpha' = g_L g_{eff}^{\uparrow\downarrow}/4\pi$, where $1/g_{eff}^{\uparrow\downarrow} = 1/g^{\uparrow\downarrow} + R_{sd}$ and $R_{sd} = \tau_{sf}\delta_{sd}/h$ is the resistance (per spin, in units of h/e^2) of the normal-metal layer of thickness λ_{sd}. It follows that the effective spin pumping out of the ferromagnet is governed by $g_{eff}^{\uparrow\downarrow}$, i.e., the conductance of the $F|N$ interface in series with diffusive normal-metal film with thickness λ_{sd}.

The second factor on the right-hand side of (13.67) suppresses the additional Gilbert damping due to the spin angular momentum that diffuses back into the ferromagnet. Because spins accumulate in the normal metal perpendicular to the ferromagnetic magnetization, the spin-accumulation-driven transport across the F|N contact, as well as the spin pumping, is governed by a mixing conductance.

13.4 Conclusions

The spin transport in the F/N-based magnetic nanostructures in magneto-static and magneto-dynamic cases has been studied in the framework of the modified Stoner model. Using the modified quantum kinetic equation for the nonlinear Green functions and the spin-dependent scattering matrix, the spin currents through and near the F|N interface are described. In the magnetostatic case, the parametric dependence of the spin current on the relative orientation of the spin polarization and magnetization is shown. In the magneto-dynamic case of the magnetization precession, the precession-induced spin pumping into the normal-metal layer is described. The accompanying effect of the spin accumulation and the spin backflow exerted via the spin torque on the magnetization precession is considered.

References

1. Akerman J (2005) Toward a Universal Memory. Science 308:508
2. Barder SD, Parkin SSP (2010) Spintronics. Ann Rev Cond Matt Phys 1:71
3. Bauer GEW (1992) Perpendicular transport through magnetic multilayers. Phys Rev Lett 9:1676

4. Binash G, Grűnberg P, Saurenbach F, Ziman W (1989) Enhanced magnetoresistance in layered magnetic structures with antiferromagnetic interlayer exchange. Phys Rev B 39:4828
5. Braganca PM, Guney BA, Wilson BA, Katine JA, Maat S, Childress JR (2010) Nanoscale magnetic field detection using a spin torque oscillator. Nanotechnology 21:235202
6. Brataas A, Nazarov YV, Inoue J, Bauer GEW (1999) Spin accumulation in small ferromagnetic double barrier junctions. Phys Rev B 59:93
7. Brataas A, Nazarov YV, Bauer GEW (2001) Spin-transport in multi-terminal normal-metal – ferromagnet systems with non-collinear magnetization. Eur Phys J B22:99
8. Bűttiker M (1986) Four-terminal Phase-Coherent Conductance. Phys Rev Lett 57:1761
9. Cheng R, Zhu J-G, Xiao D (2016) Dynamic Feedback in Ferromagnet/Spin-Hall Heterostructures. Phys Rev Lett 117:097202
10. Danilewicz P (1984) Quantum theory of Nonequilibrium Processes. Ann Phys 152:234
11. Fisher J, Gomonay O, Schlitz R, Ganzhom KN, Vestra H, Althammer M, et al (2017) Spin Hall magnetoresistance in antiferromagnet/normal metal heterostructures. arXiv:cond-mat.mes-hall
12. Gijos MAM, Bauer GEW (1997) Perpendicular giant magnetoresistance of magnetic multilayers. Adv Phys 46:285
13. Huertas D, Hemando A, Brataas A, Nazarov YV, Bauer GEW (2000) Conductance modulation by spin precessing in noncollinear ferromagnet/normal metal ferromagnetic multilayers. Phys Rev B 62:5700
14. Katine J, Fullerton EF (2008) Device implications of spin-transfer torque. J Magn Magn Mater 320:1217
15. Manchon A, Stelkova N, Ryzhanova A, Vedyaeva BA, Dienya BB, Slonczewski JC (2007) Theoretical investigation of the relationship between spin torque and magnetoresistance in spin-valves and magnetic tunnel junctions. J Magn Magn Mater 316:L977
16. Manchon A, Koo HC, Nitta J, Frolov SM, Duine RA (2015) New Perspective for Rashba Spin-Orbit Coupling. Nat Mater 36:871
17. Matsunaga S, Hiyama K, Matsumoto A, Ikeda S, Hasegawa H, Miura K, Hayakawa J, Endoh T, Ohno H, Hanyu T (2009) Standby-power-free compact ternary content-addressable memory cell chip using magnetic tunnel junction devices. Appl Phys Express 2:023004
18. Miron IM, Gaudin G, Auffer S, Rodmacq B, Schuhl A, Pizzini S, Vogel J, Gambardalla P (2010) Current-driven spin torque induced by the Rashba effect in ferromagnetic metal layer. Nat Mater 9:230
19. Myöhänem P, Stan A, Stefanucci G, Leeuwen R (2009) Kadanoff-Baym approach to quantum transport through interacting nanoscale systems: From the transient to the steady-state regime. Phys Rev B 80:115107
20. Nagasaka K (2009) CPP-GMR technology for magnetic read heads of future high-density recording systems. J Magn Magn Mater 321:508
21. Silva TJ, Rippard WH (2008) Developments in nano-transfer point-contact devices. J Magn Magn Mater 320:1260
22. Slonczewski JC (1996) Current-driven excitation of magnetic multilayers. J Magn Magn Mater 159:L1
23. Tserkovnyak Y, Brataas A, Bauer GEW (2002) Spin pumping and magnetization dynamics in metallic multilayers. Phys Rev B 66:224403
24. Tserkovnyak Y, Brataas A, Bauer GE, Halperin BI (2005) Nonlocal magnetization dynamics in ferromagnetic heterostructures. Rev Mod Phys 77:1375
25. Waintal X, Myers EB, Brouwer PW, Ralph DC (2000) Role of spin-dependent interface scattering in generating current-induced torques in magnetic multilayers. Phys Rev B 62:12317
26. Zutic L, Fabian J, Sarma SD (2004) Spintronics: fundamentals and applications. Rev Mod Phys 77:323

Part II
Nanooptics and Photonics

Chapter 14
Digital Interferometry Methods for the Surface Relief Study

Yu. Kotsiuba, H. Petrovska, V. Fitio, and Ya. Bobitski

14.1 Introduction

It is known that the control of surface relief is an important process in the study of the quality of the mechanical parts which are under the influence of external forces or factors. At the same time, information about the surface waviness and roughness is important. This information is necessary when such parameters as wear resistance [1], chemical resistance [2], and strength and density (tightness) of joints [3] are under study. Thus, the control of surface roughness is necessary in such fields as machine building and tribology and in solving fundamental problems of solid-state physics, physical chemistry, and materials science [1–4].

For its implementation, a number of methods are used, which can be divided into contact and non-contact. In contact methods, the determination of roughness is carried out with the profile line using special measuring instruments [5]. However, according to the modern world standards, the preformation of this procedure requires using the surface area of the test sample, instead of the profile line [6]. That fact reduces the relevance of contact methods and increases the need to develop alternative approaches. One of the possible solutions is the coherent optical measurement method. That allows quick and non-contact determination of surface roughness according to new measurement standards [6]. Among all of the coherent optical methods used for the surface relief study, the phase-shifting interferometry [7] and digital two-wavelength interferometry [8] are the most promising techniques.

Y. Kotsiuba · H. Petrovska (✉) · V. Fitio
Department of Photonics, Lviv Polytechnic National University, Lviv, Ukraine
e-mail: halyna.a.petrovska@lpnu.ua

Ya. Bobitski
Department of Photonics, Lviv Polytechnic National University, Lviv, Ukraine

Faculty of Mathematics and Natural Sciences, University of Rzeszow, Rzeszow, Poland

O. Fesenko, L. Yatsenko (eds.), *Nanooptics, Nanophotonics, Nanostructures, and Their Applications*, Springer Proceedings in Physics 210,
https://doi.org/10.1007/978-3-319-91083-3_14

The digital phase-shifting interferometry was successfully applied for the first time to rough objects, for which random speckle variations are observed in amplitude-phase distributions [9]. This method remain attractive due to the simplicity of the technical implementation compared with the white light interferometer [10] and the possibility of simultaneous research of a large area of the object without the scanning.

Today, the common phase-shifting methods are gradually superseded by a series of new faster and more efficient algorithms, which include a generalized phase-shifting algorithm [11], a wavelength change algorithm [12], and a two-step method with arbitrary phase shift [13]. The last method provides the simplest technical implementation, since for obtaining experimental data of the object under study, only two interferograms are used and it is not necessary to calibrate the phase shift. The experiments have confirmed the efficiency of using these methods in practice [11–13].

In the digital two-wavelength interferometry, an increase of the measured relief value range is done by use of the synthesized wavelength. Thus, the possibility of studying the relief of arbitrary objects, even small radio details [14], is realized. In [15] the method of three-color digital interferometry is used to receive three-dimensional information about the object and its deformation.

14.2 Theoretical Description of Surface Relief Study

14.2.1 The Phase-Shifting Interferometry Method

The image reconstruction process in the phase-shifting interferometry is simple and straightforward, since it is not necessary to apply the spatial frequency filtering procedure, as in other methods of digital interferometry [16]. In addition, this technology requires simpler optical setup. Let the complex amplitudes of the object and reference wave be determined as $U_o(x,y) = U_o\exp[i\varphi_o]$ and $A_r(x,y) = A_o\exp[i\varphi_o]$. Then the intensity distribution of the hologram in the detector plane can be expressed as:

$$I_H(x,y) = U_o^2(x,y) + A_r^2(x,y) + 2U_o(x,y)A_r(x,y)\cos\left[\varphi(x,y) - \varphi_r(x,y)\right] \quad (14.1)$$

Applying a phase shift with the value equal to $2\pi/\mathrm{P}$ (P – integer), we obtain:

$$I_{Hp}(x,y) = U_o^2(x,y) + A_r^2(x,y) + 2U_o(x,y)A_r(x,y)\cos\left[\varphi(x,y) - \varphi_r(x,y) + 2n\pi/P\right] \quad (14.2)$$

where $\mathrm{p} = 1, 2, \ldots, \mathrm{P}$. For a case $\mathrm{P} \geq 3$, the phase of the object field will be determined from the ratio [11]:

$$\varphi_o = \varphi_r + \arctan\left\{\frac{\sum_{n=1}^{P} I_{Hp}\sin(2\pi/P)}{\sum_{n=1}^{P} I_{Hp}\cos(2\pi/P)}\right\} \tag{14.3}$$

The amplitude can be calculated from equation:

$$U_o - \frac{1}{2A_o}\sqrt{\left(\sum_{n=1}^{P} I_{Hp}\sin(2\pi/P)\right)^2 + \left(\sum_{n=1}^{P} I_{Hp}\cos(2\pi/P)\right)^2} \tag{14.4}$$

When P = 4 we obtain a widely used method, which was proposed in [17], using four holograms with phase shift equal to π/2 [18, 19]. In this case, Eqs. (14.3) and (14.4) can be rewritten in the form:

$$\varphi_o = \varphi_r + \arctan\left\{\frac{I_{H4} - I_{H2}}{I_{H3} - I_{H1}}\right\} \tag{14.5}$$

$$U_o = \frac{1}{2A_o}\sqrt{(I_{H3} - I_{H1})^2 + (I_{H4} - I_{H2})^2} \tag{14.6}$$

The phase-shifting methods restore all information, when the amplitude and phase of the reference beam are known. In practice, these conditions are difficult to satisfy [20]. In addition, the value of the phase shift should be precisely set to obtain the correct results.

However, in [21] an algorithm that allows simplification of the phase field reconstruction was presented. This is done by recording two digital holograms with arbitrary phase shift. The corresponding phase-shift value can be defined as:

$$\alpha = \arccos\left\{\frac{\langle [I_1(x,y) - \langle I_1(x,y)\rangle][I_2(x,y) - \langle I_2(x,y)\rangle]\rangle}{\sigma_{I_1}\sigma_{I_2}}\right\} \tag{14.7}$$

where σ denotes the standard deviation and [...] stands for the arithmetic mean value. Then the phase can be determined from:

$$\varphi(x,y) = \frac{I'_{H1}(x,y)\cos\alpha - I'_{H2}(x,y)}{I'_{H1}(x,y)} \tag{14.8}$$

where $I'_{H1}(x,y)$ and $I'_{H2}(x,y)$ are the intensity distribution of two holograms after subtracting the components of the object and reference fields.

In the digital phase-shifting interferometry, the phase value is obtained modulo within the range of 2π, which corresponds to the physical value of the investigated parameter equal to λ/2. Obviously, for the study of ultra-small surface heterogeneities, it is necessary to use lasers with smaller wavelengths.

14.2.2 Digital Two-Wavelength Interferometry

However, if the phase changes are much bigger than the wavelength value, then the phase-shifting method will not restore the object surface under study. In this case, another digital holography method should be used. One possible solution is to compare a series of complex amplitudes before and after changing the object illumination conditions, namely, the wavelength.

To measure the surface shape by the two-wavelength method, the two holograms are recorded using different wavelengths λ_1 and λ_2. In conventional holographic interferometry, both holograms are recorded on the single photosensitive plate, after that they are restored using one of the wavelengths used during the recording. As a result, two object images are reconstructed. The image recorded and reconstructed with λ_1 is an exact copy of the original surface of the object. The image recorded with λ_2 but reconstructed with λ_1 moves toward the observer relative to its original position. The two reconstructed images interfere, forming a contour map of the object's surface [22].

A similar approach can be realized in digital interferometry methods [23]. Each digital hologram is reconstructed separately using the appropriate wavelength. Reconstructed from each digital hologram, the phase of the object field can be determined as:

$$\varphi_{1,2}(x,y) = \frac{2\pi}{\lambda_{1,2}} h(x,y) \tag{14.9}$$

where $h\,(x, y)$ denotes the height of the surface profile of the object at each point.

The digital interferogram of a surface in this case is calculated as the difference between two phase fields:

$$\varphi_2(x,y) - \varphi_1(x,y) = 2\pi\left(\frac{1}{\lambda_2} - \frac{1}{\lambda_1}\right) h(x,y) = \frac{2\pi}{\Lambda} h(x,y) \tag{14.10}$$

where Λ, a synthetic wavelength, is determined from the ratio:

$$\Lambda = \frac{\lambda_1 \lambda_2}{|\lambda_1 - \lambda_2|} \tag{14.11}$$

As a result, the surface profile will be determined as:

$$h(x,y) = \frac{\Delta\varphi(x,y)\Lambda}{2\pi} \tag{14.12}$$

In this case, the sensitivity of the measurements becomes proportional to the value of $\Lambda/2$ instead of $\lambda/2$. This means that it is possible to set the required sensitivity, depending on the size of the studied relief.

14.3 The Developed Software

14.3.1 The Implementation of Phase-Shifting Interferometry Method

The software for the implementation of the phase-shifting interferometry method was developed in the Labview 13.0. programming environment. It allows receiving digital interferogram with the expression (14.8). In addition, it is possible to carry out simulations of the recording of digital hologram with the ability to set arbitrary phase shift. The test surface relief used during the simulation was formed using Gwyddion software. The components of the micro- and nanorelief of the surface were separately created (see Fig. 14.1).

With the help of simulation, two holograms of the surface were formed using the phase shift $\alpha = 21\ ^\circ$. The corresponding holograms are shown in Fig. 14.2.

After that a digital interferogram was formed (see Fig. 14.3).

In the developed program, it is additionally implemented the ability to define the micro- and nanorelief, as shown in [21] and to determine the basic parameters of roughness. As a result of the program, it was possible to separately recreate the components of the micro- and nanorelief (Fig. 14.4).

Determination of roughness parameters (S_q, S_a, S_{sk}, S_{ku}) was carried out in accordance with the values of the received nanorelief.

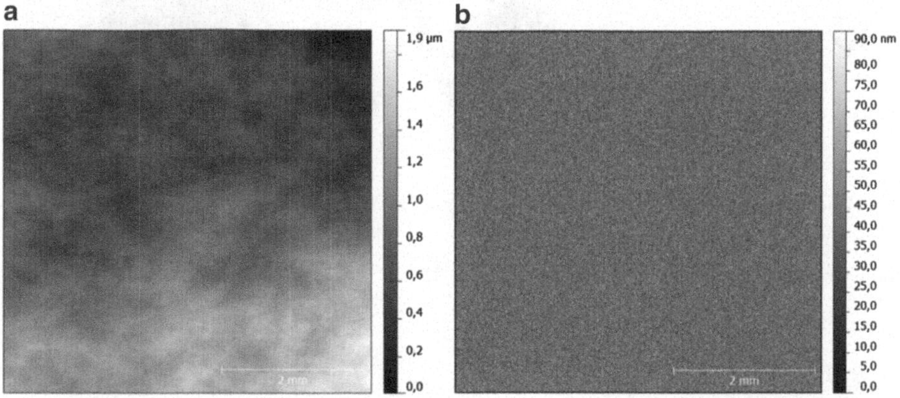

Fig. 14.1 Components of the micro- and nanorelief

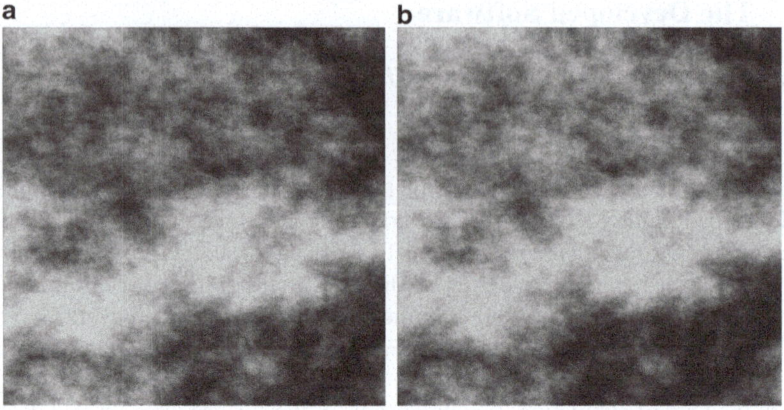

Fig. 14.2 Digital surface hologram recorded with $\alpha = 0°$ (**a**) and $\alpha = 21°$ (**b**)

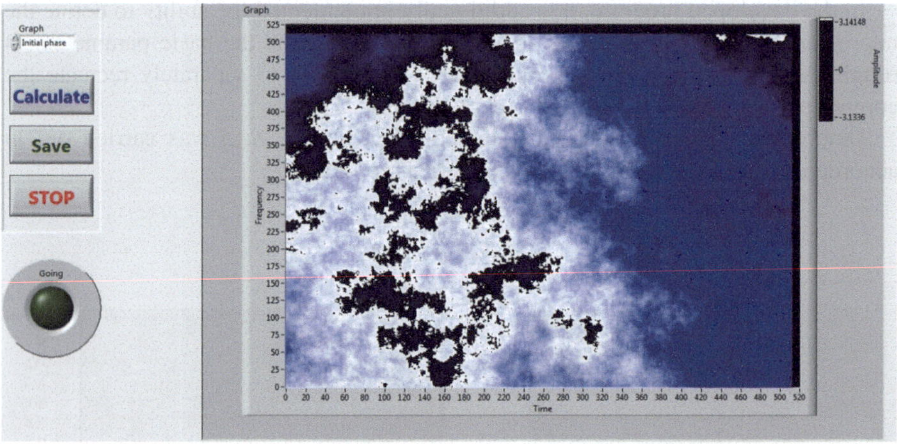

Fig. 14.3 The software interface with formed digital interferogram

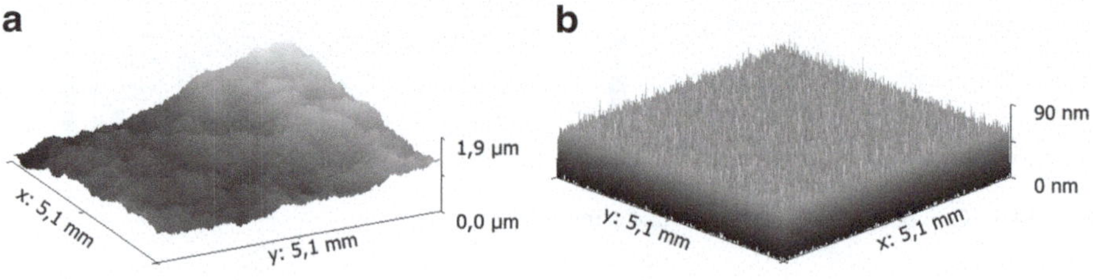

Fig. 14.4 Reconstructed micro- (**a**) and nanorelief (**b**) of the surface

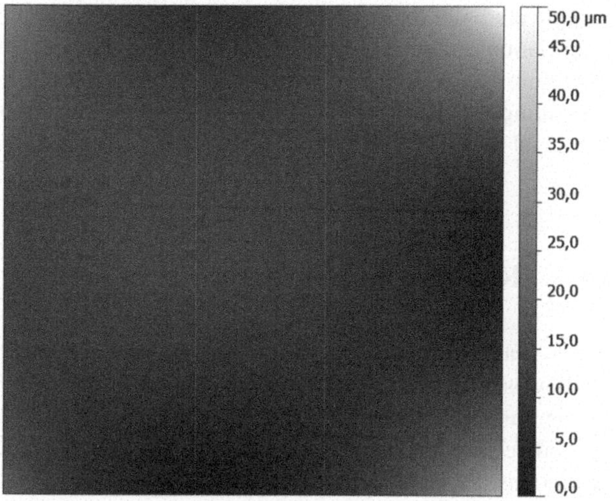

Fig. 14.5 Test surface

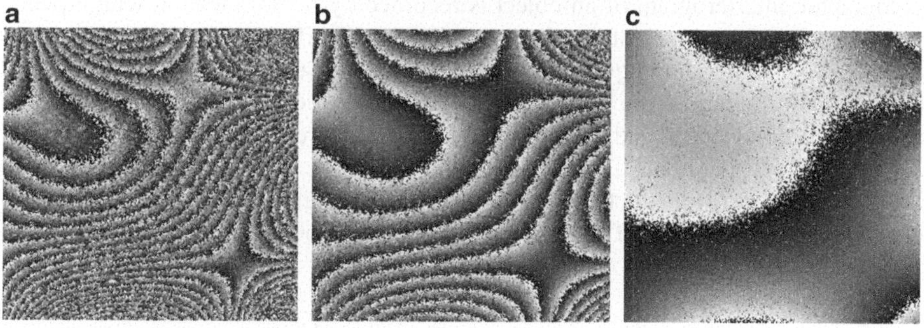

Fig. 14.6 Digital interferogram of the test surface received with $\lambda_1 = 632.8$ nm and $\lambda_2 = 450$ nm (**a**), $\lambda_1 = 632.8$ nm and $\lambda_2 = 532$ nm (**b**), and $\lambda_1 = 632.8$ nm and $\lambda_2 = 650$ nm (**c**)

14.3.2 The Implementation of the Digital Two-Wavelength Interferometry Method

When it comes to study of the surface relief that is much larger than the wavelength value, it is reasonable to use digital two-color interferometry. For the implementation of this method, the software described in [24] was used.

The reconstruction of the test surface relief (Fig. 14.5) was determined using different wavelength pairs.

During the simulation, a series of digital holograms of the test surface was sequentially written using different wavelength pairs. Digital interferograms obtained during the simulation are shown in Fig. 14.6.

Obviously, when the value of $\Delta\lambda$ is too large, the fringe density will be very high. That fact makes the process of spatial filtration [25] and phase unwrapping complicated. On the other hand, when $\Delta\lambda$ is small, some of the useful surface relief information may be lost due to the large distance between the surface levels. Therefore, the optimal wavelengths ratio should be chosen experimentally, depending on the profile value of the studied surfaces. This provides a relatively low level of noise while retaining useful information.

14.4 Results and Discussion

A two-wavelength digital holographic interferometer scheme for the surface quality study was designed (Fig. 14.7). The scheme is based on the Mach-Zehnder interferometer, and here one of the shoulders is the reference (elements 1, 2, 3, 4), and the second (elements 1, 5, 4) is the object. With this scheme, digital-focused image holograms are recorded.

For obtaining the surface relief map with a two-wavelength method, a digital holographic interferogram of an object is recorded by its illumination with a plane wave beam, using two light sources with λ_1 and λ_2 differing in $\Delta\lambda$. Helium-neon laser ($\lambda_1 = 633$ nm) and frequency-doubled Nd-YAG laser ($\lambda_2 = 532$ nm) were used as illumination sources. The beams from both lasers are combined using a 50% beam splitter 3. A condenser 4 and 5 forms the flat object and reference wavefront for both λ with a diameter of about 12 mm.

The image is reconstructed with a single wavelength. Within the illuminated space, there is a system of interference planes perpendicular to the illumination

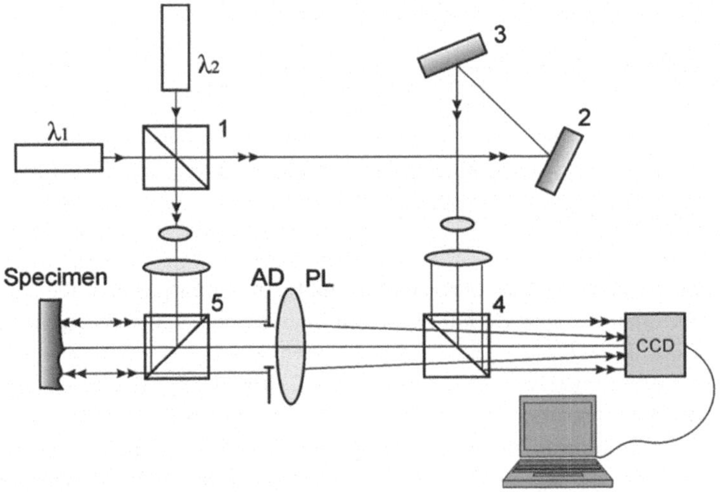

Fig. 14.7 The recording scheme

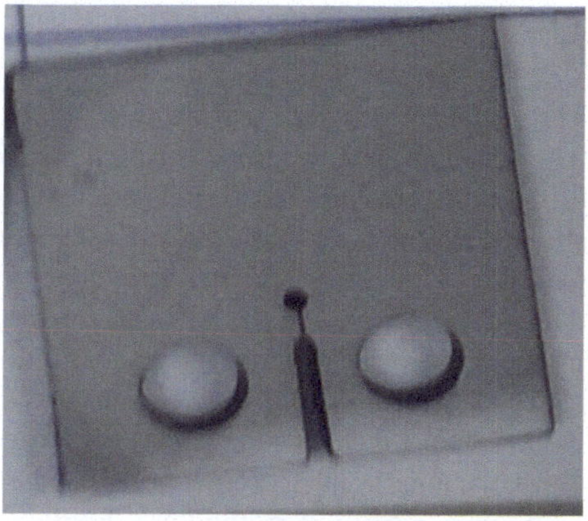

Fig. 14.8 Object of study

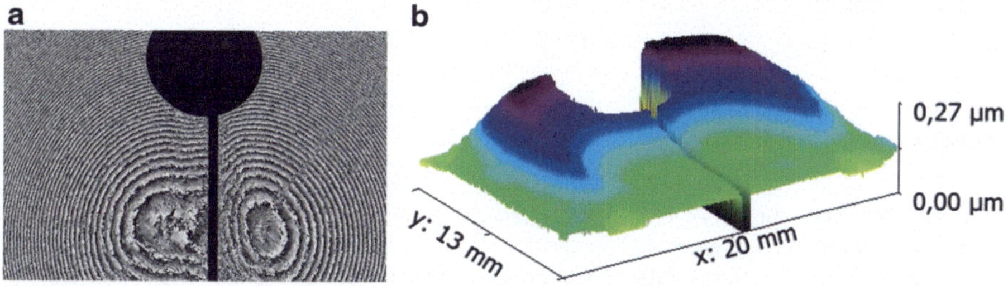

Fig. 14.9 Digital interferogram of the studied area (**a**) reconstructed surface relief (**b**)

direction. Another 50% beam splitter 6 is used for obtaining a normal illumination vector direction toward the studied surface. During the image reconstruction, one can observe a system of interference fringes representing a surface map. The distance between the two surface levels for the following wavelengths is $\lambda_1 \bullet \lambda_2 / 2\Delta\lambda = 1.667$ μm.

An objective with a focal length of 150 mm is used to form a focused image in the recording plane. The required magnification of the studied surface is achieved by selecting the mutual position of the objective and the CCD matrix.

An iris diaphragm is set before the objective and to be the aperture diaphragm. In this experiment, a CCD camera with a photosensitive area of 645 × 485 and a pixel pitch 10 μm was used.

The CT specimen, with size 62.5 × 60 mm, was used as an object of study (Fig. 14.8).

Two digital holograms were sequentially recorded with two wavelengths. The received surface interferogram is shown on Fig. 14.9a.

After that filtering and phase unwrapping were performed with developed software. The reconstructed surface relief is shown on Fig. 14.9 (b).

14.5 Conclusion

The possibilities of using digital phase-shifting interferometry and digital two-wavelength interferometry methods for studying the quality of surfaces with different macro- and nanoreliefs are investigated. The results of the performed tests showed that the phase-shifting digital interferometry method allows to effectively measure the surface micro- and nanorelief and determine the basic roughness parameters. Moreover, the software developed in this work allows separately restoring components of micro- and nanorelief.

The test and experimental results have shown that for the surface relief study, with value much greater than the wavelength, it is advisable to use digital two-wavelength interferometry. If in the experimental scheme two wavelengths with a phase-shifting method are simultaneously used, the whole relief components study will be allowed.

References

1. Lerner YS, Kingsbury GR (1997) Wear resistance properties of austempered ductile iron. J Mater Eng Perform 7(1):48–52
2. Roy DM, Arjunan P, Silsbee MR (2001) Effect of silica fume, metakaolin, and low-calcium fly ash on chemical resistance of concrete. Cem Concr Res 31(12):1809–1813
3. Syed, A. (2004, June). Accumulated creep strain and energy density based thermal fatigue life prediction models for SnAgCu solder joints. In Electronic Components and Technology Conference, 2004. Proceedings. 54th (Vol. 1, pp. 737–746). IEEE
4. Callister WD, Rethwisch DG (2011) Materials science and engineering, vol 5. Wiley, Hoboken
5. Schwider J, Zhou L (1994) Dispersive interferometric profilometer. Opt Lett 19(13):995–997
6. ISO 25178–2:2012, Geometrical product specifications (GPS) — Surface texture: Areal — Terms, definitions and surface texture parameters
7. Malacara Z, Servin M (2016) Interferogram analysis for optical testing, vol 84. CRC Press, Boca Raton
8. Schnars, U., Falldorf, C., Watson, J., & Jüptner, W. (2016). Digital holography and wavefront sensing. Springer-verlag Berlin an
9. Yamaguchi I (2005, January) Image formation and measurement of surface shape and deformation by phase-shifting digital holography. In: Information optics and photonics technology, vol 5642. International Society for Optics and Photonics, Bellingham, pp 66–78
10. Wyant JC, Creath K (1992) Advances in interferometric optical profiling. Int J Mach Tools Manuf 32(1–2):5–10
11. Greivenkamp JE (1984) Generalized data reduction for heterodyne interferometry. Opt Eng 23 (4):234350
12. Deck LL, Soobitsky JA (1999, November) Phase-shifting via wavelength tuning in very large aperture interferometers. In: Proceedings SPIE, vol 3782, No. 58, pp 432–442)
13. Cai LZ, Liu Q, Yang XL (2004) Generalized phase-shifting interferometry with arbitrary unknown phase steps for diffraction objects. Opt Lett 29(2):183–185
14. Osten W, Seebacher S, Baumbach T, Jüptner W (2001, May) Absolute shape control of microcomponents using digital holography and multiwavelength contouring. In: Proceedings SPIE, vol 4275, pp 71–84

15. Tankam P, Song Q, Karray M, Li JC, Desse JM, Picart P (2010) Real-time three-sensitivity measurements based on three-color digital Fresnel holographic interferometry. Opt Lett 35 (12):2055–2057
16. Schnars U, Jüptner WP (2002) Digital recording and numerical reconstruction of holograms. Meas Sci Technol 13(9):R85
17. Yamaguchi I (2006) Phase-shifting digital holography. In Digital holography and three-dimensional display. Springer, Boston, pp 145–171
18. Creath K (1988) V phase-measurement interferometry techniques. Prog Opt 26:349–393
19. Wyant JC (1975) Use of an ac heterodyne lateral shear interferometer with real–time wavefront correction systems. Appl Opt 14(11):2622–2626
20. Li, J., Song, Q., Tankam, P., & Picart, P. (2010, November). Eliminating zero-order diffraction in the digital holography wavefront reconstruction with adjustable magnification. In: Proceedings of SPIE, vol 7848, pp 78481Y-1
21. Muravsky LI, Kmet AB, Voronyak TI (2012) Retrieving the relief of a low-roughness surface using a two-step interferometric method with blind phase shift of a reference wave. Opt Lasers Eng 50(11):1508–1516
22. Ostrovsky YI, Shchepinov VP, Yakovlev VV (1988) Holographic interferometry methods of the deformation measurement. Nauka, Moscow
23. Kühn J, Colomb T, Montfort F, Charrière F, Emery Y, Cuche E, Depeursinge C (2007) Real-time dual-wavelength digital holographic microscopy with a single hologram acquisition. Opt Express 15(12):7231–7242
24. Kotsiuba Y, Petrovska H, Fitio VM, Bulavinets T, Bobitski YV (2016, August) Optimization of the parameters of digital holographic microscope. In: International conference on nanotechnology and nanomaterials. Springer, Cham, pp 231–247
25. Kotsiuba YM, Petrovska HA, Fitio VM, Bobitski YV (2016, September) Improving digital holographic interferogram quality by frequency filtering. In 2016 I.E. 7th International Conference on Advanced optoelectronics and lasers (CAOL), IEEE, pp 67–68

Chapter 15
Spectrum of Electron in Quantum Well with Continuous Position-Dependent Effective Mass

Julia Seti, Mykola Tкach, and Oxana Voitsekhivska

15.1 Introduction

The modern nanotechnologies allow fabrication of the nano-heterostructures with different components of semiconductor materials. The quantum cascade lasers [1, 2] and detectors [2–4] operating in the infrared range are already produced using such multicomponent structures.

The theoretical investigation of these objects in general and nano-devices at their base, in particular, means the solution of Schrodinger equation with the position-dependent effective mass of quasiparticle. The effective mass and momentum operator p do not commute in this approach, since the correct form of kinetic energy operator in Hamiltonian becomes a problem. Thus, for a long time, in theoretical papers [5–7] concerning both 3D and heterostructures, the different ways to place the position-dependent effective mass between operators p have been observed. However, the consistent theory, providing the uniformly correct form of kinetic energy operator, is still absent.

According to the abovementioned, the most realistic form to place the effective mass in the operator of kinetic energy was justified by non-direct methods. For example, in the papers [9–13], it was shown that the results obtained for the heterostructures within the Hamiltonian proposed by BenDaniel-Duke [8] correlated well with the experimental data [9] and the other theoretical models like Kronig-Penney one [10–13]. The existence of the exact solution of Schrodinger equation with position-dependent potential energy and effective mass of quasiparticle depends on their analytical expressions. Thus, in many papers [14–18], the main

J. Seti (✉) · M. Tкach · O. Voitsekhivska
Yuriy Fedkovych Chernivtsi National University, Chernivtsi, Ukraine
e-mail: j.seti@chnu.edu.ua

O. Fesenko, L. Yatsenko (eds.), *Nanooptics, Nanophotonics, Nanostructures, and Their Applications*, Springer Proceedings in Physics 210,
https://doi.org/10.1007/978-3-319-91083-3_15

attention was paid to the development of mathematical methods for the solution of the problems using the well-tested but idealized functions.

At the same time, the theory of nano-heterostructures with position-dependent potentials and quasiparticles effective masses was developed, due to the mathematical problems arising during the search of the exact solution of Schrodinger equation, using the approximated numerical methods [19, 20] or simplified abrupt model [21–24], where effective mass was skip-like at hetero-interface. However, the physical considerations confirm that in near-interface regions the effective mass of quasiparticle differs from that in the contacting materials due to the imperfect heterostructure interfaces caused by nonhomogeneous composition and different lattice parameters. In the first approximation, the effective mass can be taken as linearly dependent of coordinate in this region.

In this paper, we study the influence of near-interface region size, where the effective mass of electron is linearly dependent, on the tunnel properties of double-barrier open nano-heterostructure and on the spectral characteristics of electron sub-barrier states using the model of rectangular potential and continuous position-dependent effective mass.

15.2 Schrodinger Equation and Transmitting Coefficient in the Nano-heterostructure with Position-Dependent Effective Mass of Electron

We study the open plane double-barrier nano-heterostructure using the model of rectangular potential profile (Fig. 15.1a):

$$U(z) = \begin{cases} 0, & 0 \leq |z| \leq z_{\pm 2}, \quad z_{\pm 5} \leq |z| \leq \infty \\ U, & z_{\pm 2} \leq |z| \leq z_{\pm 5}; \end{cases} \tag{15.1}$$

and the model of position-dependent electron effective mass (Fig. 15.1b):

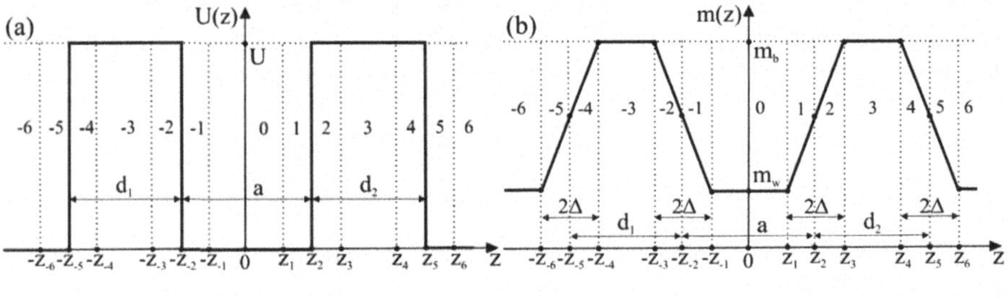

Fig. 15.1 Potential energy (**a**) and electron effective mass (**b**) as functions of z coordinate in nano-heterostructure

$$m(z) = m_w \begin{cases} 1, & 0 \leq |z| \leq z_{\pm 1}, \quad z_{\pm 6} \leq |z| \leq \infty; \\ 1 + \delta m \dfrac{|z| - z_{\pm 1}}{2\Delta}, & z_{\pm 1} \leq |z| \leq z_{\pm 3}; \\ \dfrac{m_b}{m_w} - \delta m \dfrac{|z| - z_{\pm 4}}{2\Delta}, & z_{\pm 4} \leq |z| \leq z_{\pm 6}; \\ \dfrac{m_b}{m_w}, & z_{\pm 3} \leq |z| \leq z_{\pm 4}, \end{cases} \tag{15.2}$$

which is the linear function of z coordinate in the near-interface region between wells and barriers and being constant elsewhere. Herein, U is the height of the potential barrier; a is the width of quantum well; 2Δ is the width of near-interface region; m_w and m_b are the effective masses of quasiparticle in the wells and in the barriers, respectively; and $\delta m = (m_b - m_w)/m_w$.

It is assumed that monoenergetic electronic flux with the energy E impinges at nanostructure from the left, perpendicularly to its interfaces. The one-dimensional Schrodinger equation for the electron is written in the BenDaniel-Duke [8] form:

$$\left[-\frac{\hbar^2}{2} \frac{d}{dz}\left(\frac{1}{m(z)} \frac{d}{dz}\right) + U(z)\right]\Psi(z) = E\Psi(z) \tag{15.3}$$

It is solved using the dimensionless variable $\eta = z/a_0$ (a_0 is the lattice constant of the well material). Introducing the dimensionless magnitudes: coordinates $\eta_{\pm j} = z_{\pm j}/a_0$ ($j = 1$–6), semi-width of near-interface region $d = \Delta/a_0$, the energies $\varepsilon = E/Ry$, $V = U/\mathrm{Ry}$ (Ry, Rydberg constant),

$$U(\eta) = \begin{cases} 0, & 0 \leq |\eta| \leq \eta_{\pm 2}, \eta_{\pm 5} \leq |\eta| \leq \infty \\ V, & \eta_{\pm 2} \leq |\eta| \leq \eta_{\pm 5}; \end{cases} \tag{15.4}$$

the effective mass

$$m(\eta) = \begin{cases} 1, & 0 \leq |\eta| \leq \eta_{\pm 1}, \eta_{\pm 6} \leq |\eta| \leq \infty; \\ 1 + \delta m \dfrac{|\eta| - \eta_{\pm 1}}{2d}, & \eta_{\pm 1} \leq |\eta| \leq \eta_{\pm 3}; \\ \dfrac{m_b}{m_w} - \delta m \dfrac{|\eta| - \eta_{\pm 4}}{2d}, & \eta_{\pm 4} \leq |\eta| \leq \eta_{\pm 6}; \\ \dfrac{m_b}{m_w}, & \eta_{\pm 3} \leq |\eta| \leq \eta_{\pm 4}. \end{cases} \tag{15.5}$$

and wave function $\Psi(\eta)$ by the relationship

$$\begin{aligned} &\Psi(z) = C\Psi(\eta) = C\{\Psi_0(\eta_{-1} \leq |\eta| \leq \eta_1) + \\ &\sum_{p=1}^{6}\left[\Psi_p(\eta_p \leq |\eta| \leq \eta_{p+1}) + \Psi_{-p}\left(\eta_{-p} \leq |\eta| \leq \eta_{-(p+1)}\right)\right]\}, \quad (\eta_{\pm 7} = \infty) \end{aligned} \tag{15.6}$$

instead of Eq. (15.3) we obtain the equation

$$\left[\frac{1}{m(\eta)}\frac{d^2}{d\eta^2}-\frac{1}{m(\eta)^2}\frac{dm(\eta)}{d\eta}\frac{d}{d\eta}-\frac{2a_0^2m_wRy}{\hbar^2}(U(\eta)-\varepsilon)\right]\Psi(\eta)=0. \tag{15.7}$$

Its solution is found together with the conditions of wave function and its density of current continuity at all interfaces of nanostructure [21–24], which are written, taking into account the continuity of effective mass, in the following form:

$$\begin{cases}\Psi_{\pm(p-1)}(\eta_{\pm p})=\Psi_{\pm p}(\eta_{\pm p}), & (p=1-6)\\ \left.\frac{d\Psi_{\pm(p-1)}(\eta)}{d\eta}\right|_{\eta=\eta_{\pm p}}=\left.\frac{d\Psi_{\pm p}(\eta)}{d\eta}\right|_{\eta=\eta_{\pm p}}, & \end{cases} \tag{15.8}$$

In the regions (in Fig. 15.1–0, ±3, ±6) where the electron effective mass is constant, Eq. (15.7) is equivalent to the system of equations

$$\left[\frac{d^2}{d\eta^2}+K_{\pm p}^2\right]\Psi_{\pm p}(\eta)=0,(p=0,3,6). \tag{15.9}$$

where

$$K_{\pm p}=\begin{cases}k=a_0\hbar^{-1}\sqrt{2m_w\,Ry\,\varepsilon}, & p=0,6\\ \chi=a_0\hbar^{-1}\sqrt{2m_b\,Ry\,(\varepsilon-V)}, & p=3\end{cases} \tag{15.10}$$

The solutions of Eq. (15.9) are known [23, 24]:

$$\Psi_{\pm p}(\eta)=A_{\pm p}e^{iK_{\pm p}\eta}+B_{\pm p}e^{-iK_{\pm p}\eta},(p=0,3,6). \tag{15.11}$$

In the near-interface region (in Fig. 15.1 - ±1, ±2, ±4, ±5), where the electron effective mass is linearly dependent on coordinate, Eq. (15.7) is equivalent to the equations

$$\left[\xi_{\pm p}\frac{d^2}{d\xi_{\pm p}^2}-\frac{d}{d\xi_{\pm p}}-\xi_{\pm p}^2\right]\Psi_{\pm p}(\xi_{\pm p})=0,\quad (p=1,2,4,5) \tag{15.12}$$

with new dimensionless variables

$$\xi_{\pm p}(\eta)=2m(\eta)\begin{cases}\left(-\frac{m_wa_0^2d^2Ry\,\varepsilon}{\delta m^2\hbar^2}\right)^{1/3}, & p=1,5\\ \left(\frac{m_wa_0^2d^2Ry\,(V-\varepsilon)}{\delta m^2\hbar^2}\right)^{1/3}, & p=2,4\end{cases} \tag{15.13}$$

The system of Eq. (15.12) also has the exact solutions [25], as linear superposition of the derivatives from Airy functions

$$\Psi_{\pm p}(\xi_{\pm p}) = A_{\pm p}\mathrm{Ai}'(\xi_{\pm p}) + B_{\pm p}\mathrm{Bi}'(\xi_{\pm p}), (p = 1, 2, 4, 5). \tag{15.14}$$

All unknown coefficients $A_{\pm p}$, $B_{\pm p}$ are obtained from the system of fitting conditions (8), in its turn, completely defining the electron wave function $\Psi(z)$.

Now calculating, according to the quantum mechanics [24, 26], the densities of input ($z = -z_{-6}$) and output ($z = z_6$) electronic currents

$$j(z, E) = \frac{i\hbar}{2m_w}\left[\Psi(z, E)\frac{\partial\Psi^*(z, E)}{\partial z} - \Psi^*(z, E)\frac{\partial\Psi(z, E)}{\partial z}\right], \tag{15.15}$$

and taking their ratio, we obtain the transmission coefficient

$$D(E) = \left|\frac{A_6(E)}{A_{-6}(E)}\right|^2. \tag{15.16}$$

Using the function $D(E)$ [27], one can obtain the resonance energies (E_n) and resonance widths (Γ_n) of electron quasi-stationary states in nano-heterostructure and study their dependences on the width of near-interface region between the wells and barriers where the electron effective mass is linear function of coordinate.

15.3 Analysis of the Results

At the base of the developed theory, we calculated the transmission coefficient together with the resonance energies and widths of electron quasi-stationary states in the sub-barrier region of energies ($E_n < U$). For the comparison, we also calculated the influence of the width of near-barrier region on the transmission coefficient and spectral characteristics of electron quasi-stationary states: the resonance energies (E_n^{abr}) and widths (Γ_n^{abr}) in the wide spread model of abrupt effective mass [24].

We observed the symmetric A ($d_1 = d_2$) and asymmetric B ($d_1 \neq d_2$) structures with $In_{0.53}Ga_{0.47}As$ wells and $In_{0.52}Al_{0.48}As$ barriers, being the operating elements of modern quantum cascade lasers [2]. Their physical parameters are known [26]: $m_w = 0.047m_e$, $m_b = 0.089m_e$, U = 516 meV, and geometrical ones are taken typical for the completing materials of devices cascades.

Figure 15.2 shows the transmission coefficient D as function of the energy E calculated in the abrupt mass model ($\Delta = 0$) and in the model of position-dependent electron effective mass at two fixed widths of near-interface region ($\Delta = 0.25\ d_1$, $0.5\ d_1$) in the symmetric A ($d_1 = d_2 = 2$ nm) and asymmetric B ($d_1 = 1.5$ nm, $d_2 = 2.5$ nm) structure with the potential well of the width $a = 5.6$ nm. The widths of the barriers in both cases (A and B) are taken such that their sum width is the same (4 nm).

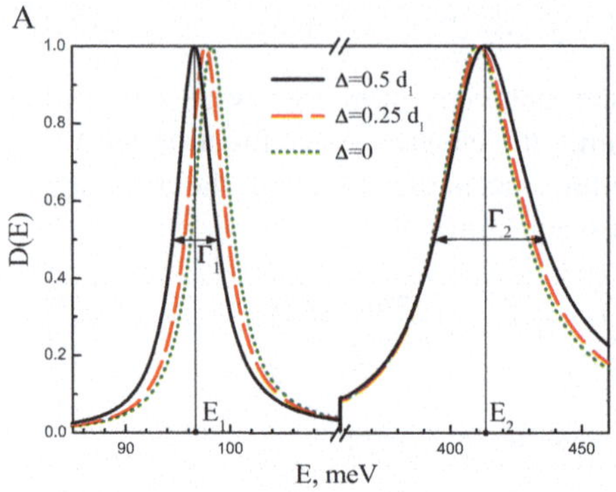

Fig. 15.2 The transmission coefficient D as function of the energy E in the symmetric (**a**) and asymmetric (**b**) structures calculated at two fixed widths of near-interface region ($\Delta = 0.25d_1$, $0.5d_1$) and in the abrupt effective mass model ($\Delta = 0$)

Figure proves that at any Δ in both structures one can see the Lorentz-like peaks at the functions $D(E)$. Their position, according to the general theory [27], determines the resonance energies (E_n) of electron quasi-stationary states and their widths at the half of the height – the resonance widths (Γ_n) of these states. The transmission coefficient of the symmetric structure A is maximal, $D(E_n) \approx 1$, in the small vicinity of resonance energies E_n. The asymmetrical widths of the barriers in the structure B bring to the decreases of transmission in the vicinity of E_n, herein $D(E_1) < D(E_2) < 1$. The bigger widths of near-interface region (Δ) in both structures A and B cause the increasing distance between two sub-barrier peaks, shifting them into the opposite directions and broaden them a little. However, it does not change the magnitudes of the maxima of $D(E_n)$ peaks and weakly deforms their Lorentz-like shape. Such behavior of the transmission peaks functions of Δ determines the corresponding dependences of spectral parameters (E_n, Γ_n) of electron quasi-stationary states.

In Fig. 15.3 , the resonance energies E_n and widths Γ_n of electron sub-barrier states are shown as functions of Δ for the symmetric A and asymmetric B structure. The resonance energies (E_n^{abr}) and widths (Γ_n^{abr}) calculated in the abrupt effective mass model are presented too, for the comparison. It is clear that the asymmetric widths of the barriers cause the small shift of E_1 into the low-energy region while E_2 shifts into the high-energy one. Also, the resonance widths (Γ_1, Γ_2) of the both states - increase. The functions $E_n(\Delta)$ and $\Gamma_n(\Delta)$ are qualitatively similar for the both structures (A and B).

The resonance energies of the first ($n = 1$) and second ($n = 2$) states of electron in A and B structures differently depend on Δ. The energy E_1 only decreases while E_2 - increases. The resonance widths Γ_n of the both states increase at bigger Δ and the respective life times ($\tau_n = \hbar/\Gamma_n$) decrease due to the smaller "power" of the potential barriers. At $\Delta \to 0$, as it should be, the resonance energies E_n and widths Γ_n, calculated within the model of continuous position-dependent effective mass, tend to the respective resonance energies (E_n^{abr}) and widths (Γ_n^{abr}) obtained in abrupt mass model.

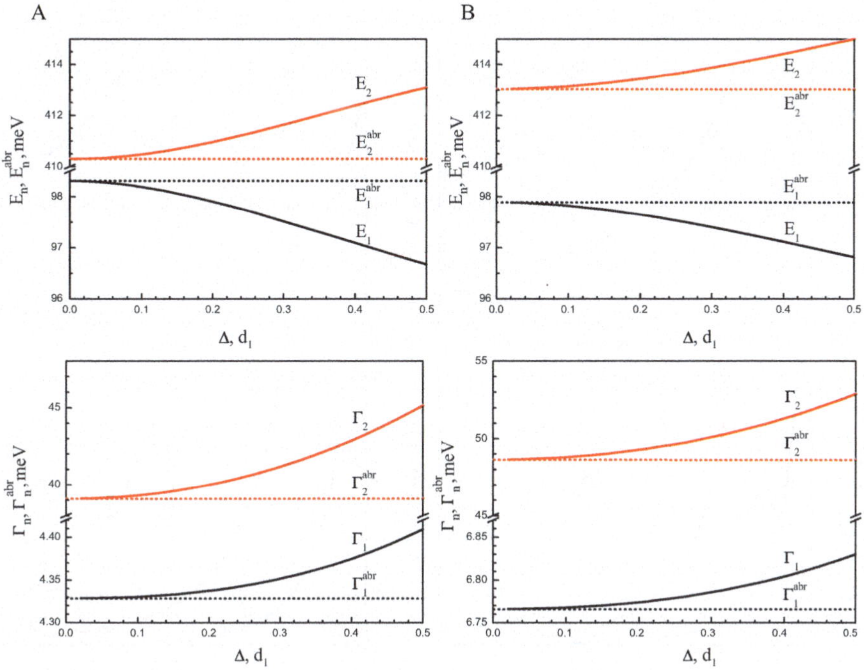

Fig. 15.3 Dependences of resonance energies E_n, E_n^{abr} and widths Γ_n, Γ_n^{abr} on the width of near-interface region for the symmetric A and asymmetric B structure at $a = 5.6$nm

The resonance energies E_n, E_n^{abr} and widths Γ_n, Γ_n^{abr} of sub-barrier states as functions of well width a for the symmetric A and asymmetric B structure are presented in Fig. 15.4 at $\Delta = 0.1d_1$, $0.25d_1$, $0.5d_1$, and $\Delta = 0$, respectively.

It is clear that when the width of the well (a) increases, the resonance energies and widths of electron states in the structures A and B shift into the low-energy region in the both models of effective mass and the number of levels in the sub-barrier region becomes bigger. In the scale of Fig. 15.4, the differences between the spectral characteristics calculated in the abrupt model and position-dependent one are almost not visible. Thus, in Fig. 15.5, one can see the differences of resonance energies $\Delta E_n = E_n - E_n^{abr}$ and widths $\Delta\Gamma_n = \Gamma_n - \Gamma_n^{abr}$ as functions of well width a for the structures A and B.

Figure 15.5 proves that in both structures the differences of energies ΔE_n as functions of a for all quasi-stationary states (n) are the complicated sign-alternating functions. Such behavior of ΔE_n(a) is caused by the corresponding functions of resonance energies: $E_n(a)$ and $E_n^{abr}(a)$. Really, when the well width is such that the certain n-th level with the energy E_n only begins to involve into the sub-barrier

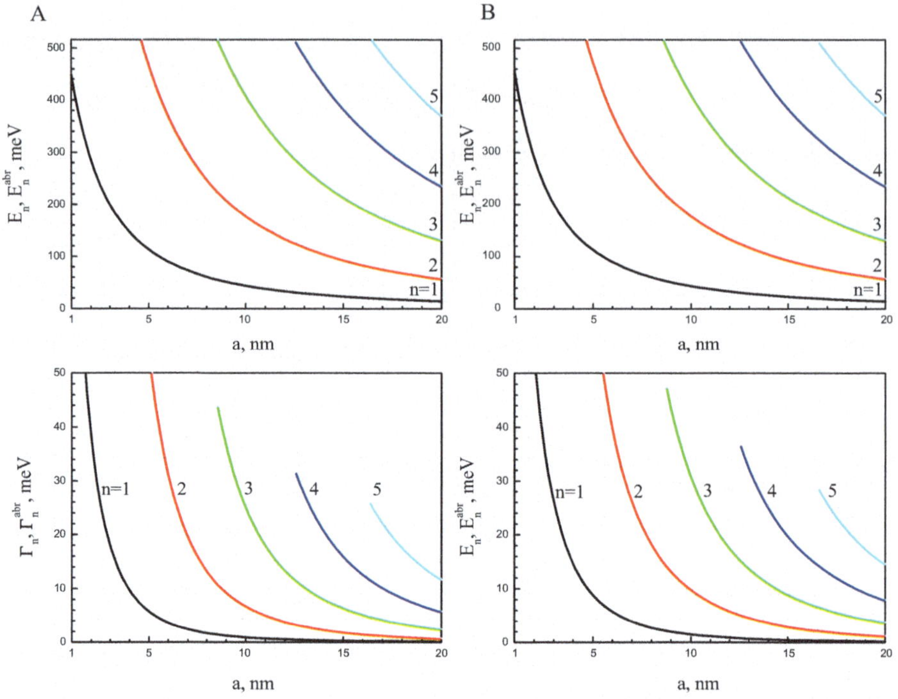

Fig. 15.4 The resonance energies E_n, E_n^{abr} and widths Γ_n, Γ_n^{abr} of sub-barrier electron states as functions of well width a for the symmetric A and asymmetric B structure at $\Delta = 0.1d_1$, $0.25d_1$, $0.5d_1$, and $\Delta = 0$, respectively

region ($E_n < U$), that of the energy E_n^{abr} (in the abrupt model) is already involved into the potential well at certain magnitude, as a result, $\Delta E_n > 0$. The increasing well width, as it is clear from Fig. 15.4, though, only shifts E_n and E_n^{abr} into the low-energy region, but the energy E_n^{abr} decays slower than E_n. Thus, at certain a, the both energies become equal and $\Delta E_n = 0$. At further increasing of well width, the difference ΔE_n becomes negative, reaches its minimum, and tends to zero at $a \to \infty$. Figure 15.5 also shows that the differences between the resonance widths $\Delta \Gamma_n$ as functions of a are sharply decaying functions in both structures A and B.

So, it is clear that the biggest difference between the energies E_n, E_n^{abr} and widths Γ_n, Γ_n^{abr} in the both models for the electron effective mass is observed at such width of the well when the n-th level with E_n only appears in the sub-barrier region of the potential well. The calculations prove that at $\Delta = 0.5d_1$ for the arbitrary n-th quasi-stationary state of electron, the relative maximal difference of energies $\Delta E_n / E_n$ does not exceed 1% and that of widths $\Delta \Gamma_n / \Gamma_n$ - 12%.

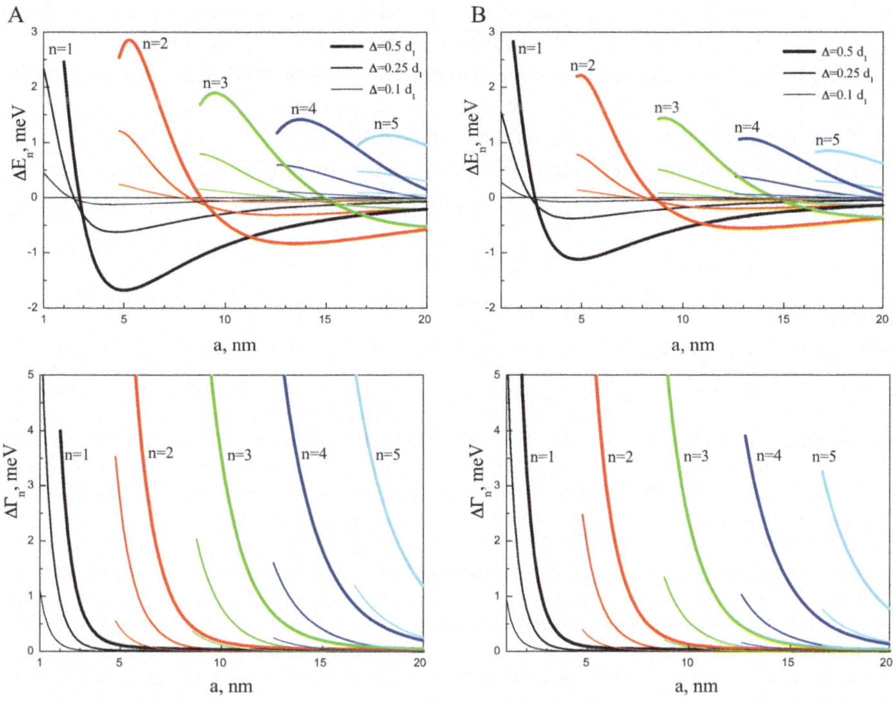

Fig. 15.5 The differences of resonance energies $\Delta E_n = E_n - E_n^{abr}$ and widths $\Delta\Gamma_n = \Gamma_n - \Gamma_n^{abr}$ of electron sub-barrier states as functions of well width a for the symmetric A and asymmetric structure B at $\Delta = 0.1\text{d}_1, 0.25\text{d}_1, 0.5\text{d}_1$

15.4 Conclusions

1. The exact solutions of Schrodinger equation for the electron are obtained using the approximation of the rectangular potential profile and position-dependent effective mass, which is the linear function of coordinate in the near-interface region between the wells and barriers of the open plane double-barrier nanostructure.
2. Using the exact set of wave functions, the transmission coefficient, resonance energies and widths of sub-barrier electron quasi-stationary states are calculated for the structure with $In_{0.53}Ga_{0.47}As$ wells and $In_{0.52}Al_{0.48}As$ barriers. It is shown that the increasing near-interface region changes the resonance energies and widths of electron states but does not effect on the transmission coefficient respectively to the monoenergetic electronic flux.

3. It is revealed that the difference of energies ΔE_n of arbitrary n-th level calculated as functions of quantum well width within the models of position-dependent and abrupt effective mass are the sign-alternating strongly nonlinear functions while the differences of widths $\Delta \Gamma_n$ are positive and sharply decaying ones.
4. It is established that when the width of near-interface region is of the order of the half of the smallest potentials barrier, the relative differences of energies $\Delta E_n / E_n$ and widths $\Delta \Gamma_n / \Gamma_n$ do not exceed 1 and 12%, respectively.

References

1. Chastanet D, Bousseksou A, Lollia G, Bahriz M, Julien FH, Baranov AN, Teissier R, Colombelli R (2014) High temperature, single mode, long infrared (λ = 17.8 μm) InAs-based quantum cascade lasers. Appl Phys Lett 105:111118. https://doi.org/10.1063/1.4895763
2. Schwarz B, Ristanic D, Reininger P, Zederbauer T, MacFarland D, Detz H, Maxwell Andrews A, Schrenk W, Strasser G (2015) High performance bi-functional quantum cascade laser and detector. Appl Phys Lett 107:071104. https://doi.org/10.1063/1.4927851
3. Sakr S, Crozat P, Gacemi D, Kotsar Y, Pesach A, Quach P, Isac N, Tchernycheva M, Vivien L, Bahir G, Monroy E (2013) F. H. Julien. GaN/AlGaN waveguide quantum cascade photodetectors at 1.55m with enhanced responsivity and 40GHz frequency bandwidth. Appl Phys Lett 102:011135. https://doi.org/10.1063/1.4775374
4. Reininger P, Zederbauer T, Schwarz B, Detz H, MacFarland D, Maxwell Andrews A, Schrenk W, Strasser G (2015) InAs/AlAsSb based quantum cascade detector. Appl Phys Lett 107:081107. https://doi.org/10.1063/1.4929501
5. Bastard G (1981) Superlattice band structure in the envelope-function approximation. Phys Rev B 24:5693. https://doi.org/10.1103/PhysRevB.24.5693
6. Zhu Q-G, Kroemer H (1983) Interface connection rules for effective-mass wave functions at an abrupt heterojunction between two different semiconductors. Phys Rev B 27:3519. https://doi.org/10.1103/PhysRevB.27.3519
7. von Roos O (1983) Position-dependent effective masses in semiconductor theory. Phys Rev B 27:7547. https://doi.org/10.1103/PhysRevB.27.7547
8. BenDaniel DJ, Duke CB (1966) Space-charge effects on electron tunneling. Phys Rev 152:683. https://doi.org/10.1103/PhysRev.152.683
9. Galbraith G (1988) Duggan. Envelope-function matching conditions for GaAs/(Al, Ga)As heterojunctions. Phys Rev B 38:10057. https://doi.org/10.1103/PhysRevB.38.10057
10. Einevoll GT (1988) P C Hemmer. The effective-mass Hamiltonian for abrupt heterostructures. J Phys C Solid State Phys 21:L1193. https://doi.org/10.1088/0022-3719/21/36/001
11. Einevoll GT, Hemmer PC, Thomsen J (1990) Operator ordering in effective-mass theory for heterostructures. I. Comparison with exact results for superlattices, quantum wells, and localized potentials. Phys Rev B 42:3485. https://doi.org/10.1103/PhysRevB.42.3485
12. Einevoll GT (1990) Operator ordering in effective-mass theory for heterostructures. II. Strained systems. Phys Rev B 42:3497. https://doi.org/10.1103/PhysRevB.42.3497
13. Einevoll GT, Hemmer PC (1991) Superlattice minibands-explicit formulae for band gaps and effective masses. Semicond Sci Technol 6:590. https://doi.org/10.1088/0268-1242/6/7/004
14. Roy B, Roy P (2002) A Lie algebraic approach to effective mass Schrödinger equations. J Phys A Math Gen 35(3961). https://doi.org/10.1088/0305-4470/35/17/310
15. Gonul B, Ozer O, Gonul B, Uzgun F (2002) Exact solutions of effective-mass Schrodinger equations. Modern Phys Lett A 17:2453. https://doi.org/10.1142/S0217732302008514
16. Koc R, Koca M, Sahinoglu G (2005) Scattering in abrupt heterostructures using a position dependent mass Hamiltonian. Eur Phys J B 48:583. https://doi.org/10.1140/epjb/e2005-00422-x

17. Ganguly A, Kuru S, Negro J, Nieto LM (2006) A study of the bound states for square potential wells with position-dependent mass. Phys Lett A 360:228. https://doi.org/10.1016/j.physleta.2006.08.032
18. Hassanabadi H, Chung WS, Zare S, Alimohammadi M (2017) Scattering of position-dependent mass Schrodinger equation with delta potential. Eur Phys J Plus 132:135. https://doi.org/10.1140/epjp/i2017-11422-0
19. Wang H, Farias GA, Freire VN (1999) Graded interface effects on the carriers confinement in single GaN/$Al_xGa_{1-x}N$ wurtzite quantum wells. Solid State Commun 110:587. https://doi.org/10.1016/S0038-1098(99)00112-X
20. Lima FMS, Enders BG, Fonseca ALA, Nunes OAC, Freire VN, Freire JAK, Farias GA, da Silva EF, Jr. (2004) Effect of non-abrupt doping profiles on the carrier sheet density in one-side modulation-doped GaAs/AlGaAs quantum wells. Phys Status Solidi C 1:S2–S215. https://doi.org/10.1002/pssc.200405143
21. Tkach MV, Seti JO, Boyko IV (2013) O.M.Voitsekhivska. Optimization of quantum cascade laser operation by geometric design of cascade active band in open and closed models. Condens Matter Phys 16:33701. https://doi.org/10.5488/CMP.16.33701
22. Tkach MV, Seti JO, Boyko IV, Voitsekhivska OM (2013) Dynamic conductivity of resonance tunnel structures in the models of open cascades in nanolasers. Rom Rep Phys 65:1443
23. Jirauschek C, Kubis T (2014) Modeling techniques for quantum cascade lasers. Appl Phys Rev **1**:011307. https://doi.org/10.1063/1.4863665
24. P. Harrison, A. Valavanis. Quantum wells, wires and dots: theoretical and computational physics of semiconductor nanostructures, 4th edn. (Wiley, Hoboken/Chichester, 2016).
25. Abramowitz M, Stegun IA (1964) Handbook of mathematical functions with formulas, graphs, and mathematical tables. Courier Corporation, New York
26. Davies JH (1998) The physics of low-dimensional semiconductors. Cambridge University Press, Cambridge
27. Tkach NV, Seti YA (2009) Evolution and collapse of quasistationary states of an electron in planar symmetric three-barrier resonance-tunneling structures. Low Temp Phys 35:556. https://doi.org/10.1063/1.3170931

Chapter 16
Design and Synthesis of Silver Nanoparticles with Different Shapes Under the Influence of Photon Flows

T. Bulavinets, V. Varyshchuk, I. Yaremchuk, and Ya Bobitski

16.1 Introduction

Recently noble metal nanoparticles are actively investigated and attract a lot of attention from researchers. Modern achievements in nanomaterial synthesis and their unique features are of great interest and are widely used in almost all areas of human activity. Noble metal nanoparticles are actively used for solving problems of electronics, energy, photonics, biomedicine, and ecology due to their interesting physical, chemical, biological, catalytic, and bactericidal properties [1–3].

There are many modern methods of silver nanoparticle synthesis, particularly chemical reduction, ablation, photochemical methods, microwave processing, and different biological synthetic methods [4, 5]. However, the problem of the controlled process of nanoparticle formation and improvement of the stabilization methods of their colloids were studied not enough. The general approach to size control is based on testing different parameters via attempts and errors. Real nanoparticle synthesis mechanisms and the influence of synthesis parameters on the formation process remain unstudied. Also, the actual tasks of modern physical chemistry are improving conditions and nanoparticle synthesis methods and stabilizing their colloids.

One of the perspective methods of synthesis is a photostimulated recovery of silver ions from a silver salt solution ($AgNO_3$) [6–8]. The ability to use light for the controlled formation and synthesis of nanoparticles is very perspective. Light can significantly affect on the synthesis of nanoparticles. Using photon flows is an

---

T. Bulavinets (✉) · V. Varyshchuk · I. Yaremchuk
Lviv Polytechnic National University, Lviv, Ukraine
e-mail: tetiana.o.bulavinets@lpnu.ua

Y. Bobitski
Lviv Polytechnic National University, Lviv, Ukraine

Faculty of Mathematics and Natural Sciences, University of Rzeszow, Rzeszow, Poland

O. Fesenko, L. Yatsenko (eds.), *Nanooptics, Nanophotonics, Nanostructures, and Their Applications*, Springer Proceedings in Physics 210,
https://doi.org/10.1007/978-3-319-91083-3_16

effective way to change the nanoparticle shape, since nanoparticles have the ability to absorb light with a certain wavelength [9, 10]. Photostimulated recovery has several advantages in comparison with other chemical methods. This method gives a possibility of synthesizing nanoparticles at low temperatures in liquid and solid environments. The resulting colloidal solutions also differ by higher purity of the synthesized nanoparticles and don't need cooling.

This work presents the results of research process of formation and synthesis of silver nanoparticles using the photostimulated recovery method. Various substances are used to recover nanoparticles from the salt solution, including sodium citrate [11, 12], sodium borohydride $NaBH_4$ [13], hydroquinone $C_6H_6O_2$ [14], and others. The crystalline sodium citrate ($Na_3C_6H_5O_7$) was used to recover Ag by anions of citrate acid. It is also used as a stabilizer due to adsorption on the surface of silver nanoparticles and the creation of an electrostatic barrier that prevents aggregation of silver nanoparticles. Since sodium citrate refers to organic substances, it is not toxic and used in the food industry, in the production of pharmaceutical preparations and for synthesis of biocompatible stable colloids of silver. This opens up new perspectives of using such nanomaterials in medicine as components of systems of diagnosis, prevention, and therapy [15–17].

16.2 Theoretical Model

It is well known that the surface plasmon resonance in metallic nanoparticles is very sensitive to the size of the nanoparticle, its shape, and properties of the environment. Thus, the optical properties of nanoparticles can easily be adjusted by changing their shape and size [18–20]. In order to fully use this phenomenon in practical tasks, it is important to understand well the interaction of light with metallic nanoparticles on a very small scale. For solving this problem, theoretical methods based on the Mie theory and the discrete dipole approximation (DDA) are widely used to describe the optical properties of nanoparticles of different sizes, shapes, and composition [21]. Exact analytical solutions to the problems of scattering and absorption of electromagnetic waves based on the theory of Mie are known only for simple geometric forms [22]. For theoretical study of objects with complex form, it is often necessary to apply approximate methods, one of which is the DDA method, which is a flexible and powerful tool for determining the interaction of electromagnetic radiation with nanoparticles and calculating the scattering and absorption of radiation by nanoparticles with arbitrary forms. The basis of the method is that the nanoparticle is discretized into an array of dipoles or polarizable points. After that this method solves the Maxwell equation for the incident electromagnetic wave that interacts with each of the dipoles separately. The resulting solutions then can be used to calculate the scattering and absorption properties of an array of dipoles and hence nanoparticle. DDA method relies on the assumption that the dielectric properties of nanoparticle, as well as its interaction with the incident electromagnetic field, are directly related to the polarization of the arrays of dipoles [23] (Fig. 16.1).

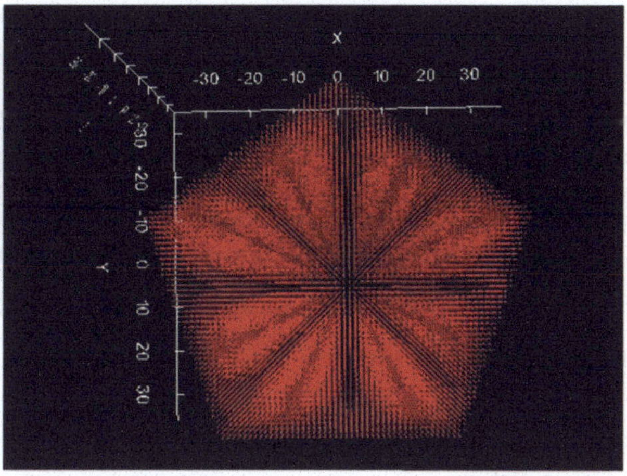

Fig. 16.1 The image of decahedra splitted into an array of dipoles, each one has 1 nm size

The software implementation of the DDA method, presented in free access, is called DDSCAT [24]. This program allows calculating the absorption and scattering spectra of nanoparticles with arbitrary shapes and sizes and provides initial data of calculations in the form of radiation absorption and scattering spectra at different wavelengths of incident light [19]. The corresponding quantitative characteristics that are calculated when using DDSCAT are the absorption efficiency factor Q_{abs} and the scattering efficiency factor Q_{sca}, which are calculated, respectively, on the basis of expressions [22]:

$$Q_{abs} = C_{abs}/\pi a_{eff}^2 \tag{16.1}$$

$$Q_{sca} = C_{sca}/\pi a_{eff}^2, \tag{16.2}$$

where C_{abs} and C_{sca} are the absorption and scattering cross sections for an object, respectively, and aeff is the effective radius, which is calculated as $a_{\mathrm{eff}} = ((3\ V)/(4\pi))^{1/3}$, where V is the volume of the object.

A group of researchers from the University of Illinois has developed software that implements a simple and intuitive interface for the DDSCAT code. It provides experimental researchers, who are not familiar with computational methods, the ability to use the visual user interface for modeling the interaction of radiation with nanoparticles of arbitrary shape [25].

In this work, a simulation of the interaction of radiation with metallic nanoparticles of various forms is carried out. In particular, the properties of silver nanoparticles of spherical, decahedral, and triangular prism shapes are considered. All presented results of the simulation of optical properties of metallic nanoparticles of arbitrary geometry and size are obtained using DDSCAT 7.3. As mentioned above, DDSCAT uses an array of polarized points to approximate the geometric shape of an object. In the case of a decahedra and prism, its shapes are not available in the standard implementation of the program it was set and imported into the DDSCAT (Fig. 16.2).

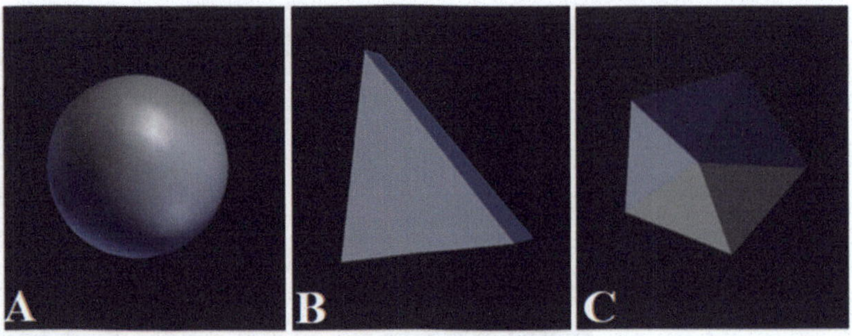

Fig. 16.2 Simulated 3D shapes of spherical (**a**), prismatic (**b**), and decahedral (**c**) nanoparticles

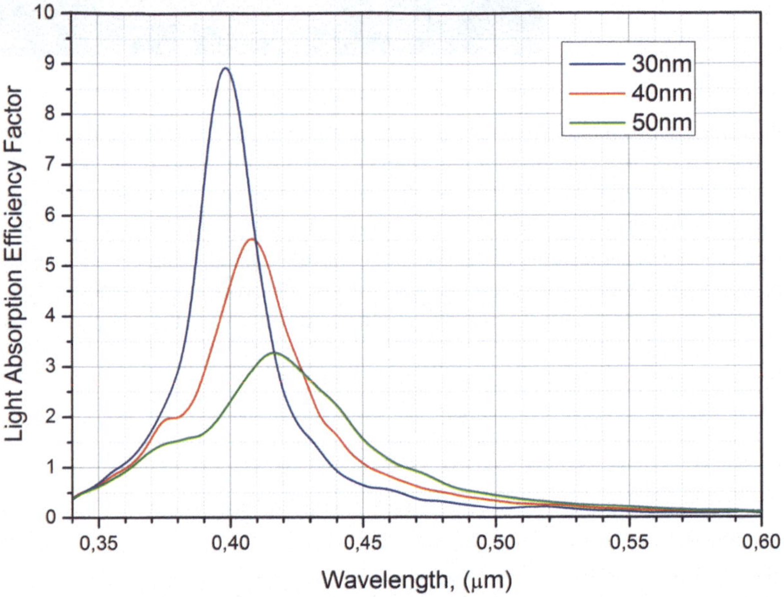

Fig. 16.3 Absorption spectra of spherical silver nanoparticles with different diameters

In order to illustrate the effect of size on the optical properties of spherical silver nanoparticles, absorption spectra for nanoparticles with different diameters were simulated. The simulation results for silver nanospheres with diameters 30 nm, 40 nm, and 50 nm are depicted in Fig. 16.3. It is noticeable that smaller nanospheres mainly absorb light, while larger spheres demonstrate increased scattering and have more extended peaks slightly shifted to a longer wavelength region.

For the theoretical prediction of the size of nanoparticles obtained experimentally, the simulation of absorption spectra for spherical, prismatic, and decahedral nanoparticles was carried out. Appropriate sizes were selected in accordance with the optimal absorption wavelength for each of the shapes obtained experimentally. It should be noted that for silver decahedra, a strong dependence of the position of the absorption peak of the nanoparticle and its intensity on the orientation of the incident electromagnetic wave can be observed.

16.3 Materials and Methods

Silver nitrate ($AgNO_3$) (Sfera Sim, UA) was used as a source for ion recovery of Ag + to Ag^0. Sodium citrate ($Na_3C_6H_5O_7$) (Sfera Sim, UA) was additionally used for acceleration of the recovery reaction. Citric acid anions also act as a stabilizer due to their adsorption on the surface of formed silver nanoparticles and create an electrostatic barrier that prevents nanoparticles from aggregating with each other [26]. The use of sodium citrate in the nanoparticle synthesis also allows obtaining stable biocompatible silver colloids. These materials were used as initial source without additional filtering. All solutions of reacting materials were prepared in distilled water.

Synthesis of silver colloids was carried out by photostimulated recovery of silver ions from a solution of silver salt. In this case, the nanoparticle formation occurs due to irradiation of the silver salt solution by visible light to excite plasmon oscillations and restore Ag + using citrate. Thus, the forming process of silver nanoparticles can be represented as a sequence of stages: the recovery of Ag + → accumulation of critical concentrations of metal atoms → the formation of particle germs → growth (Fig. 16.4).

The initial assessment of formed silver nanoparticles and their geometric parameters is carried out by determining the color of colloid. Ordinary silver has a light gray color, and the color of nanosilver solution, depending on the nanoparticle size, changes from yellow to reddish (Fig. 16.5, used from open source from the Internet).

During the experiment, 10 ml of a solution of $AgNO_3$ (4 mM) was irradiated by semiconductor laser with a wavelength $\lambda = 445$ nm and a power of 1 W. To this solution was dropwise added 50 ml of sodium citrate solution (4 mM). During the photorecovery process, the solution was continuously mixed. The investigated solution changed the color under the influence of laser irradiation from colorless to yellow, indicating the formation of spherical silver nanoparticles in the solution [27].

The photon flows were used for research of silver nanoparticle growth process and in changing their geometrical parameters. Obtained silver colloids were exposed to irradiation by power RGB diodes with a power of 1 W under the same conditions during 120 h.

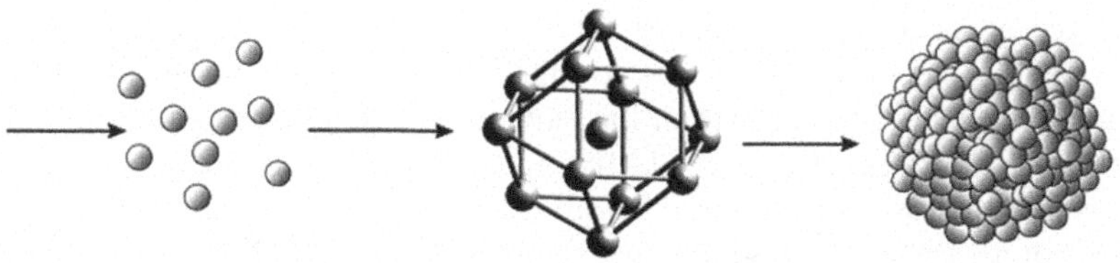

Fig. 16.4 Formation stages of silver nanoparticles

Fig. 16.5 Color dependence on the size of silver nanoparticles (size increases from left to right)

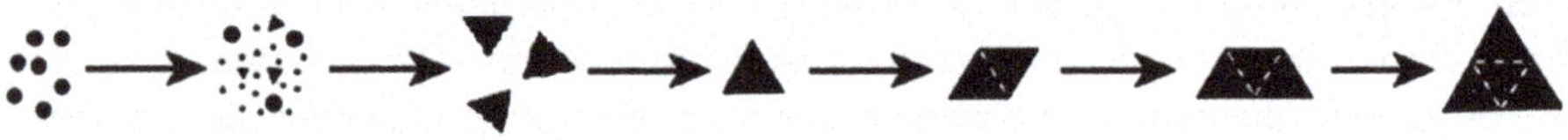

Fig. 16.6 Nanoprism growth process

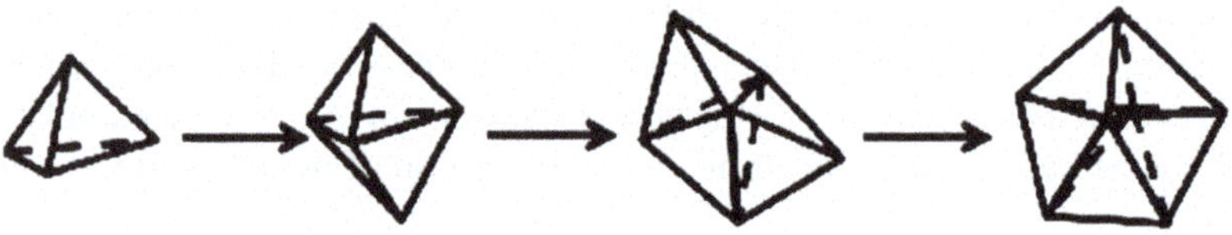

Fig. 16.7 Mechanism of decahedron growth

Since nanoparticles have the ability to absorb light with a certain wavelength, it is possible to obtain nanoparticles of different shapes by lighting them with separate wavelengths. The choice of the wavelength of excitation for a photostimulated growth process can greatly affect the size and morphology of the obtained nanoparticles. Triangular nanoparticles are usually obtained by irradiating of colloidal spherical nanoparticles by green or red light. Decahedra structures are usually formed under the influence of blue light [28].

The change of nanoparticle shape is due to the rocking of the plasmon to amplitude, commensurate with the light wavelength, which creates a temporary dipole, causing the addition of free ions Ag + to the nanosphere. Thus, nanoparticles of various shapes (decahedron, nanoprisms) are gradually formed. The growth process stops when the possibility of forming a temporary dipole under the action of blue (red or green) light is exhausted [10].

Changes of the initial Ag seeds to small clusters with the small nanoprism formation and nanoprism growth at the expense of this clusters occur only due to a photoinduced process (Fig. 16.6) [29].

Silver decahedron also evolves from silver seeds under light influence. Formation of decahedral structure might be via symmetric growth of the twinning seeds (Fig. 16.7).

16.4 Results and Discussion

Measurement of the absorption coefficient obtained by photostimulated recovery silver colloids was conducted using the spectrophotometer UV-3600. The position of the plasmon resonance peak on the spectral range also was determined (Fig. 16.8).

The absorption peak position indicates that the average size of formed nanoparticles is about 47 nm. Figure 16.8 shows that the obtained experimental result correlates well with the theoretically calculated absorption curve of silver nanoparticles. The absorption peak of Ag nanoparticles is in wavelength range 380–440 nm. This is typical for spherical nanosilver.

The next step was to study the influence of electromagnetic radiation on changing the geometric parameters of silver nanoparticles. For this, obtained silver colloids were irradiated by power RGB diodes. The color of the irradiated samples visually differs from the initial solution (Fig. 16.9).

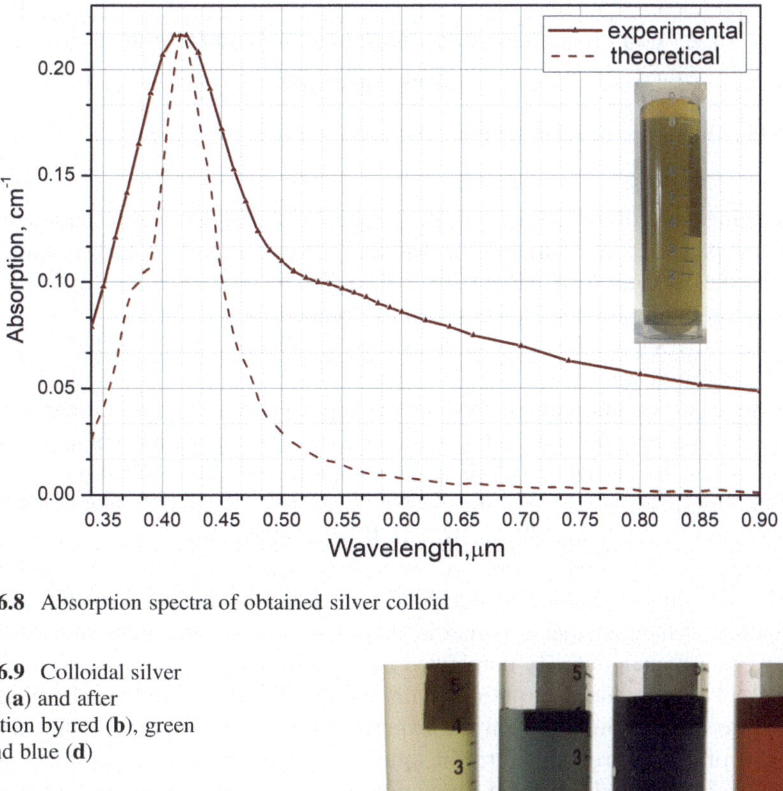

Fig. 16.8 Absorption spectra of obtained silver colloid

Fig. 16.9 Colloidal silver before (**a**) and after irradiation by red (**b**), green (**b**), and blue (**d**)

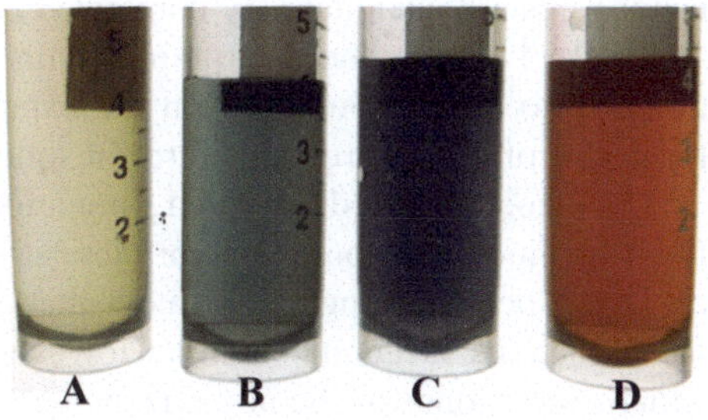

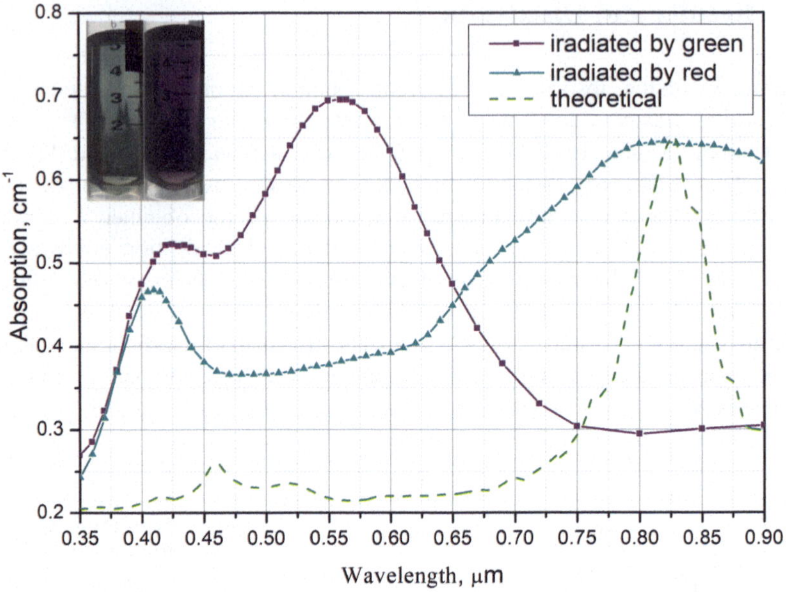

Fig. 16.10 Absorption spectra of silver colloid irradiated by red and green

Changing the colloid color indicates a change in the geometric parameters of silver nanoparticles. A change in the samples color from yellow to green and blue indicates the formation of nanoprism with different size. A color changing from yellow to orange indicates the formation of decahedron [29, 30].

Figure 16.10 shows the absorption spectra of colloidal silver, irradiated by green and red photon flows.

On the experimental curves, besides the typical absorption peak near 420 nm, additional peaks near 560 nm and 825 nm appeared. The appearance of an additional peak approves the formation of prismatic nanoparticles in the colloid. For example, for absorption peak on 825 nm, the average size of the obtained nanoprisms by the DDA method was calculated (Fig. 16.10). It is about 76 nm.

Intensity of appeared additional absorption peaks (Fig. 16.10) significantly exceeds the initial peak on $\lambda = 420$ nm. This indicates that most of the spherical nanoparticles changed their geometric parameters under the light influence and nanoprisms are quantitatively dominated in the solution.

Fig. 16.11 shows the absorption spectra of colloidal silver, irradiated by blue light flows. Two absorption peaks on the experimental curve (Fig. 16.11) indicate the formation of decahedron structures in solution with spherical nanoparticles.

The theoretical prediction and simulation of size and absorption peak position by DDA method were conducted for these structures. Second absorption peak position on the spectral range indicates that average size of formed decahedron nanostructures is about 64 nm. However, the intensity of the absorption peak, which corresponds to the initial ($\lambda = 420$ nm) received for spherical nanoparticles,

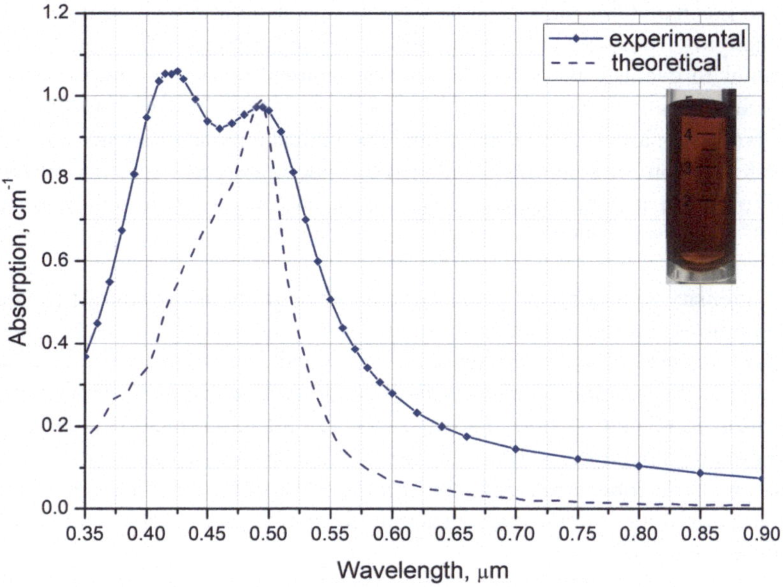

Fig. 16.11 Absorption spectra of silver colloid irradiated by blue

remains higher. This indicates that not all silver nanoparticles in the solution changed their geometry. It is can be assumed that such an effect is associated with high nanoparticle concentrations in the solution.

16.5 Conclusions

This paper presents the study results of the influence of light flows on the change in the geometric parameters of spherical silver nanoparticles, obtained by photostimulated recovery of silver ions from the salt solution ($AgNO_3$).

The theoretical prediction of the plasmon resonance peak position on the spectral scale and size of silver nanoparticles with different shapes (nanoprism, sphere, and decahedron) were carried out using the DDA method.

The synthesis of silver nanoparticles by photostimulated recovery of silver ions from a salt solution was conducted. The absorption peak of obtained Ag nanoparticles is on wavelength 420 nm, which is typical for spherical nanosilver. The plasmon resonance peak position indicates that the average size of formed nanoparticles is about 47 nm.

To research the change in the geometric parameters of obtained colloids, they were irradiated by photon flows during 5 days. It is shown that color of the irradiated samples visually differs from the initial solution. A change in the samples color from yellow to green and blue indicates the formation of nanoprism with different size. A color changing from yellow to orange indicates the formation of decahedron.

In order to evaluate the reliability of the obtained results, a comparison of the calculated absorption coefficient with the experimental dependences for the synthesized silver nanoparticles was carried out. Theoretical simulation of size and absorption peak position for solutions, irradiated by blue, green, and red, was conducted. For the obtained nanoprisms, the average size calculated by the DDA method is about 76 nm. For decahedron nanostructures, the calculated average size is about 64 nm.

The intensity of absorption peaks in experimental curves, irradiated by red and green solutions, indicates that most of the spherical nanoparticles changed their geometric parameters under the light influence and nanoprisms are quantitatively dominated in the solution. The intensity of absorption peaks in experimental curve, for solution irradiated by blue, indicates that not all initial nanoparticles changed their geometrical parameters. It can be assumed that such an effect is associated with high nanoparticle concentrations in the solution. The influence of concentration and the irradiation time on the nanostructure formation in the solution should be researched additionally.

Acknowledgment Financial support of the Ministry of Education and Science of Ukraine should be acknowledged (grant DB/Fotonika № 0117U007176).

References

1. Kim JS, Kuk E et al (2007) Antimicrobial effects of silver nanoparticles. Nanomed Nanotechnol Biol Med 3(1):95–101
2. Chow PE (2010) Gold nanoparticles: properties, characterization and fabrication. Nova Science Publishers, 343 p
3. Bulavinets, T., Yaremchuk, I., Bobitski, Y. (2017, February). Spectral properties of TiO2-Ag nanoshells with different shapes for biomedical applications. In: 2017 14th international conference the Experience of Designing and Application of CAD Systems in Microelectronics (CADSM), IEEE, pp 441–444
4. Burda C, Chen X et al (2005) Chemistry and properties of nanocrystals of different shapes. Chem Rev 105(4):1025–1102
5. Iravani S et al (2014) Synthesis of silver nanoparticles: chemical, physical and biological methods. Res Pharm Sci 9(6):385
6. Popov AK et al (2006) Laser-stimulated synthesis of large fractal silver nanoaggregates. Nanotechnology 17(8):1901
7. Ma J, Guo X et al (2014) Catalytic performance of TiO 2@ Ag composites prepared by modified photodeposition method. Chem Eng J 258:247–253
8. Tang B, Zhang M et al (2014) Photoinduced reversible shape conversion of silver nanoparticles assisted by TiO_2. Phys Chem Chem Phys 16(40):21999–22007
9. Maier SA (2007) Plasmonics: fundamentals and applications. Springer, New York
10. Grigel VA et al (2014) Dependence of silver nanoparticles sensory properties on their form. Scientific Journal of the Chernivtsi University, 683: Chemistry, 42–48 (in Ukrainian)
11. Wan Y, Guo Z et al (2013) Quasi-spherical silver nanoparticles: aqueous synthesis and size control by the seed-mediated Lee–Meisel method. J Colloid Interface Sci 394:263–268

12. Qin Y, Ji X et al (2010) Size control over spherical silver nanoparticles by ascorbic acid reduction. Colloids Surf A Physicochem Eng Asp 372(1):172–176
13. Mitas AW, Bugdol MD (2010) An idea of human voice reaction measurement system under the aspect of behavioral biometric. In: Information technologies in biomedicine. Springer, Berlin/Heidelberg, pp 219–228
14. Illika AI et al (2011) Photostimulated recovery of Ag ions with the formation of decahedron nanoparticles. Scientific Journal of the Chernivtsi University, 555: Chemistry, 40–43 (in Ukrainian)
15. Shafer-Peltier KE et al (2003) Toward a glucose biosensor based on surface-enhanced Raman scattering. J Am Chem Soc 125(2):588–593
16. Loo C, Lin A et al (2004) Nanoshell-enabled photonics-based imaging and therapy of cancer. Technol Cancer Res Treat 3(1):33–40
17. Baptista PV et al (2006) Gold-nanoparticle-probe–based assay for rapid and direct detection of Mycobacterium tuberculosis DNA in clinical samples. Clin Chem 52(7):1433–1434
18. Kelly KL et al (2003) The optical properties of metal nanoparticles: the influence of size, shape, and dielectric environment. The Journal of Physical Chemistry B 107(3):668–677
19. Bulavinets TO, Yaremchuk IYa, et al (2016, September) Modeling absorption and scattering cross sections of the multilayer nanoshells in the near infrared spectrum region. In Laser and Fiber-Optical Networks Modeling (LFNM), 2016 I.E. 13th International Conference on IEEE, pp 35–37
20. Bulavinets T, Yaremchuk I, Bobitski Y (2016) Modeling optical characteristics of multilayer nanoparticles of different sizes for applications in biomedicine. In: Nanophysics, nanophotonics, surface studies, and applications. Springer, Cham, pp 101–115
21. Logsdail AJ et al (2010) Theoretical and experimental studies of the optical properties of conjoined gold– palladium nanospheres. J Phys Chem C 114(49):21247–21251
22. Yurkin MA, Hoekstra AG (2007) The discrete dipole approximation: an overview and recent developments. J Quant Spectrosc Radiat Transf 106(1):558–589
23. Kokhanovsky A (2009) Light scattering reviews 4: single light scattering and radiative transfer. Springer, Berlin/Heidelberg
24. Draine BT, Flatau PJ (2013) User guide for the discrete dipole approximation code DDSCAT 7.3. arXiv preprint arXiv:1305.6497
25. Jain PK, et al (2015) "nanoDDSCAT"
26. Nishioka M et al (2011) Continuous synthesis of monodispersed silver nanoparticles using a homogeneous heating microwave reactor system. Nanoscale 3(6):2621–2626
27. Sönnichsen C (2001) Plasmons in metal nanostructures. Doctoral dissertation, Lmu
28. Zheng X et al (2009) Photochemical formation of silver nanodecahedra: structural selection by the excitation wavelength. Langmuir 25(6):3802–3807
29. Pastoriza-Santos I, Liz-Marzán LM (2008) Colloidal silver nanoplates. State of the art and future challenges. J Mater Chem 18(15):1724–1737
30. Zheng X, Zhao X, Guo et al (2009) Photochemical formation of silver nanodecahedra: structural selection by the excitation wavelength. Langmuir 25(6):3802–3807

Chapter 17
Internal Reflection of the Surface of a Plasmonic Substrate Covered by Active Nanoparticles

Eugene Bortchagovsky and Yurii Demydenko

17.1 Introduction

Surface plasmon polaritons (SPPs) are surface electromagnetic waves propagating along metal-dielectric [1, 2] or doped semiconductor-dielectric [2, 3] interfaces. Under phase-matching or resonance conditions, the energy of the exciting light wave is coupled to the collective oscillation of conducting electrons on the metal or semiconductor surface. There are a number of standard experimental methods for the realization of such an energy transfer [4, 5]. In the visible part of the spectrum, the Kretschmann configuration [5] based on a prism coupler with an active plasmonic film on its base is mainly used. Experimentally, the energy transferred at the excitation of a SPP wave is observed as a dip of a reflectance spectrum. As the excitation of the SPP wave has a resonant nature, the shape and the position of this minimum are highly sensitive to any changes in the refractive index at the surface of a plasmonic film. So, this configuration has been extensively applied to various chemical and biochemical sensing applications [6–9].

In the case when the surface of the plasmonic film is covered with a layer of active nanoparticles, SPP can act as a source of excitation of so-called localized plasmon polaritons (LPPs) [10–12] which, similar to conventional planar SPP in an extended system, are charge density oscillations confined within the nanoparticles. LPP is also resonant excitation, and the electromagnetic field of LPP near the surface of a nanoparticle may be strongly enhanced. Spectral properties of LPP depend on the material of nanoparticles, their structure, size, shape, and on the refractive index of local dielectric environment [12–17].

E. Bortchagovsky · Y. Demydenko (✉)
V. Lashkaryov Institute of Semiconductor Physics of National Academy of Sciences of Ukraine, Kyiv, Ukraine

O. Fesenko, L. Yatsenko (eds.), *Nanooptics, Nanophotonics, Nanostructures, and Their Applications*, Springer Proceedings in Physics 210,
https://doi.org/10.1007/978-3-319-91083-3_17

The interaction of LPPs excited on the nanoparticles with SPP waves leads to a modification of optical properties of SPPs which results, depending on the coupling of two excitations, in changing of the position of the minimum of reflectance (or the position of maximum of an absorption/extinction spectra of SPPs) or in the splitting of that minimum corresponding to the hybridization of two plasmonic modes. For the cases of high surface concentration of active nanoparticles on the surface of the plasmonic film, interactions between LPPs of individual nanoparticles lead to formation of a new collective long-range electromagnetic surface mode. In this case due to the interaction of this new mode with the SPP, we can definitely expect the mode hybridization and the occurrence of new minimum in reflectance (maximum in absorption/extinction) spectra of SPPs as well as to splitting of existed extremum [16–19].

For practical applications of plasmonic nanosystems, in particular in sensor applications, it is important to understand how one can govern their optical properties most efficiently by means of optimizing their structure and design [20–22]. In this connection the studying of the influence of parameters of the layer of active nanoparticles placed on the surface of an active plasmonic film on optical properties of such a system has both fundamental and practical interest.

In this paper we present general theoretical considerations of the influence of parameters of a layer of cylinder-like active nanoparticles placed on the surface of a plasmonic film on the reflectance spectrum measured in the Kretschmann configuration. Calculations of the optical properties of such a system are based on the Green function method [23, 24] taking into account of the local field effects in the system with uniformly distributed nanoparticles on the surface of the plasmonic film.

The paper is organized as follows: In Sect. 17.2 we give general remarks on the model and outline the geometry of the task. Section 17.3 deals with the common analytical method of the calculation of reflection coefficients for the chosen model. In Sect. 17.4 we apply the developed formalism for the numerical calculations of the reflection coefficient for the case of p-polarized excitation. These calculations are performed for the Kretschmann configuration assuming the uniform distributed cylindrical gold nanoparticles for the case when n-doped semiconductor $F{:}SnO_2$ is considered as the plasmonic substrate. The "mirror" case of exchanged materials of the nanoparticles and the substrate is considered as well. The shape and particle's concentration influence on the reflectance spectrum are scrutinized. Section 17.5 presents some conclusions for the results obtained in the presented work. In Appendix A we present the analytical calculation of the susceptibility of a single nanoparticle on a surface in the near-field approximation based on the effective susceptibility concept developed in [16, 17, 25–27]. In Appendix B the scheme of the calculation of the susceptibility of a nanoparticle layer within the Green function method is presented in brief [23, 24].

17.2 Problem Definition

Let us consider the Kretschmann scheme of the frustrated total internal reflection [5] with a glass prism covered by a thin film of a plasmonic material; on the surface of the film, a layer of active nanoparticles is placed, as it is depicted in Fig. 17.1. The incident radiation illuminates the plasmonic film in the Kretschmann scheme from the glass side at the angle of θ, and, for angles θ bigger than the Brewster angle, SPP wave can be excited at the interface of plasmonic film with environment.

The layer of active nanoparticles consists of identical cylindrical nanoparticles randomly distributed on the surface of the plasmonic film with the thickness h (Fig. 17.2). We will consider nanoparticles with diameter d and the height h_p, which are much less than the wavelength of light λ_0 (d, $h_p << \lambda_0$). Supposing that

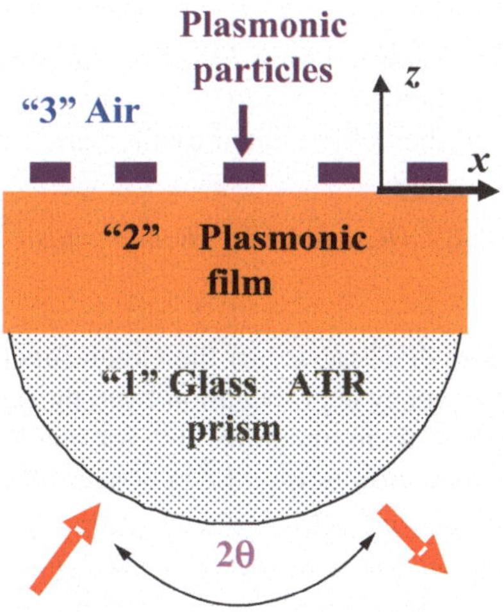

Fig. 17.1 Experimental scheme: Layer of cylindrical plasmonic nanoparticles is placed on a plasmonic film deposited on the base of a glass prism. The structure is illuminated in the Kretschmann configuration

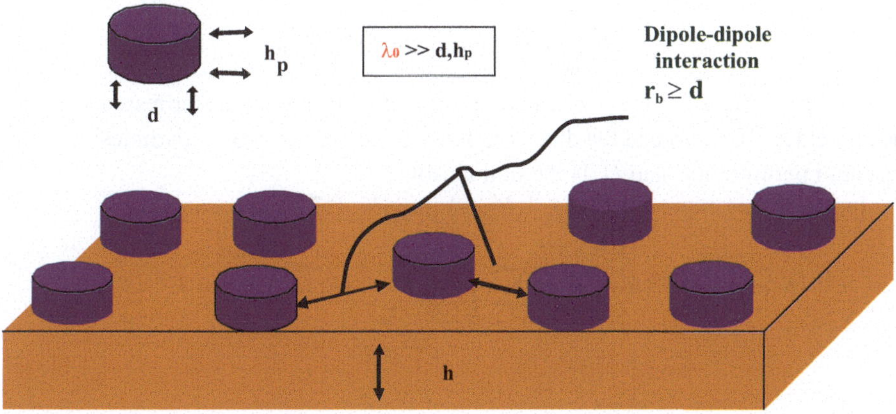

Fig. 17.2 Illustration of a particle layer placed on the surface of a plasmonic film. The particles are distributed randomly and uniformly with surface concentration n

the minimal distance between neighboring nanoparticles r_b is bigger than their diameter ($r_b \geq d$), one can treat interparticle interaction as dipole-dipole one [28].

The SPP wave interacts with localized plasmons of the active nanoparticles in the layer. As the properties of localized plasmons of nanoparticles strongly depend on the shape, size of nanoparticles, and the interaction between them, these properties depend on the concentration of particles on the surface. Thus, altering system parameters such as shape and particle concentration, one can change the optical properties of the whole system that results in modification of corresponding reflectance spectra.

17.3 Reflected Field

The reflected field $E^{(1)}(R, \omega)$ at the point **R** can be represented as superposition of two fields – the field reflected by the system without the layer of particles$E^{(01)}(R, \omega)$ and the field radiated by the particles

$$E^{(1)}(R,\omega) = E^{(01)}(R,\omega) - i\mu_0\omega \sum_{\upsilon=1}^{N} \int_{V_\upsilon} dR'_\upsilon \widetilde{G}^{(13)}(R, R'_\upsilon, \omega) j^{(3)}\left(R'_\upsilon, \omega\right), \qquad (17.1)$$

with $\widetilde{G}^{(ab)}(R, R', \omega)$ the Green function (the photon propagator) describing field of frequency ω propagation from the source point **R**′ in medium "a" to the observation point **R** in medium "b" for the system with two smooth interfaces [29]. The summation in Eq. (17.1) is over all positions υ which are occupied by the particles, $j^{(3)}\left(R'_\upsilon, \omega\right)$ is the induced current density within the υth particle, V_υ is the volume of the υth particle, and μ_0 is the vacuum permeability.

Supposing of a local relation between the current density $j^{(3)}\left(R'_\upsilon, \omega\right)$ and the local field $E^{(3)}\left(R'_\upsilon, \omega\right)$ at the point R'_υ, one can write

$$j^{(3)}\left(R'_\upsilon, \omega\right) = -i\omega\varepsilon_0 \widetilde{\chi}(\omega) E^{(3)}\left(R'_\upsilon, \omega\right), \qquad (17.2)$$

with $\widetilde{\chi}(\omega) = \left(\widetilde{\varepsilon}(\omega) - \widetilde{U}\right)/4\pi$, where $\widetilde{\chi}(\omega)$ and $\widetilde{\varepsilon}(\omega)$ are the linear response to the local field $E^{(3)}\left(R'_\upsilon, \omega\right)$ and the dielectric function of the material of particles, ε_0 is the vacuum permittivity, and $\widetilde{U}$ is the unit tensor.

Inserting this expression for $j^{(3)}\left(R'_\upsilon, \omega\right)$ into Eq. (17.1), one can write

$$E^{(1)}(R,\omega) = E^{(01)}(R,\omega) - k_0^2 \sum_{\upsilon=1}^{N} \int_{V_\upsilon} dR'_\upsilon \widetilde{G}^{(13)}(R, R'_\upsilon, \omega) \widetilde{\chi}(\omega) E^{(3)}\left(R'_\upsilon, \omega\right), \qquad (17.3)$$

where $k_0^2 = \omega/c$ and c is the light velocity. Considering each particle as the point-like dipole located at its center of mass with the polarizability of the cylindrical particle (the so-called quasi-point-like approximation [24–26]), the second term of Eq. (17.3) can be rewritten in the form

$$\sum_{\upsilon=1}^{N} \int_{V_\upsilon} dR'_\upsilon \widetilde{G}^{(13)}(R, R'_\upsilon, \omega)\widetilde{\chi}(\omega)E^{(3)}\left(R'_\upsilon, \omega\right)$$

$$\approx V \sum_{\upsilon=1}^{N} \widetilde{G}^{(13)}\left(r - r'_\upsilon, z, z_c, \omega\right)\widetilde{\chi}^{(S)}(\omega)E^{(3)}\left(r'_\upsilon, z_c, \omega\right), \qquad (17.4)$$

where vectors r'_υ define the positions of the particles in the z_c-plane and z_c denotes the z-coordinate of the center of mass of a particle. The tensor,$\widetilde{\chi}^{(s)}(\omega)$, introduced in Eq. (17.4) determines the linear response of a cylindrical particle on the surface of a substrate to the local field acting inside the particle volume V (the dimensionless polarizability). The implicit expression of $\widetilde{\chi}^{(s)}(\omega)$is given in Appendix A.

Taking into account that the considered system is homogeneous in the XOY plane, it is possible to perform the Fourier transformation of Eq. (17.4) in the plane and to average the obtained result with the function of the homogeneous distribution [25–27]. Then Eq. (17.3) for the reflected field $E^{(1)}(k, z, \omega)$ at the observation point z is written in the **k**-z representation as

$$E^{(1)}(k, z, \omega) = E^{(01)}(k, z, \omega) - \rho\widetilde{G}^{(13)}(k, z, z_c, \omega)\widetilde{\chi}^{(S)}(\omega)E^{(3)}(k, z_c, \omega), \qquad (17.5)$$

with $\rho = Vk_0^2 n$, where n is a concentration of nanoparticles on the surface of the plasmonic film. $E^{(01)}(k, z, \omega)$ is the field reflected in Kretschmann configuration from the structure without the layer of particles

$$E^{(10)}(k, z', \omega) = -i\mu_0\omega \int_{l_0} dz'' \widetilde{G}_I^{(11)}(k, z', z'', \omega)j^{1ext}(k, z'', \omega), \qquad (17.6)$$

where $\widetilde{G}_I^{(11)}$ is the indirect part of the Green's function [29] and $j^{1ext}(k, z'', \omega)$ is the current density of a source of the external to the system field.

Using connection between local field $E^{(3)}(k, z_c, \omega)$ and the external field acting on a nanoparticle in the layer $E^{(30)}(k, z_c, \omega)$, which is given by Eq. (17.B6), Eq. (17.5) takes the form

$$E^{(1)}(k, z, \omega) = E^{(01)}(k, z, \omega)$$
$$- \rho\widetilde{G}^{(13)}(k, z, z_c, \omega)\widetilde{\chi}^{(S)}(\omega)\widetilde{\Omega}(k, z_c, z_c, \omega)E^{(30)}(k, z_c, \omega) \qquad (17.7)$$

with

$$E^{(30)}(k, z_c, \omega) = -i\mu_0\omega \int_{l_0} dz'' \widetilde{G}^{(31)}(k, z_c, z''\omega)j^{1ext}(k, z'', \omega). \qquad (17.8)$$

Supposing that the source of the external electric field is a point dipole located at z_0, we can write

$$j^{1ext}(k, z'', \omega) = j^{1ext}(k, \omega)\delta(z'' - z_0). \tag{17.9}$$

Substituting Eq. (17.9) into Eq. ((17.7), we receive

$$E^{(1)}(k, z, \omega) = -i\mu_0\omega\Big(\widetilde{G}_I^{(11)}(k, z, z_0, \omega) \\ -\rho\widetilde{G}^{(13)}(k, z, z_c, \omega)\widetilde{\chi}(\omega)\widetilde{\Omega}(k, z_c, z_c, \omega)\widetilde{G}^{(31)}(k, z_c, z_0, \omega)\Big)j^{1ext}(k, \omega). \tag{17.10}$$

Taking into account the explicit expressions of the Green function [29], Eq. (17.10) may be rewritten in the form

$$E^{(1)}(k, z, \omega) = -i\mu_0\omega e^{\text{æ}_1 z}\Big(\widetilde{G}_I^{(11)}(k, \omega) - \rho\widetilde{T}^{(11)}(k, \omega)\Big)j^{(1ext)}(k, z_0, \omega), \tag{17.11}$$

where the next designations are introduced

$$\widetilde{T}^{(11)}(k, \omega) = \widetilde{G}^{(13)}(k, \omega)\widetilde{\chi}(\omega)\widetilde{\Omega}(k, z_c, z_c, \omega)\widetilde{G}^{(31)}(k, \omega)\exp(-2\text{æ}_3 z_c), \tag{17.12}$$

$$j^{(1ext)}(k, z_0, \omega) = j^{(1ext)}(k, \omega)\exp(-\text{æ}_1|z_0|), \tag{17.13}$$

$$\text{æ}_u = \sqrt{|k|^2 - \varepsilon_u k_0^2}, \quad (u = 1, 2, 3) \tag{17.14}$$

with ε_u denoting dielectric constants of media constituting the system.

To calculate the reflectance of the system, the expression for the field illuminating the system $\mathbf{E}^{(1ext)}$ is necessary, which can be obtained via the direct part of the Green's function, $\widetilde{G}_D^{(11)}$ [29]

$$E^{(1ext)}(k, z, \omega) = -i\mu_0\omega e^{-\text{æ}_1 z}\widetilde{G}_D^{(11)}(k, \omega)j^{1ext}(k, z_0, \omega). \tag{17.15}$$

Dividing the field on s- and p-polarized components, we can calculate corresponding reflectivities R_s and R_p from Eqs. (17.11) and (17.15):

$$R_s(k, \omega) = \left|G_{Iyy}^{(11)}(k, \omega) - \rho T_{yy}^{(11)}(k, \omega)\right|^2 / \left|G_{Dyy}^{(11)}(k, \omega)\right|^2, \tag{17.16}$$

$$R_p(k, \omega) \\ = \begin{pmatrix} \left|\left(G_{Ixx}^{(11)}(k, \omega) - \rho T_{xx}^{(11)}(k, \omega)\right)\cos\theta + \left(G_{Ixz}^{(11)}(k, \omega) - \rho T_{xz}^{(11)}(k, \omega)\right)\sin\theta\right|^2 \\ + \left|\left(G_{Izx}^{(11)}(k, \omega) - \rho T_{zx}^{(11)}(k, \omega)\right)\cos\theta + \left(G_{Izz}^{(11)}(k, \omega) - \rho T_{zz}^{(11)}(k, \omega)\right)\sin\theta\right|^2 \end{pmatrix} \\ / \left(\left|G_{Dxx}^{(11)}(k, \omega)\cos\theta + G_{Dxz}^{(11)}(k, \omega)\sin\theta\right|^2 + \left|G_{Dzx}^{(11)}(k, \omega)\cos\theta + G_{Dzz}^{(11)}(k, \omega)\sin\theta\right|^2\right). \tag{17.17}$$

Eqs. (17.16) and (17.17) were written under assumption that $Re\ae_1 = 0$ (a nonabsorbing glass prism).

17.4 Numerical Calculation of Reflectance Spectra and Discussion

We will consider two "mirror" cases with respect to active materials of nanoparticles and the plasmonic film. In the first case, metallic cylindrical particles are placed on the surface of a thin film of a n-doped semiconductor transparent in the visible, as it is shown in Figs.17.1 and 17.2 and the second case when the materials of the particles and the substrate are replaced by each other.

Let us suppose that optical properties of the considered system can be described rather adequately taking the dielectric functions of a metal in Drude form:

$$\widetilde{\varepsilon}(\omega) = \left[\varepsilon_{1\infty} - \omega_p^2/\left(\omega^2 + i\omega\gamma_p\right)\right]\widetilde{U}, \tag{17.18}$$

where $\varepsilon_{1\infty}$ is the optical high-frequency dielectric constant of a metal and ω_p and γ_p are the plasma frequency and the rate constant for the metal, respectively. We shall take the dielectric function of a n-doped semiconductor in the form

$$\widetilde{\varepsilon}(\omega) = \varepsilon_{2\infty}\left[1 + \left(\omega_{LO}^2 - \omega_{TO}^2\right)/\left(\omega_{TO}^2 - \omega^2 - i\omega\gamma_{ph}\right) - \widetilde{\omega}_e^2/\left(\omega^2 + i\omega\gamma_e\right)\right]\widetilde{U}, \tag{17.19}$$

with $\varepsilon_{2\infty}$ denoting the optical high-frequency dielectric constant of a semiconductor; ω_{LO}, ω_{TO}, and γ_{ph} are the transverse and longitudinal frequency of the optical phonons and the phonon damping constant, respectively; and $\widetilde{\omega}_e = \omega_e/\sqrt{\varepsilon_{2\infty}}$, ω_e, and γ_e are the plasma frequency and the damping rate constant of free electrons for the doped semiconductor, respectively. The following effective parameters were taken for the dielectric function of metal (gold) [30]: $\varepsilon_{1\infty} = 9{,}84$, $\omega_p = 1{,}38\cdot10^{16}\ s^{-1}$, and $\gamma_p = 1{,}09\cdot10^{14}\ s^{-1}$. Fluorinated tin oxide (F:SnO_2) is considered as a n-doped semiconductor with the following effective parameters [31]: $\varepsilon_{2\infty} = 3{,}19$, $\omega_{LO} = 9{,}03\cdot10^{15}\ s^{-1}$, $\omega_{TO} = 7{,}99\cdot10^{15}\ s^{-1}$, $\omega_e = 2{,}24\cdot10^{15}\ s^{-1}$, $\gamma_{ph} = 1{,}94\cdot10^{14}\ s^{-1}$, and $\gamma_e = 2{,}54\cdot10^{14}\ s^{-1}$.

We will consider the case when p-polarized radiation excites the system coming from the medium "1" at the angle of incidence $\theta = 60^0$ in the (XOZ) plane (Fig. 17.1). The BK7 glass prism with the refraction coefficient $n_{pr} = 1{,}47$ is considered as the medium "1", and the wave vector of the incident radiation is $k = \left(k_0 n_{pr} \sin\theta e_x, 0\right)$. The thickness of the active layer $h = 50$ nm is chosen the same for both considered cases.

Sizes of cylindrical gold nanoparticles were taken as $h_p = 25$ nm and $d = 100$ nm, and their dielectric function was obtained by Eq. (17.18). Then, considering F:SnO2 as the active film with dielectric function defined by Eq. (17.19), numerical

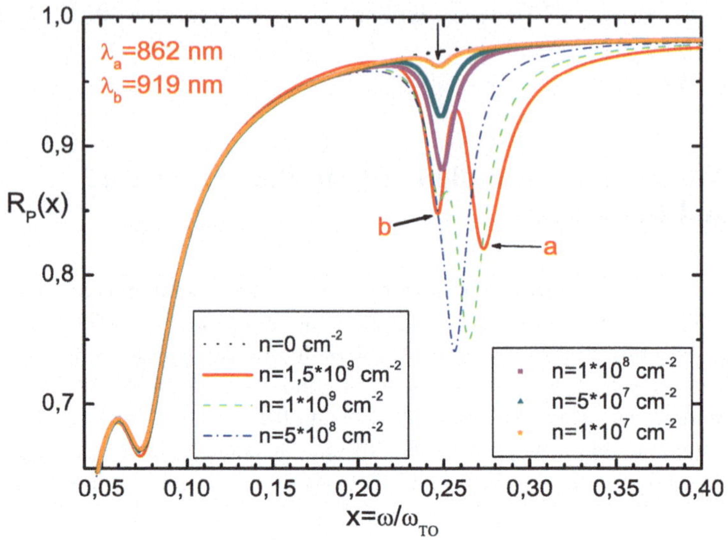

Fig. 17.3 Influence of the particle concentration *n* on the reflectance spectra. The spectra calculated for the case with F:SnO_2 as a substrate for gold particles. The angle of incidence $\theta = 60^0$

calculations of an influence of the surface concentration *n* on the reflection spectrum of the system in the Kretschmann geometry were performed using Eq. (17.17). The results are presented in Fig. 17.3.

As it is seen from Fig. 17.3, the reflection spectra for selected parameters remarkably depend on the particle concentration. Without gold particles on the surface of F:SnO_2, there is only one dip in the reflectance spectra around normalized frequency x = 0,07 connected with the excitation of the plasmon of a subsystem of conducting electrons. Deposition of gold particles results in the arising of a new dip close to x = 0,25 corresponding to the excitation of out of plane localized plasmons. The spectral position of this localized plasmon for one separated nanoparticle on the surface of F:SnO_2 film is x = 0,248, which is shown in figures by arrow. At small particle concentration of about $n \leq 10^7$ cm^{-2}, this dip is not pronounced but further rises with the increase of the number of particles. This behavior is obvious as the absorbed energy should correlate with the number of absorbers. However, the depth of the minimum rises not linear with the number of particles became slower at bigger concentrations. Along with the increase of the depth of that minimum, its blue shift is observed (spectral positions of sequential minima are given in Fig. 17.3 and shown in Fig. 17.4). It is the exhibition of interparticle interactions that results in the creation of common mode of the ensemble of degenerated localized plasmons of individual particles. The blue shift also indicates that we observe out of plane plasmonic resonance of nanoparticles [32]. The blue shift becomes more pronounced for concentrations of particles more than $1 \cdot 10^8$ cm^{-2}.

However, this behavior of the minimum changes remarkably after the concentration exceeds about $5 \cdot 10^8$ cm^{-2}. One minimum splits demonstrating at $x \approx 0,245$

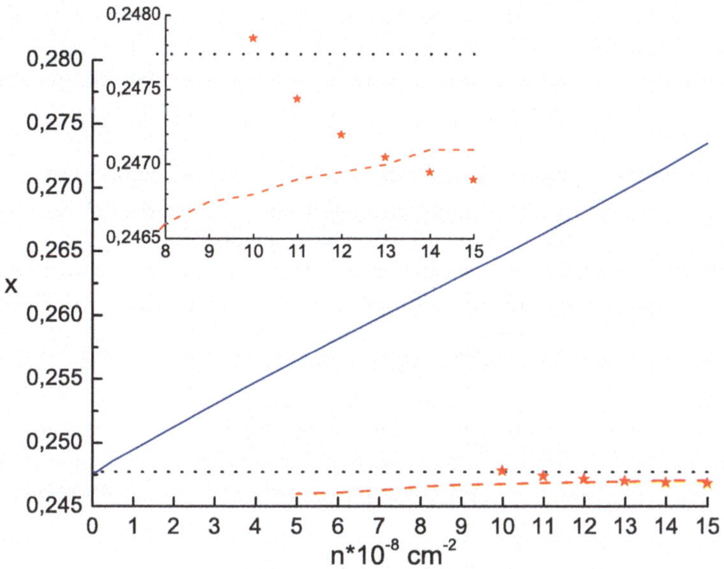

Fig. 17.4 Position of minima in reflectivity spectra versus the concentration of nanoparticles on the surface. Dashed line gives the position of the resonance of one separated particle as blue and red ones of two blue and red shifted minima noticed in Fig. 17.3 by a and b correspondingly. The insert demonstrates the position of red shifted minimum in more details. The line represents the position defined from the second derivative as stars by visually visible minimum positions

small minimum as the deep one moves to the blue side to about $x = 0{,}26$. Further increase of the concentration makes both minima equally pronounced with smaller but comparable depths indicated in Fig. 17.3 by letters "a" and "b." The spectral position of the second dip gets about $x \approx 0{,}275$ at the highest concentration $1{,}5 \cdot 10^9$ cm^{-2} what is about the limit to which we can restrict ourself by dipole-dipole interactions. This behavior demonstrates the transfer from weak coupling of SPP with collective mode of nanoparticles to strong one with their hybridization and splitting. It happens when the energy of coupling exceeds the damping [33] and modes can exchange energy of excitation many times before decay.

Fig. 17.4 exhibits the normalized spectral position of minima of reflectivity spectra versus the surface concentration of nanoparticles on the surface. Two branches, which correspond to two minima shown in Fig. 17.3, are visible. It is interesting that the behavior of the blue shifted minimum noticed in Fig. 17.3 by "a" is practically linear with the concentration. The second minimum becomes noticeable only at concentrations higher than $5 \cdot 10^8$ cm^{-2} and visually recognizable after $1 \cdot 10^9$ cm^{-2}. The positions of that red shifted minimum were defined either from the second derivative revealing flex points of the curve what is shown by solid line or from the visually recognizable minima position (first derivative) at higher concentrations what is shown by stars. The difference behavior of stars can be explained by the influence of not well-separated blue shifted minimum and the exhibition of the red shifted minimum as a shoulder on the dip of the blue shifted minimum. It gives visually shifted minimum position to the direction of the main minimum (in our case to blue side) according to the real one. It is obvious that the flex point should be closer to the minimum position just when it is very shallow and not revealed by the first derivative. Finite numerical precision explains not smooth behavior of the flex point.

Despite shown dependences are not dispersion curves, demonstrated behavior proves that we have different coupling conditions for surface and localized resonances at different concentration of deposited nanoparticles in the system gold nanoparticles – n-doped semiconductor F:SnO_2 and transfer between these two regimes. To demonstrate this transfer more pronouncedly, reflectivity of our system is shown in more details in Fig. 17.5 for three ranges of the particle concentration: small up to $1 \cdot 10^8$ cm^{-2}, intermediate from $1 \cdot 10^8$ cm^{-2} to $5 \cdot 10^8$ cm^{-2}, and high up to $1{,}5 \cdot 10^9$ cm^{-2}.

Panel 4a shows sequential increasing of the depth of the absorption dip with rather small blue shift, as the average distances between particles in our system at the concentration $1 \cdot 10^8$ cm^{-2} are still rather large of about 1 μm remarkably exceeding

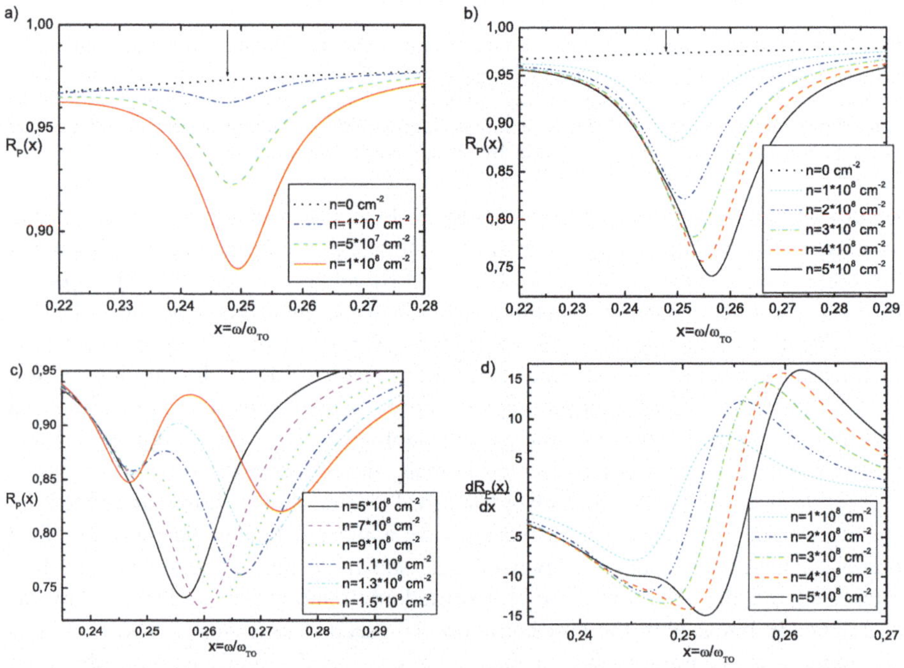

Fig. 17.5 Detailed presentation of the reflectivity from Fig. 17.3 for different ranges of concentrations: (**a**) small ones up to $1 \cdot 10^8$ cm^{-2}, (**b**) intermediate ones for the range $1 \div 5 \cdot 10^8$ cm^{-2}, (**c**) high ones over $5 \cdot 10^8$ cm^{-2}, and (**d**) the first derivatives of curves from the panel b

the lateral size of the particle. Panel 4b exhibits the range of intermediate concentrations of $1 \div 5 \cdot 10^8$ cm^{-2}. For this range pronounced blue shift is visible up to about x ≈ 0,255 what is far from the position of the resonance of the only particle indicated by arrow. Panel 4c exhibits the case of higher concentrations when the splitting of minima is clearly visible. It may be seen that the position of the resonance at x ≈ 0,245 is practically nonshifted from the resonance of one separated particle. So this resonance corresponds to the localized-like branch of two hybridized modes.

It is necessary to note that careful consideration of the reflectance curve for the concentration of $5 \cdot 10^8$ cm^{-2} reveals its slight asymmetry. To clarify it, first derivatives of curves from panel 4b are shown in panel 4d. It is clearly seen that already at the concentration of $5 \cdot 10^8$ cm^{-2} and less pronounced at $4 \cdot 10^8$ cm^{-2}, there is the trace of the second minimum corresponding to the localized-like mode. It means that already at the distance between edges of particles exceeding their size in $1.5 \div 2$ times, there is the creation of the common mode of the layer and their hybridization with the surface plasmon of the substrate resulting in the splitting of two modes. The hybridization of modes at smaller concentration is probably absent. At least even the second derivative does not demonstrate any additional features beside of the one resonance for the concentration of $3 \cdot 10^8$ cm^{-2}.

So we suppose that presented three diapasons of the particle concentration correspond to three different coupling regimes for surface and localized plasmons in the considered system. At small concentrations of nanoparticles when interparticle interactions are practically unobservable, there is weak coupling of modes. Rising the concentration of deposited nanoparticles, we come to the case of noticeable interparticle interactions resulting in the shift of the common resonance of the layer of nanoparticles, but still intermediate coupling of surface and localized modes when the strength of the coupling is comparable with the mode dumping [33] and splitting is not yet pronounced. Only the trace of the splitting is visible in this case as the shoulders of the reflectivity spectrum. The third regime of strong coupling with pronounced Rabi splitting is realized at high particle concentration.

As it is well known [34, 35], the positions of minima of the reflectance spectrum at the excitation of SPPs are determined by the positions of zeroes of the pole part of the susceptibility of the considered system. These zeroes define points on the plane (θ,ω) where SPPs can be excited under given parameters of the system. The number of these points determines curves, which are dispersion characteristics of excited SPPs. In the case considered in this work, the appropriate zero points can be obtained from the equation

$$\mathrm{Re}\left\{\det\left[\left(\widetilde{\Omega}\left(k, z_c, z_c, \omega\right)\right)^{-1}\right]\right\} = 0, \quad \left(k = k_0 n_{pr} \sin\theta\right). \tag{17.20}$$

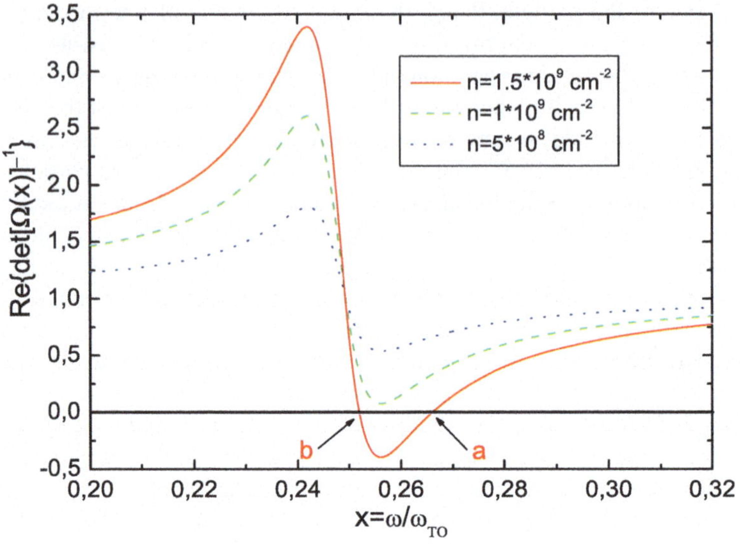

Fig. 17.6 The illustration of the appearance of zeroes of the pole part of $\widetilde{\Omega}\,(k, z_c, z_c, \omega)$ when the surface nanoparticle concentration, n, increases

Let us consider the plot of $\mathrm{Re}\left\{\det\left[\left(\widetilde{\Omega}\,(k, z_c, z_c, \omega)\right)^{-1}\right]\right\}$ as the function of the normalized frequency $x = \omega / \omega_{TO}$ for the case of high concentration of nanoparticles with parameters used in previous calculations. These curves are shown in Fig. 17.6 for different concentrations of nanoparticles.

As it is seen from Fig. 17.6, for the nanoparticle concentrations for which $n \geq 1 \cdot 10^9$ cm^{-2}, one can say that requirement of the excitation of SPP waves in the considered system is satisfied. The discrepancy of this value with the value $n \geq 5 \cdot 10^8$ cm^{-2}, obtained in the analysis of reflectance spectra, may be explained by damping. Only at $n \geq 1 \cdot 10^9$ cm^{-2}, two minima are well seen, before the red one is seen only as the shoulder on the curve. The account of damping would correspond to the spreading of linear curves in Fig. 17.6 to stripes, so even at smaller concentrations, there would be touching of zero line corresponding to the excitation of SPP in the system.

As it was already mentioned in the introduction [12–17], optical properties of LPP strongly depend on the shape of the nanoparticle. Thus, we can expect effective control of the optical properties and the reflectance spectra of the system modifying the shape of deposited nanoparticles. Let us fix the particle concentration at $n = 1{,}5 \cdot 10^9$ cm^{-2} and the value of the particle volume at $V = 1{,}96 \cdot 10^{-22}$ m^3 (the value of the volume corresponds to $h_p = 25$ nm and $d = 100$ nm; these geometrical parameters will be used as initial ones) and study numerically the influence of the change of the particle shape on the reflectance spectrum using Eq. (17.17) with the

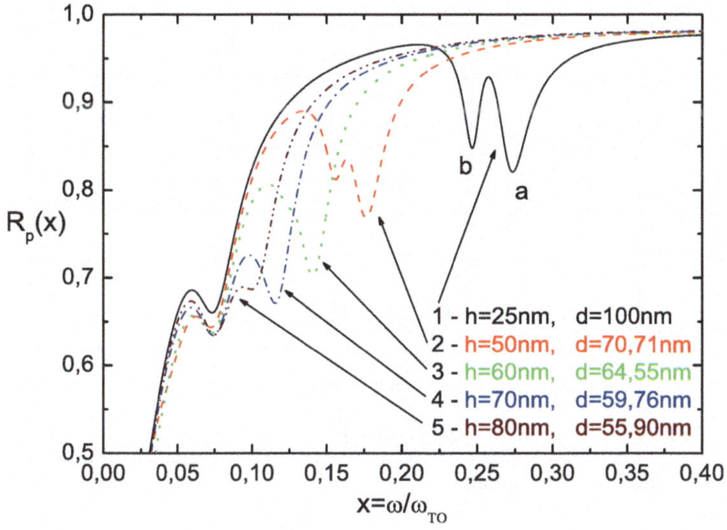

Fig. 17.7 The influence of the modification of the particle shape on the reflectance spectrum for the case of high particle concentration. For all curves the particle volume *V* is taken constant

same other parameters of the system as it was used in previous calculations. The result of these calculations is presented in Fig. 17.7.

As it is seen from Fig. 17.7, the change of the particle form from the disk-like to pillar-like one leads to quick disappearance of the "red" (b) localized-like mode. This happens along with rapid displacement of the remaining "blue" (a) toward low frequencies that results in the disappearance of this minimum at the frequency of SPP mode of the F:SnO_2 substrate at given angle of incidence $\theta = 60^0$. The local field interactions of localized plasmons of nanoparticles with surface plasmonic excitation of conducting electrons on the F:SnO2 substrate lead to slight improvement of SPP generation condition exhibited by a bit dipper minimum at $x = 0{,}07$.

Let us exchange the materials of deposited nanoparticles and the active film assuming semiconducting particles on gold and using Eq. (17.17) consider the influence of particle concentration on the reflectance spectrum of such a modified system in the same way as it was done for the previous system presented in Fig. 17.3. The incident angle of external radiation is now $\theta = 44^0$, as for angles of incidence bigger than 50^0, the minimum in reflectivity spectra is not pronounced. The height and diameter of nanoparticles are the same as for Fig. 17.3. At these conditions the well-known SPP is generated on the surface of gold layer. The field of SPP interacts with the disk-like nanoparticles of F:SnO_2 that leads to worsening of the generation condition of the SPP wave at high concentrations, and, as the result of the interaction, the minimum of SPP becomes shallower and a bit shifted than for the clean surface of a gold film. The result of calculations is shown in Fig. 17.8. So, no

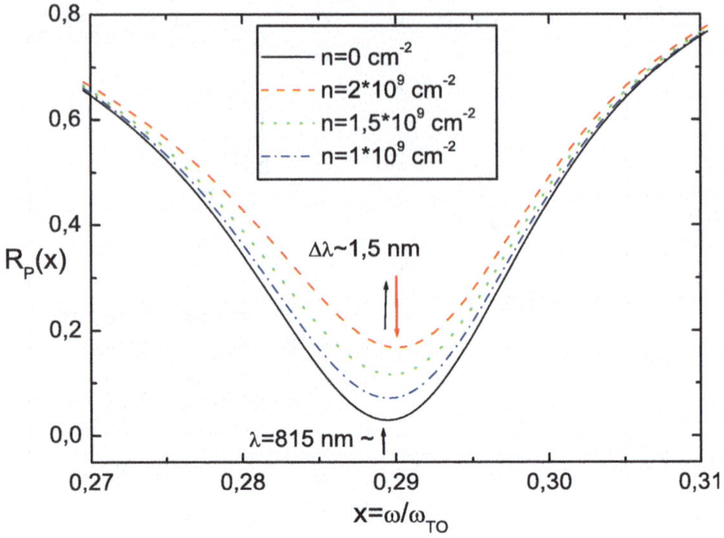

Fig. 17.8 The influence of particle concentration *n* on the reflectance spectra of "reversed" structure with semiconducting particles on the gold film

distinctive influence on the reflectance spectrum is found but the fact that loading of active semiconducting particles of F:SnO_2 gives slight blue shift of the plasmonic resonance in contrast to deposition of any passive material on the gold surface.

17.5 Conclusion

Within the frame of the Green functions method with the use of the concept of effective susceptibility the approach for the calculation of reflectance spectra of the layer of possessing LP cylindrical nanoparticles situated on the surface of a plasmonic film was developed. The modeling was performed for the system of gold disks on the surface of the n-doped semiconductor F:SnO_2 for the illumination of the system in the Kretschmann configuration with the excitation of SPP. The dependences of the reflectivity on the particle concentration and on their shape were investigated. Obtained results clearly demonstrate the possibility of effective control of optical properties of the system and its reflectance spectrum changing parameters of the layer of particles. The spectral position of resonances of such a system can be tuned by changing the particle shape.

Investigation of the concentration dependence of the optical properties of the considered system reveals the existence of regions of weak and strong coupling of surface and localized modes and clear transition from one to the other regime. At higher concentration of deposited nanoparticles, damping of the common mode of nanoparticle layer, which is created from degenerated localized resonance by interparticle interactions, drops, and we obtain the strong coupling regime with the hybridization of modes and Rabi splitting. This transition between two regimes was analyzed in detail.

The system considered in this study can be used as active element of sensors.

Appendices

Appendix A

Susceptibility of a Nanoparticle on the Surface in the Near-Field Approximation

Let us consider a cylindrical nanoparticle placed on the surface of an isotropic active plasmonic substrate (metal or n-doped semiconductor) with the dielectric constant $\varepsilon_2(\omega)$ illuminated by long-range electrical field $\mathbf{E}^{(0)}(\mathbf{R},\omega)$ as it is illustrated in Fig. 17.2. As linear dimension of the particle is supposed less than the wavelength, all electrodynamic interactions between the particle and the substrate as well as with a surrounding medium can be taken into account by using the Green function in near-field approximation [16, 17]. Then the local field $\mathbf{E}(\mathbf{R},\omega)$ at any point $\mathbf{R}$ in the system can be calculated by the equation

$$E(R,\omega) = E^{(0)}(R,\omega) - i\omega\mu_0 \lim_{\delta\to 0} \int\limits_{V-V_\delta} dR' \widetilde{G}^{NF}\left(R,R',\omega\right) j\left(R',\omega\right) - \frac{1}{i\omega\varepsilon_0}\widetilde{L} j(R,\omega), \tag{17.A1}$$

where V_δ is the so-called exclusion volume which is used to remove the singularity of $\widetilde{G}^{NF}$ at $\mathbf{R} = \mathbf{R}'$ and $\widetilde{L}$ is the source dyadic accounting depolarizing properties of V_δ, which depends solely on the shape of the exclusion volume [36, 37].

Choosing the origin of a Cartesian coordinate system on the surface of the plasmonic film with the z-axis directed along the axis of the cylindrical particle, as shown in Fig. 17.2, for our case $\widetilde{L}$ has the next simple diagonal form:

$$\widetilde{L} = 0.5a(\xi)\widetilde{U}_t + (1 - a(\xi))\left(e_z \bigotimes e_z\right), \tag{17.A2}$$

where $a(\xi) = 1/\sqrt{1+\xi^2}$ and $\xi = d/h_p$ with $\widetilde{U}_t = \left(e_i \bigotimes e_i\right)$ the transversal unit tensor $\mathbf{e}_i$ ($i = x, y$) and $\mathbf{e}_z$ denote unit vectors in a Cartesian xyz-coordinate system.

According to the method developed in [26], we suppose that similar to Eq. (17.2) for the local current density $\mathbf{j}(\mathbf{R},\omega)$, one can introduce the relation connecting $\mathbf{j}(\mathbf{R},\omega)$ with the external field $\mathbf{E}^{(0)}(\mathbf{R},\omega)$ illuminating the system:

$$j(R,\omega) = -i\omega\varepsilon_0\tilde{\chi}^{(S)}(R,\omega)E^{(0)}(R,\omega), \tag{17.A3}$$

where$\tilde{\chi}^{(S)}(R,\omega)$ is the tensor of linear response of the particle on the surface to the external field $\mathbf{E}^{(0)}(\mathbf{R},\omega)$. Then using the constitutive relation Eq. (17.2) in the form

$$j(R,\omega) = -i\omega\varepsilon_0\tilde{\chi}(\omega)E(R,\omega), \tag{17.A4}$$

and Eq. (17.A3), the external field can be expressed as

$$E^{(0)}(R,\omega) = \left(\tilde{\chi}^{(S)}(R,\omega\right)^{-1}\tilde{\chi}(\omega)E(R,\omega). \tag{17.A5}$$

The reverse matrix $\left(\tilde{\chi}^{(S)}(R,\omega\right)^{-1}$ exists because there is dissipation in the system.

Substituting Eq. (17.A4) into Eq. (17.A1), one obtains analogously to [26]

$$\begin{aligned} E(R,\omega) &= \left(\tilde{\chi}^{(S)}(R,\omega)\right)^{-1}\tilde{\chi}(\omega)E(R,\omega) \\ &-k_0^2\lim_{\delta\to 0}\int\limits_{V-V_\delta} dR'\tilde{G}^{NF}\left(R,R',\omega\right)\tilde{\chi}(\omega)E\left(R',\omega\right)+\tilde{L}\tilde{\chi}(\omega)E(R,\omega). \end{aligned} \tag{17.A6}$$

Since Eq. (17.A6) fulfills for all points of the system including points within the volume V - V_δ, we can act on Eq. (17.A6) by the operator $\lim\limits_{\delta\to 0}\int\limits_{V-V_\delta} dR$. Then, supposing that local field can be represented in the form [26]

$$E(R,\omega) = \sum_K E(K,\omega)\exp(i\mathrm{KR}), \tag{17.A7}$$

after interchanging of the dummy variables $R' \leftrightarrow R$ and, correspondingly, the order of integration, one obtains from Eq. (17.A6)

$$\begin{aligned} &\sum_K \lim_{\delta\to 0}\int\limits_{V-V\delta} dR\Bigg\{\tilde{U}-\tilde{L}\tilde{\chi}(\omega)+k_0^2\lim_{\delta\to\infty}\int\limits_{V-V_\delta} dR'\tilde{G}^{NF}(R',R,\omega)\tilde{\chi}(\omega) \\ &\quad-\left(\tilde{\chi}^{(S)}(R,\omega)\right)^{-1}\tilde{\chi}(\omega)\Bigg\}E(K,\omega)\exp(i\mathrm{KR})=0. \end{aligned} \tag{17.A8}$$

Using the fact that exponents form the complete set of orthonormal functions, we obtain from Eq. (17.A8)

$$\tilde{\chi}^{(S)}(R,\omega) = \tilde{\chi}(\omega)\left[\tilde{U}+\left(k_0^2\lim_{\delta\to 0}\int\limits_{V-V_\delta} dR'\tilde{G}^{NF}\left(R',R,\omega\right)-\tilde{L}\right)\tilde{\chi}(\omega)\right]^{-1}. \tag{17.A9}$$

Using Eqs. (17.A3), (17.A4), and (17.A9), one can receive the relation between the local and external fields existing within the volume $V - V_\delta$:

$$E(R,\omega) = \widetilde{\Theta}(R,\omega)E^{(0)}(R,\omega), \tag{17.A10}$$

$$\widetilde{\Theta}(R,\omega) = \left[\widetilde{U} + \left(k_0^2 \lim_{\delta\to 0} \int\limits_{V-V_\delta} dR' \widetilde{G}^{NF}(R',R,\omega) - \widetilde{L}\right)\widetilde{\chi}(\omega)\right]^{-1}. \tag{17.A11}$$

The electric dipole moment of the current density distribution $\mathbf{j}(\mathbf{R},\omega)$

$$p(\omega) = (i/\omega)\int\limits_V d\mathrm{R}\mathrm{j}(R,\omega). \tag{17.A10}$$

Taking into account Eq. (17.A4), it can be rewritten in the form

$$p(\omega) = \varepsilon_0\widetilde{\chi}(\omega)\int\limits_V d\mathrm{R}\mathrm{E}(R,\omega). \tag{17.A11}$$

Replacing the local field $\mathbf{E}(\mathbf{R},\omega)$ in Eq. (17.A11) by the external field $\mathbf{E}^{(0)}(\mathbf{R},\omega)$, we obtain

$$p(\omega) = \varepsilon_0\widetilde{\chi}(\omega)\int\limits_V dR\widetilde{\Theta}(R,\omega)E^{(0)}(R,\omega). \tag{17.A12}$$

Since we consider nanoparticles with sizes less than the wavelength of the external illuminating field, one can omit the variation of the external field across the particle, i.e., in Eq. (17.A12), one can replace $\mathbf{E}^{(0)}(\mathbf{R},\omega)$ by $\mathbf{E}^{(0)}(\mathbf{R}_c,\omega)$, then

$$p(\omega) = \varepsilon_0\widetilde{\chi}(\omega)\left[\int\limits_V dR\widetilde{\Theta}(R,\omega)\right]E^{(0)}(R_c,\omega), R_c = (\mathbf{0}, z_c). \tag{17.A13}$$

Evaluating the integral in square brackets of Eq. (17.A13) as

$$\left[\int\limits_V dR\widetilde{\Theta}(R,\omega)\right] \approx V\widetilde{\Theta}(R_c,\omega) \tag{17.A14}$$

we can introduce the dyadic dipole-dipole polarizability of the particle via

$$\widetilde{\alpha}(\omega) = \varepsilon_0 V\widetilde{\chi}^{(S)}(\omega). \tag{17.A15}$$

The tensor $\widetilde{\chi}^{(S)}(\omega)$ for the considered case has the diagonal form

$$\widetilde{\chi}^{(S)}(\omega) = \chi_{\parallel}^{(S)}(\omega)U_t + \chi_{\perp}^{(S)}(\omega)\left(e_z \bigotimes e_z\right) \quad (i = x, y) \tag{17.A16}$$

with the components of the polarizability written as

$$\chi_q^{(S)}(\omega) = \chi(\omega)/\left(1 + A_q(\omega;\xi)\right) \ (q = \parallel, \perp), \tag{17.A17}$$

where

$$A_{\parallel}(\omega;\xi) = 0,25(-1 + a(\xi)(3 + 0,5\eta(\omega)) - 1,5b(\xi)\eta(\omega)), \tag{17.A18}$$

$$A_{\perp}(\omega;\xi) = -0,5 \\ + 0,25\left(-a(\xi)\left(6 - \eta(\omega)\left[3 + 4/\xi^2\right]\right) + b(\xi)\eta(\omega)\left[1 + 12/\xi^2\right]\right), \tag{17.A19}$$

$$b(\xi) = 1/\sqrt{9 + \xi^2}, \eta(\omega) = (\varepsilon_2(\omega) - \varepsilon_3)/(\varepsilon_2(\omega) + \varepsilon_3), \tag{17.A20}$$

ε_3 is a dielectric function of surrounding medium.

Calculation of the Susceptibility of a Nanoparticle Layer

Let us suppose that the tensor of linear response of the cylindrical particle on the surface of an active medium $\widetilde{\chi}^{(S)}(\omega)$ is known; then the electric field in the medium "3" where the particles are located can be written in the form similar to Eq. (17.5):

$$E^{(3)}(k,z,\omega) = E^{(03)}(k,z,\omega) - \rho\widetilde{G}^{(33)}(k,z,z_c,\omega)\widetilde{\chi}^{(S)}(\omega)E^{(3)}(k,z_c,\omega), \tag{17.B1}$$

where $E^{(03)}(k,z,\omega)$ is the external long-range electrical field acting on the particle in the layer

$$E^{(03)}(k,z,\omega) = -i\mu_0\omega \int_{l_0} dz''\widetilde{G}^{(31)}(k,z,z'',\omega)j^{1ext}(k,z'',\omega). \tag{17.B2}$$

Let us assume that the Green function of the considered system $\widetilde{F}^{(31)}$ is known, then

$$E^{(3)}(k,z,\omega) = -i\mu_0\omega \int_{l_0} dz''\widetilde{F}^{(31)}(k,z,z'',\omega)j^{1ext}(k,z'',\omega). \tag{17.B3}$$

Substituting Eq. (17.B3) into Eq. (17.B1), one can write

$$\begin{aligned}&\widetilde{F}^{(31)}(k,z,z'',\omega)+\rho\widetilde{G}^{(33)}(k,z,z_c,\omega)\widetilde{\chi}^{(S)}(\omega)\widetilde{F}^{(31)}(k,z_c,z'',\omega)\\&\quad-\widetilde{G}^{(03)}(k,z,z'',\omega)=0.\end{aligned} \tag{17.B4}$$

Taking Eq. (17.B4) at the point $z = z_c$ and using Eq. (17.B3), we obtain the simple relation between the local field $\mathbf{E}^{(3)}$ and the external field $\mathbf{E}^{(03)}$ at the point $z = z_c$:

$$E^{(3)}(k,z_c,\omega)=\widetilde{\Omega}(k,z_c,z_c,\omega)E^{(03)}(k,z_c,\omega), \tag{17.B5}$$

$$\widetilde{\Omega}(k,z_c,z_c,\omega)=\left[\widetilde{U}+\rho\widetilde{G}^{(33)}(k,z_c,z_c,\omega)\widetilde{\chi}^{(S)}(\omega)\right]^{-1}. \tag{17.B6}$$

The tensor $\widetilde{\Omega}(k,z_c,z_c,\omega)$ is the dimensionless effective susceptibility of the layer of cylindrical nanoparticles on a surface [16, 17], which accounts both near- and far-field electromagnetic interactions within the layer and with the substrate.

References

1. Agranovich VM, Mills DL (eds) (1982) Surface polaritons: electromagnetic waves at surfaces and interfaces. North Holland, Amsterdam
2. Raether H (1988) Surface plasmons on smooth and rough surfaces and on gratings. Springer, Berlin. https://doi.org/10.1007/BFb0048317
3. Franzen S (2008) Surface plasmon polaritons and screened plasma absorption in indium tin oxide compared to silver and gold. J Phys Chem C 112(15):6027–6032. https://doi.org/10.1021/jp7097813
4. Harrick NJ (1967) Internal reflection spectroscopy. Wiley, New-York
5. Kretschmann E, Raether H (1968) Radiative decay of non-radiative surface plasmons excited by of non-radiative surface plasmons excited by light. Z Naturforsch A23:2135–2136. https://doi.org/10.1515/zna-1968-1247
6. Homola J, Yee SS, Gauglitz G (1999) Surface plasmon resonance sensors: review. J Sensors Actuators B: Chem 54(1–2):3–15. https://doi.org/10.1016/S0925-4005(98)00321-9
7. Gwon HR, Lee SH (2010) Spectral and angular responses of surface Plasmon resonance based on the Kretschmann prism configuration. Mater Trans 51(6):1150–1155. https://doi.org/10.2320/matertrans.M2010003
8. Devanarayanan VP, Manjuladevi V, Gupta RK (2016) Surface plasmon resonance sensor based on a new opto-mechanical scanning mechanism. Sensors Actuators B: Chem 227:643–648. https://doi.org/10.1016/j.snb.2016.01.027
9. Ilchenko SG, Lymarenko RA, Taranenko VB (2017) Using metal-multilayer-dielectric structure to increase sensitivity of surface Plasmon resonance sensor. Nanoscale Res Lett 12:295. https://doi.org/10.1186/s11671-017-2073-1
10. Boardman AD (1982) Electromagnetic surface modes. Wiley, New York
11. Kreibig U, Vollmer M (1995) Optical properties of metal clusters. Springer, Berlin. https://doi.org/10.1007/978-3-662-09109-8
12. Lance KL, Coronado E, Zhao LL, Schatz GC (2003) The optical properties of metal nanoparticles: the influence of size, shape, and dielectric environment. J Phys Chem B 107(3):668–677. https://doi.org/10.1021/jp026731y
13. Hutter E, Fendler JH (2004) Exploitation of localized surface plasmon resonance. Adv Mater 16(19):1685–1706. https://doi.org/10.1002/adma.200400271

14. Chegel V, Demidenko Y, Lozovski V, Tsykhonya A (2008) Influence of the shape of the particles covering the metal surface on dispersion relations of surface plasmons. Surf Sci 602 (8):1540–1546. https://doi.org/10.1016/j.susc.2008.02.015
15. Chen W-Y, Lin C-H (2010) A standing-wave interpretation of plasmon resonance excitation in split-ring resonators. Opt Express 18(13):14280–14292. https://doi.org/10.1364/OE.18.014280
16. Baryakhtar I, Demidenko Y, Lozovski V (2013) Interaction between localized-on-nanoparticles plasmon polaritons and surface plasmon polaritons. J Opt Soc Am B 30(4):10200–11026. https://doi.org/10.1364/JOSAB.30.001022
17. Demydenko Y, Juodkazis S, Lozovski V (2014) Composite Au-on-SiC nanorods for sensing. J Opt Soc Am B 31(11):2893–2900. https://doi.org/10.1364/JOSAB.31.002893
18. Malynych S, Chumanov G (2003) Light-induced coherent interactions between silver nanoparticles in two-dimensional arrays. J Am Chem Soc 125(10):2896–2898. https://doi.org/10.1021/ja029453p
19. Balci S, Karademir E, Kocabas C (2015) Strong coupling between localized and propagating plasmon polaritons. Opt Lett 40(13):3177–3180. https://doi.org/10.1364/OL.40.003177
20. Guo L, Jackman JA, Yang H–H, Chen P, Cho N–J, Kim D–H (2015) Strategies for enhancing the sensitivity of plasmonic nanosensors. J Nanotoday 10(2):213–239. https://doi.org/10.1016/j.nantod.2015.02.007
21. Stewart ME, Anderton CR, Thompson LB, Maria J, Gray SK, Rogers JA, Nuzzo RG (2008) Nanostructured plasmonic sensors. Chem Rev 108(2):494–521. https://doi.org/10.1021/cr068126n
22. Li M, Cushing SK, Wu N (2015) Plasmon-enhanced optical sensors: a review. Analyst 140 (2):386–406. https://doi.org/10.1039/C4AN01079E
23. Khudik, B.I., Lozovskii, V.Z., and Baryakhtar, I.V.: Macroscopic electrodynamics of ultra-thin films. Phys Status Solidi B 153(1), 167–177 (1989). doi: https://doi.org/10.1002/pssb.2221530117
24. Keller O (1996) Local fields in electrodynamics of mesoscopic media. Phys Rep 268:85–262. https://doi.org/10.1016/0370-1573(95)00059-3
25. Demidenko Y, Makarov D, Krone P, Lozovski V (2009) Effect of a magnetic field on the optical properties of nonmagnetic nanorods in a dielectric matrix. J Opt A Pure Appl Opt 11 (12):125001. https://doi.org/10.1088/1464-4258/11/12/125001
26. Lozovski V (2010) The effective susceptibility concept in the electrodynamics of nano-system. J comput Theor Nanosci 7(7):1–17. https://doi.org/10.1166/jctn.2010.1588
27. Demidenko YV, Makarov D, Lozovski V (2010) Local-field effects in magneto-plasmonic nanocomposites. J Opt Soc Am B 27(12):2700–2706. https://doi.org/10.1364/JOSAB.27.002700
28. Haarmans MT, Bedeaux D (1993) The polarizability and the optical properties of lattices and random distributions of small metal spheres on a substrate. Thin Solid Films 224(1):117–131. https://doi.org/10.1016/0040-6090(93)90468-5
29. Bah ML, Akjouj A, Dobrzynski L (1992) Response functions in layered dielectric media. Surf Sci Rep 16(3):97–131. https://doi.org/10.1016/0167-5729(92)90010-9
30. Kolwas K, Derkachova A (2013) Damping rates of surface plasmons for particles of size nano- to micrometers; reduction of the nonradiative decay. JQSRT 114:45–55. https://doi.org/10.1016/j.jqsrt.2012.08.007
31. von Rottkay, K., Rubin, M.: Optical indices of pyrolytic tin-oxide glass. Mater Res Soc Symp Proc 426, 449–456 (1996). doi: https://doi.org/10.1557/PROC-426-449
32. Persson BNJ (1983) Lateral interactions in small particle systems. J de Physique 44(C10) C10-409-C10-419. https://doi.org/10.1051/jphyscol:19831084
33. Törmä P, Barnes WL (2015) Strong coupling between surface plasmon polaritons and emitters: a review. Rep Prog Phys 78, 013901-1-34. https://doi.org/10.1088/0034-4885/78/1/013901
34. Abrickosov AA, Gor'kov LP, Dzyaloshinskii IY (1965) Quantum field theoretical methods in statistical physics. Pergamon, Oxford

35. Lifshitz EM, Pitaevskii LP (1980) Statistical physics (Part.2), course of theoretical physics, vol 9. Pergamon, Oxford
36. Yaghjian AD (1980) Electric dyadic green's functions in the source region. Proc IEEE 68 (2):248–263. https://doi.org/10.1109/PROC.1980.11620
37. van Bladel, J.: Singular electromagnetic fields and sources. Clarendon, Oxford (1991). doi: https://doi.org/10.1137/1034110

Chapter 18
Temperature Studies of Luminescence in Nanosize SnO_2 Films

V. Grinevych, L. Filevska, V. Smyntyna, and B. Ulug

18.1 Introduction

Tin dioxide, widely used in various electronics industries, is not known as a luminophore material; however, low-temperature luminescence of crystalline tin dioxide has been described in the literature back to 1979 [1]. The band of intrinsic SnO_2 luminescence is located in the near-ultraviolet spectrum region. In the bulk tin dioxide samples in the visible range, wide photoluminescence (PL) bands are observed at low temperatures in the regions of 2 and 2.5 eV, which are associated with interstitial tin/oxygen vacancies electronic transitions [2, 3]. However, with temperature increasing, the intensity of such a photoluminescence decreases, and at a room temperature, the PL is practically invisible. The PL spectra of nanoscale tin dioxide samples differ from the bulk material spectra, which were shown by a number of researches [2–4]. The nanoscale forms of this material have recently been increasingly characterized as luminescing at the room temperature [5–8].

Photoluminescence of tin dioxide nanosized particles at the liquid nitrogen temperature was studied by Meier and colleagues [4]. PL was excited by GaN LED at a wavelength of 400 nm. The researchers detected a PL peak at a wavelength of 625 nm ($E = 2$ eV) independent of particles' size, which witnesses, according to the authors, about the group defect responsible for it, associated with oxygen vacancies. At the same time, the PL intensity increases with nanoparticle size growth from 5 to 20 nm and with the samples' composition approaching to stoichiometry

V. Grinevych (✉) · L. Filevska · V. Smyntyna
Department of Physics, Odessa I.I. Mechnikov National University, Dvoryanskaya str., 2, Odessa 65082, Ukraine
e-mail: grinevich@onu.edu.ua

B. Ulug
Department of Physics, Faculty of Science, Akdeniz University, Antalya, Turkey

O. Fesenko, L. Yatsenko (eds.), *Nanooptics, Nanophotonics, Nanostructures, and Their Applications*, Springer Proceedings in Physics 210,
https://doi.org/10.1007/978-3-319-91083-3_18

from $SnO_{1.5}$ to $SnO_{1.7}$. This fact correlates with a high density of electrons' states within the gap width obtained from the absorption measurements presented at the same work. The authors associate this intensity increase with the low rate of nonradiative recombinations in $SnO_{1.7}$ due to the low total defect density. Obviously, in an ideal, in a defect-free SnO_2 crystal, there should be no defects responsible for PL. If the number of oxygen vacancies increases, the PL intensity will also increase until the processes of nonradiative recombination will begin to dominate, as in the case of $SnO_{1.5}$.

Korean researchers also explained the influence of oxygen vacancies on PL observed by them in the region of 2.5 eV for thin SnO_2 films deposited by the CVD method. Bonu et al. [7] also explained the observed luminescence at 2.54 eV and 2.42 eV by oxygen defects, namely, in-plane and bridging "O" vacancies. The broad luminescence peak, observed by [7], at about 1.96 eV in NPs SnO_2 was associated with the hydroxyl groups OH^- on the surface of particles. In [8], the broad luminescence bands observed in the 350–550 nm region were associated with defect states on the surface of SnO_2 nanoparticles. As it may be seen, the surface states play an important role in the luminescence of tin dioxide nanoforms.

This paper presents the luminescence low-temperature studies of nanostructured tin dioxide films obtained using polymers in order to establish the mechanisms of radiation and its temperature quenching.

18.2 Film Preparation and Research Methods

SnO_2 films were obtained by a sol-gel method using polyvinyl acetate (PVA) to improve structuring. In samples' series fabrication, the content of the precursor (bis (acetylacetonato) dichlorotin (BADCT)) in the initial solution varied from 1% to 10%. After deposition on the glass substrate, the samples were annealed until the organic components were removed and a transparent layer of tin dioxide was formed.

The experiment was carried out according to the standard procedure [9, 10]. A nitrogen laser with a radiation wavelength of 337 nm was used to excite the luminescence. The photoluminescence of the samples was studied in the temperature range 9–300 K.

18.3 Experimental Results and Discussion

The films of tin dioxide obtained with the use of polymers showed photoluminescence at room temperature in the orange-red spectrum region. The results of photoluminescence at room temperature on SnO_2 layers, depending on the amount of precursor, are shown in Fig. 18.1.

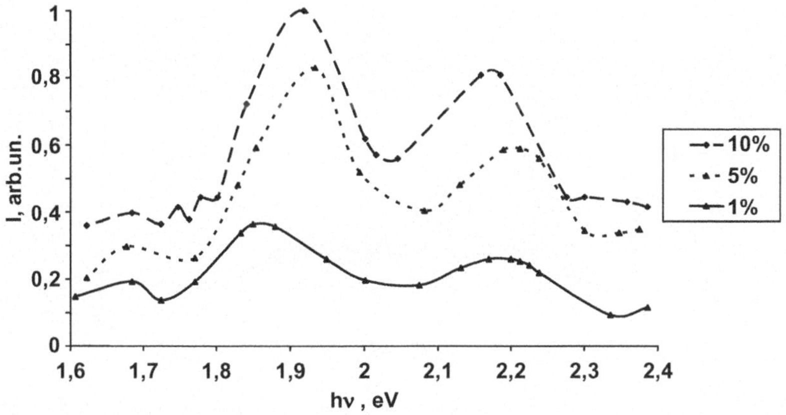

Fig. 18.1 The photoluminescence spectrum of SnO_2 films with different precursor concentrations (0.05% PVA) at room temperature

Table 18.1 Energy of maxima of PL bands for SnO_2 films with different precursor concentrations in the initial solution

Precursor concentrations	1%	5%	10%
The band in the 647 nm region	1.85 eV	1,88 eV	1,92 eV
The band in the 579 nm region	2,18 eV	2,2 eV	2,17 eV

As it can be seen (Fig. 18.1), in the emission spectrum, there are two intensive wide bands (1.92 eV for 10%, 1.94 eV for 5%, and 1.86 eV for 1% of BADCT and 2.17 eV for 10%, 2.2 eV for 5%, and 2.18 eV for 1% of BADCT). Moreover, the energy positions of the maxima of these bands are nonlinearly dependent on the concentration of tin dioxide precursor in the original gel. The bands themselves are quite wide and complex. Thus, the peak in the region 1.86 eV (for 1% of the precursor) is considered as a broadening of the bands 1.92 eV and 1.94 eV for films with high precursor content. The figure also shows an increase in the radiation intensity with an increase in the precursor concentration. This may indicate an improvement in the crystallinity of the SnO_2 film resulted in a decrease of nonradiative transitions. The values of both peaks in films of different compositions are presented in the Table 18.1.

Meier and colleagues [4] associate the detected photoluminescence peak at 625 nm (2 eV) at a liquid nitrogen temperature in nanoscale SnO_x particles with a group defect formed by oxygen vacancies. The PL was established in the present work in nanostructured tin dioxide films at room temperature. It was supposed that the peak at 647 nm (1.9 eV) is also associated with oxygen vacancies in the samples studied. The difference in the peaks' energies may be due to the different energy of the PL excitation and the differences in the samples' structure. At the same time, according to Bonu's data [7] on hydroxyl groups on the surfaces of SnO_2 nanoscale particles, this peak may also be attributed to them. This supposition can be confirmed

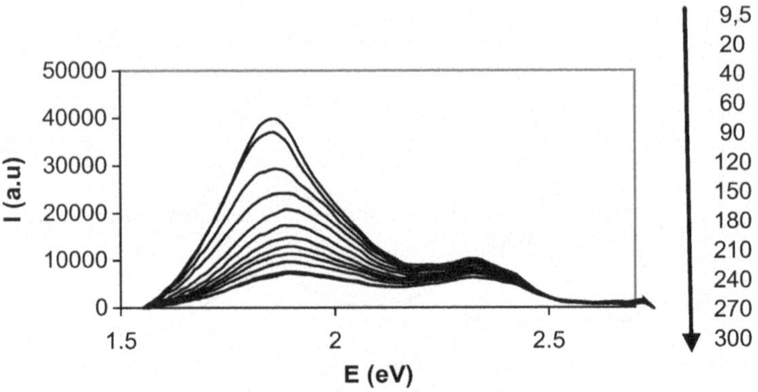

Fig. 18.2 Luminescence spectra of a tin dioxide film (produced from a solution containing 0.05% PVA and 1% precursor of tin), measured at temperatures from 9.5 to 300 K

by additional studies of PL in vacuum at elevated temperatures after additional annealing, but this was not the task of this paper.

It is known from the reference data [11] that a singly charged Sn has a band of 579 nm (2.15 eV) in its spectrum. Comparing these facts, it may be supposed the correspondence of the group of bands 2.17–2.2 eV in the films to the centers of luminescence, which are the interstitial tin atoms or clusters of these atoms. The presence of metal clusters in SnO_2 films was established earlier by the authors [12]. Their notable contribution was marked in the electrical conductivity and adsorption activity of SnO_2 layers. In addition, it was shown in [13] that at $T = 723$ K, there were at least three phases in tin dioxide films: Sn, SnO, and SnO_2. On the other hand, the presence of excess tin indicates the presence of vacancies in the oxygen structure. Having the reports of luminescence on oxygen vacancies in this region [2, 3, 5, 6], it is possible to associate the observed peak with them.

Figure 18.2 shows the temperature dependences of the luminescence of tin dioxide films studied in the temperature range from 9.5 to 300 K.

As it can be seen from the figure, two peaks are observed in the photoluminescence spectra: the first is in the region 1.85–1.9 eV, and the second is in the region 2.32 eV. The value of the second peak slightly differs from the previously measured value (2.18 eV). This may indicate the structural changes that occur in the layer in the process of research procedure, namely, the surface oxygen influence and its gradual incorporation into the crystal structure.

The results of studies in the forms of the peaks' energy, intensity, and half-width temperature dependencies are shown in Figs. 18.3 and 18.4.

The temperature dependences show a decrease in the peaks' luminescence intensity with temperature growth. The observed effect is explained by temperature quenching of luminescence as a result of nonradiative recombination. This is a consequence of a significant density of electron states in the forbidden band, specific for tin dioxide [14], especially for nanoscale structures and thin films with a

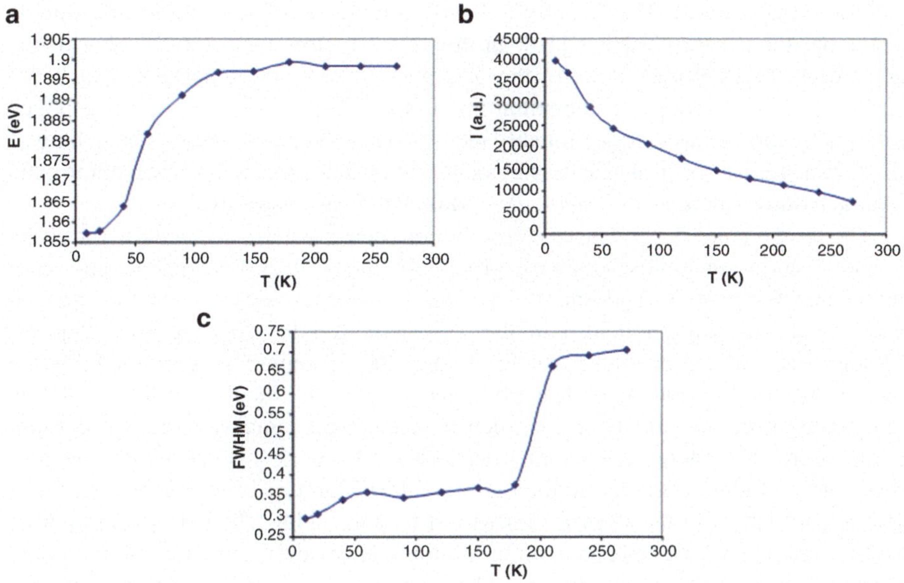

Fig. 18.3 Temperature dependences of the peak' energy 1 (**a**), its intensity (**b**), and half-width (**c**)

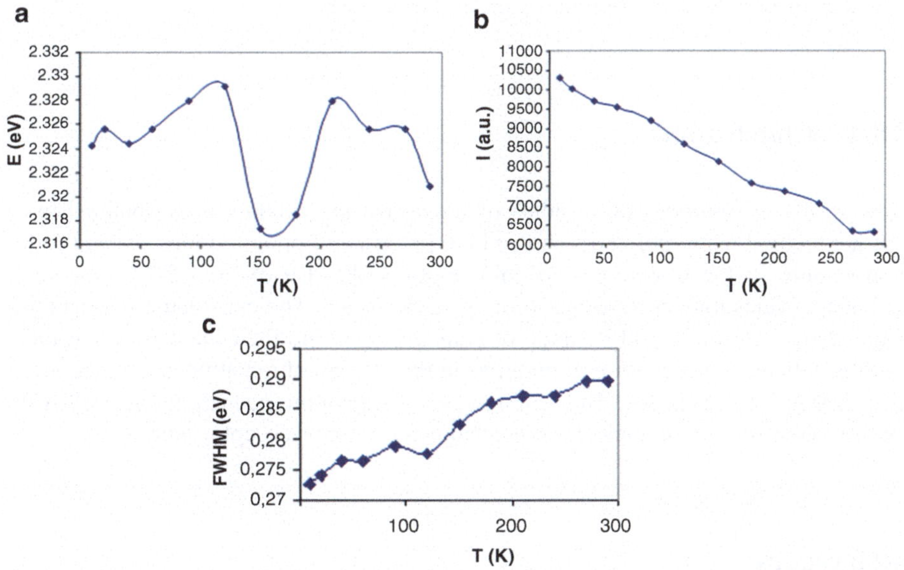

Fig. 18.4 Temperature dependences of the peak' energy 2 (**a**), its intensity (**b**), and half-width (**c**)

significant disruption of long-range order [2, 4]. The change in the peak' 1 temperature dependence of intensity built as a graph of semilogarithmic dependence on the reciprocal temperature showed the presence of two regions with activation energies (E_a) ~ 0.0114 eV (at a lower temperature up to ~ 50 K) and ~ 0.0156 eV (from 120 K to room temperature). This usually indicates a temperature quenching of luminescence on donor-acceptor pairs [10]. In this case, the low-temperature value of the activation energy corresponds to the activation of donors, which can match the surface hydroxyl groups or interstitial tin atoms. A higher value of E_a corresponds to the activation of acceptors, probably identified as oxygen vacancies. Quenching of the luminescence for peak 2 is less evident and also occurs on donor-acceptor pairs with activation energies of 0.0001 eV and 0.00545 eV, respectively.

The temperature dependences of the luminescence bands' half-widths for peaks 1 and 2 show an almost stepwise growth in the 180–200 K region and more noticeable for peak 1. The energy of peak 1 increases with temperature growth and reaches its saturation also in the region of 200 K. This indicates that the luminescence at temperature growth in the 200 K region is supported by the increasing contribution of optical phonons, which are usually not active at low temperatures (up to 200 K) [15]. Such a mechanism remains basic up to room temperature. The energy temperature dependence for peak 2 is not monotonic. The peak energy failure observed in the region of 150 K and its growth in the region of 200 K correlate with the stepwise growth of its half-width. The latter fact indicates both a change in the mechanism of luminescence and activation of phonons' participation in it. As it may be seen, the participation of phonons in the emission process is specific for both peaks. In favor of this is also the fact that tin dioxide is an indirect-gap semiconductor [16], and the participation of phonons in such a case is specific for optical absorption and emission processes.

18.4 Conclusion

The results of luminescence studies of nanoscale tin dioxide films obtained by a sol-gel method using polymers showed the presence of photoluminescence at room temperature in the orange-red spectral region (1.85–1.9 eV and 2.32 eV). Such radiation values may correspond to oxygen vacancies. The considered temperature dependences in the 9–300 K range of peak energies and their intensities and half-widths indicate the activation of phonons in the process of radiation at a temperature of ~ 200 K for both peaks. The luminescence temperature quenching of tin dioxide occurs according to the quenching mechanism on donor-acceptor pairs.

References

1. Agekyan VT (1979) Slojnyj spektr exitinno-primesnych komplexov v defektnych kristallach dvuokisi olova. Complex spectrum of exciton-impurity complexes in defective crystals of tin dioxide. Letters in the JETP 29(8):475–479
2. Ryabtsev SV (2011) Thesises of the doctor of sciences dissertation. Voronezh State University, Voronezh

3. Agyekyan VF, Yu Serov A, Filosofov NG (2014) Izluchenije sveta kristallami dvuokisi olova (Radiation of light by tin dioxide crystals). FTP (Semiconductors) 48(4):458–461
4. Meier C, Luttjohann S, Kravets VG, Nienhaus H, Lorke A, Ifeacho P, Wiggers H, Schulz C, Kennedy MK, Kruis FE (2006) Vibrational and defect states in SnO_x nanoparticles. J Appl Phys 99:113108
5. Gu F, Wang SF, Song CF, Lü MK, Qi YX, Zhou GJ, Xu D, Yuan DR (2003) Synthesis and luminescence properties of SnO_2 nanoparticles. Chem Phys Lett 372(3–4):451–454
6. Jeong J, Choi SP, Hong KJ, Song HJ, Park JS (2006) Structural and optical properties of SnO_2 thin films deposited by using CVD techniques. J Korean Phys Soc 48(5):960–963
7. Bonu V, Das A, Amirthapandian S, Dhara S, Tyagi AK (2015) Photoluminescence of oxygen vacancies and hydroxyl group surface functionalized SnO_2 nanoparticles. Phys Chem Chem Phys 17(15):9794–9801. https://doi.org/10.1039/c5cp00060b
8. Lin T, Wan N, Xu J, Xu L, Chen KJ (2010) Size-dependent optical properties of SnO_2 nanoparticles prepared by soft chemical technique. J Nanosci Nanotechnol 10(7):4357–4362
9. Emmanuel NM, Kuzmina MG (eds) (1985) Experimentalnye metody chimicheskoj kinetiki (Experimental methods of chemical kinetics). Publishing House of Moscow University, Moscow
10. Serdyuk VV, Vaksman YF (1988) Luminescencija poluprovodnicov (Luminescence of semiconductors). Vishcha shkola, Kiev-Odessa
11. Seidel AN, Prokofiev VK, Raiskij SM, Slavny VA, Ya E (1977) Shreider. Tablitscy spektralnych linij (Tables of spectral lines). Publishing House Science, Moscow, p 679
12. Viter RV, Smyntyna VA, Evtushenko NG, Filevskaya LN, Kurkov VV (2002) Issledovanije adsorbtscionno-kineticheskich characteristic tonkich plenok SnO_2 (Investigation of the adsorption-kinetic characteristics of thin SnO_2 films). Photoelectronics 11:c.109–113. Golovanov V, Rantala T, Lantto V (2001) Rehybrydization at (110) faces of SnO_2. Photoelectronics 10:80–83
13. Khoviv AM, Logacheva VA, Novikov OV (2004) Features of oxidation of tin films under conditions of reduced and atmospheric pressure of oxygen under the influence of IR radiation. VESTNIK VSU Series: Chemistry Biology Pharmacy 1:101–106. (in Russian)
14. Kilic C, Zunger A (2002) Origins of coexistence of conductivity and transparency in SnO_2. Phys Rev Lett 88(9–4):095501–095505
15. Canestraro CD, Roman LS, Persson C (2009) Polarization dependence of the optical response in SnO_2 and the effects from heavily F doping. Thin Solid Films 517:6301–6304
16. Anastasescu M, Gartner M, Mihaiu S, Anastasescu C, Purica M, Manea E, Zaharescu M (2006) Optical and structural properties of SnO_2-based sol-gel thin films. In: Proceedings of the International Semiconductor Conference (Sinaia, 2006) IEEE, vol 1, pp 163–166

Part III
Nanostructured Interfaces and Surfaces

Chapter 19
Fabrication and Characterization of High-Performance Anti-reflecting Nanotextured Si Surfaces for Solar Cells

Stepan Nichkalo, Anatoly Druzhinin, Valeriy Yerokhov, and Oleksandr Ostapiv

19.1 Introduction

In the past decades, great efforts have been undertaken to develop various photovoltaic (PV) devices [1–11]. Today Si-based PV devices remain the basis of the current PV industry due to the abundance of Si materials and the high efficiency of Si solar cells [12]. However, the high reflective index of Si causes the reflection of more than 30% of incident light. As a result, the photoconversion efficiency of the Si-PV device greatly reduces. For traditional Si wafer-based solar cells, the pyramidal or inverted pyramidal structures were generally constructed on Si surface to reduce the reflection loss for incident light and to increase the light absorption [13, 14]. However, such microstructured surfaces are not able to reduce reflectance below 10% [15–18]. At the same time, the anti-reflecting coating layers, such as Si_3N_4, MgF_2, and Si_3N_4/MgF_2, are also applied to suppress light reflection of microstructured surface, but they suffer from destructive interference of reflected light and lead to a narrowing of the solar spectrum of absorbed photons, as a result [19–21]. In this regard, the nanotexturing of Si surface is widely studied to reach maximum values of the absorption coefficient and to reduce reflection in a wide spectral range, improving the efficiency of solar cells.

Among microstructures, Si nanowires (SiNWs) are also widely considered as an important class of nanoscale building blocks for high-performance devices due to their unique structural, electrical, and thermoelectric properties in addition to their compatibility with current Si-based microelectronics [22–33]. Recent studies on the optical characteristics of SiNWs demonstrate their promising applications in solar cell [34–40]. These nanostructures are long enough to absorb most of incident light, and their small diameters provide a short collection length for excited carriers in a

S. Nichkalo (✉) · A. Druzhinin · V. Yerokhov · O. Ostapiv
Lviv Polytechnic National University, Lviv, Ukraine

O. Fesenko, L. Yatsenko (eds.), *Nanooptics, Nanophotonics, Nanostructures, and Their Applications*, Springer Proceedings in Physics 210,
https://doi.org/10.1007/978-3-319-91083-3_19

direction normal to the light absorption [38, 41]. Additionally, unlike bulk Si with its indirect optical band gap of 1.12 eV and low absorption coefficient of 10^4/cm, the optical band gap of SiNWs can be varied between 1.1 eV and 3.5 eV by decreasing SiNW diameter between 7 nm and 1.3 nm, respectively [41, 42]. Several research groups showed that in the near ultraviolet-visible-near-infrared region, optical reflectance of SiNWs is one to two orders of magnitude lower than Si [43–47]. However, such structures are effective the best under direct sunlight illumination and could not face the sun at the appropriate angle from morning to evening [48, 49]. The use of sun-tracking systems for the purpose of receiving direct light illumination is not cost-effective for practical applications. The concept of a complex structure combining the advantages of pyramids and SiNWs might help to achieve the omnidirectional light absorption and overcome the directional dependence of PV performance [49]. Realization of this concept predicts utilizing of traditional alkaline texturing of Si wafer to produce the pyramidal Si and the metal-assisted chemical etching (MACE) to fabricate SiNWs on the side faces of pyramids. The use of MACE method for fabricating of SiNWs was motivated by its simplicity, good cost-efficiency, versatility, and ability to control various parameters, e.g. cross section, shape, diameter, length, and orientation [50, 51]. In contrast, the crystallographic orientation of SiNWs grown by vapour-liquid-solid (VLS) technique depends upon the diameter of nanowire [52, 53]. Moreover, the crystalline quality of SiNWs produced by MACE method generally is high, and their surfaces are typically rougher than VLS-grown NWs [50, 54]. In general, SiNWs prepared by MACE demonstrate higher solar cell performance over VLS-grown SiNWs [55].

Here we present experimental results on application of MACE method for fabricating different morphologies on the surface of Si wafer and the comparison of their optical properties in terms to develop photosensitive structures with low reflectivity and high light absorption over an extended solar spectral range.

19.2 Experimental Details

Single-crystalline *p*-type Si (100) and Si (111) wafers with resistivity of 0.3–1 $\Omega \times$ cm were chosen for experiments. The wafers were cut into square samples of 2×2 cm^2 and then sequentially cleaned in acetone, isopropyl alcohol (IPA), and deionized (DI) water for 20 min at room temperature. Afterwards, samples were dipped in the mixed solution of ammonium hydroxide NH_4OH (30%), hydrogen peroxide H_2O_2 (35%), and DI water H_2O with volumetric ratio v/v/v = 1/1/5 at 70 °C for 10 min to remove any organic residues. In order to obtain an efficient anti-reflecting surface with a maximum absorption and the least possible reflection for possible use in photovoltaics, three types of morphology of Si surface were made and examined, namely, (1) rough nanograss, (2) random pyramids, and (3) pyramids with nanotextured side faces.

Fabrication of the nanograss morphology was carried out through metal-assisted chemical etching of Si (111) wafer, accompanied by the following steps. At the first stage, clean square Si wafer samples were dipped in a solution of 5 M HF and 2 mM $AgNO_3$ for 2 min at room temperature that resulted in a deposition of Ag particles on the Si surface. Then, samples were rinsed by DI water for 10 s and immersed in the etchant solution HF(40%) + H_2O_2(35%) + H_2O (v/v/v = 4/1/4) at room temperature for 10 min. After etching process, samples were immersed in a mixed solution of NH_4OH and H_2O_2 (v/v = 3/1) at room temperature for 10 min to remove Ag residues. Prepared samples were rinsed with DI water for 10 min and dried.

Random pyramids were fabricated by anisotropic etching of Si (100) wafer in a mixed solution of potassium hydroxide (KOH) (9 wt %) and acetic acid (5 vol %) at 75 °C for 10 min. The textured Si samples were then immersed in dilute hydrochloric acid (HCl) for 10 min and hydrofluoric acid (HF) for 5 min to remove any residue KOH and silicon dioxide, respectively. At the next, pyramid-textured Si samples were rinsed with DI water for 20 min at room temperature and dried.

For the nanotexturing of the side faces of Si pyramids, the MACE method was used. First, Ag particles were deposited on precleaned pyramid-textured Si wafers by immersing samples in a solution of 0.14 M HF and 0.5 mM $AgNO_3$ for 7 min at room temperature and then rinsed by DI water for 10 s. Low molar concentrations of HF and $AgNO_3$ were used in order to obtain well-defined and small seeds of Ag on the side faces of Si micropyramids. Subsequently, pyramid-textured Si samples covered by Ag nanoseeds were immersed in the mixed etching solution of HF (40%), H_2O_2(35%), and DI water (v/v/v = 4/1/4) at room temperature for 12 min. Finally, all the samples were immersed in a mixed solution NH_4OH + H_2O_2 (v/v = 3/1) at room temperature for 10 min to remove the residual Ag on the surface of the samples.

The surface of Si wafer was characterized by using a SELMI 106I scanning electron microscope (SEM). Optical spectra of the samples were measured by spectrophotometer at the wavelength ranging from 300 to 1100 nm.

19.3 Results and Discussion

Figure 19.1 shows the cross-section view SEM images of Si wafer after 10 min of Ag-assisted chemical etching in HF + H_2O_2 + H_2O solution at room temperature. As a result, vertically aligned SiNWs were formed on Si surface. As can be seen in Fig. 19.1a–b, the NWs are preferentially cone-shaped reminding a grass, with diameters ranging from 100 nm to 500 nm and an average height of about 4 μm.

Shown in Fig. 19.2a–b are the top-view and cross-section view SEM images of random Si pyramids formed as a result of etching of Si wafer in KOH-based solution at 75 °C for 10 min. As we see, the average size of pyramids does not exceed 5 μm.

Fig. 19.3 shows the top-view SEM image of pyramid-textured Si wafers with predeposited Ag nanoseeds after immersing samples in a solution of 0.14 M HF/0.5 mM $AgNO_3$ for 7 min at room temperature. Consequently, chemical treatment of pyramidal samples covered by Ag nanoseeds in etching solution of HF (40%)/H_2O_2(35%)/H_2O (v/v/v = 4/1/4) at room temperature for 12 min revealed to

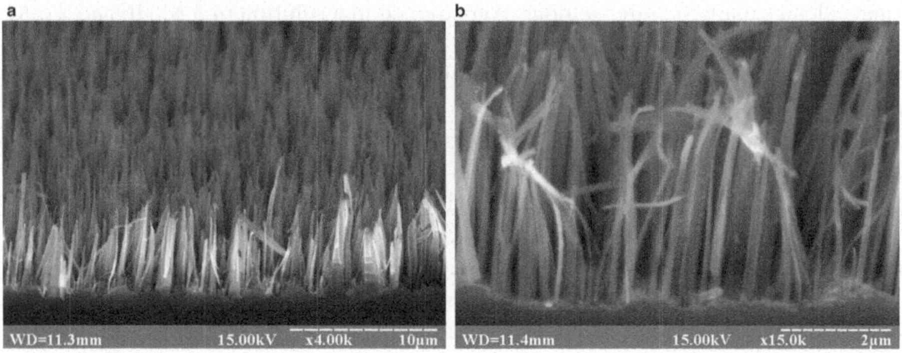

Fig. 19.1 Cross-section view SEM images of Si wafer after 10 min of Ag-assisted chemical etching in $HF/H_2O_2/H_2O$ (v/v/v = 4/1/4) solution at room temperature at different scale. (**a**) The scale 10 μm. (**b**) The scale 2 μm

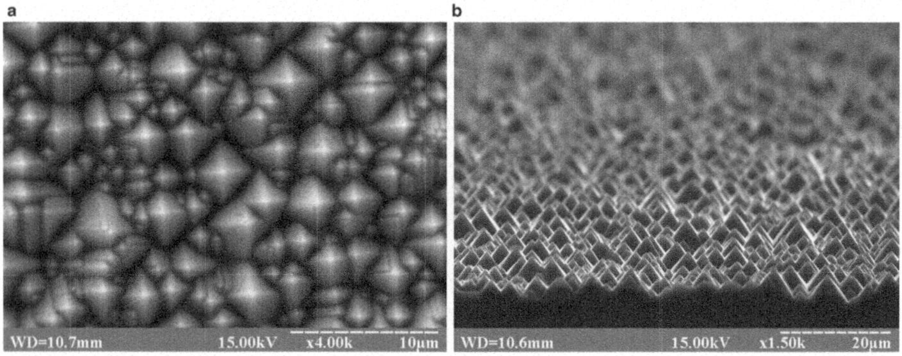

Fig. 19.2 SEM images of random Si pyramids formed as a result of etching of Si wafer in a solution of KOH (9%) and acetic acid (5%) at 75 °C for 10 min. (**a**) Top view. (**b**) Cross-section view

the formation of <111> – oriented SiNWs which were etched normally to the surface planes of Si pyramids (see Fig. 19.4a). As can be seen in Fig. 19.4b, the NWs are 325 nm in height with mean diameter about 130 nm.

The rough nanograss on micropyramids predicted to have a strong absorption and extremely low reflection ability. This suggestion is confirmed by experimental results of measurement of absorbance (Fig. 19.5a) and reflectance spectra (Fig. 19.5b) of Si wafer samples with various surface morphologies. It was found that in comparison to other textures, the pyramidal Si samples textured by Si nanograss demonstrate a highest value of light absorption, more than 98%, whereas for a non-textured Si wafer and a pyramid-textured Si, this value is found to be typical, 60% and 85%, respectively (Fig. 19.5a). The high absorption of Si nanograss, about 95%, could be explained in terms of specific nanowire geometry, which is conical (see Fig. 19.1b). According to [16], compared with planar thin film

Fig. 19.3 Top-view SEM image of pyramidal Si wafer with deposited Ag nanoseeds after immersing samples in a solution of 0.14 M HF/0.5 mM $AgNO_3$ for 7 min at room temperature

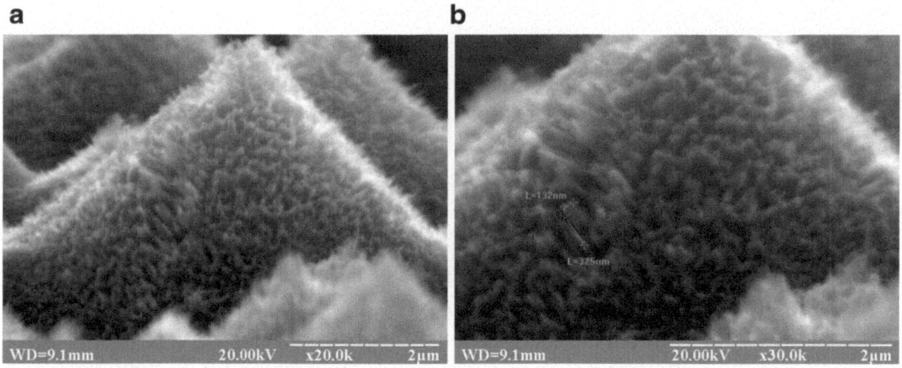

Fig. 19.4 Cross-section view SEM images of Si micropyramids nanotextured by Si nanograss after etching in HF(40%)/H_2O_2(35%)/H_2O (v/v/v = 4/1/4) at room temperature for 12 min. (**a**) General view of the pyramid. (**b**) Scaled-up view of the side face of a pyramid with SiNWs

and NWs with uniform diameters, the cone-shaped Si nanostructures show improved absorption because of a gradual increase of the effective refractive index from the Si surface to air. From another hand, in aperiodic SiNWs, absorption at long-wavelength regime is not degraded, unlike periodic NWs, in which calculated absorptance for early band edge photons dwarfs due to the significant transmission loss [56]. In their work [56], authors showed that optimization of aperiodic NWs by using a random walk algorithm reveals to an over 100% enhancement factor in PV efficiency compared to the periodic counterparts.

Figure 19.5b compares the optical reflection between pyramid-textured Si, Si nanograss, and Si pyramids with a nanograss texture of side faces of pyramids. In contrast to pyramid-textured Si and Si nanograss, the Si pyramids with a nanograss texture possess a lowest reflection ability (< 1%) in all optical spectral range of wavelength. For the record, the low reflection value (about 4%) was observed for Si

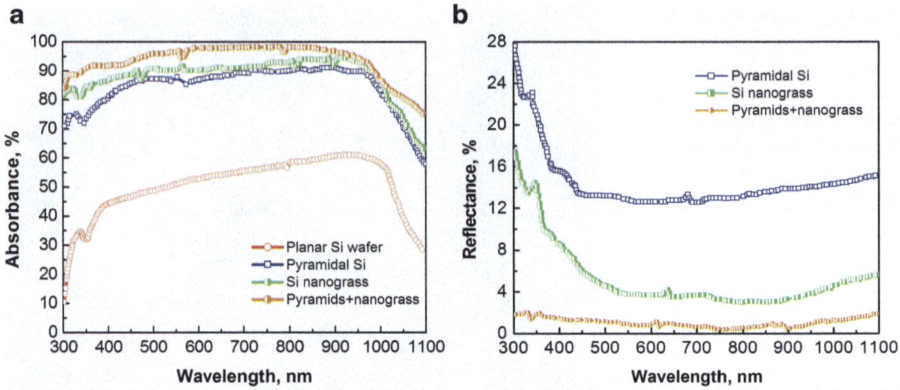

Fig. 19.5 (**a**) Absorbance spectra for pyramid-textured Si, Si nanograss, complex structure Si pyramids+Si nanograss, and planar Si wafer (serve as reference). (**b**) Reflectance spectra for pyramid-textured Si, Si nanograss, complex structure Si pyramids+Si nanograss

nanograss as well. These findings confirm the accusation that vertically aligned NWs with random position or diameter can also achieve similar low reflectance and high absorption as the ordered NWs with uniform diameter [57]. In their work [57], authors showed that the random diameters of the nanowire arrays can lead to different broadened resonant frequencies which give rise to absorption enhancement. While the absorption at the original off-resonance frequency is enhanced, the original resonance frequency maintains high, caused by multiple scattering induced by the random arrangement or diameter of NWs. In other words, aperiodic Si NWs demonstrate suppression of light reflection similar to the random pyramid texture. Thus, almost the same behaviour of a reflection dependence for Si nanograss and pyramidal Si samples is observed (see Fig. 19.5b).

19.4 Conclusions

It was demonstrated that the metal-assisted chemical etching is a simple, versatile, and flexible method, which can be successfully utilized for fabrication of Si nanostructures on the surface of Si wafer with good anti-reflecting properties. By using this method, an array of SiNWs and a complex Si structure composed of Si pyramids, obtained by conventional alkali chemical etching, and SiNWs prepared on the side faces of pyramids have been produced on Si wafers. It was found that Si micropyramids textured by a rough SiNWs with an average diameter of 130 nm, and 325 nm in height, show a strong absorption and extremely low reflection ability. In particular, the absorbance of pyramidal Si samples textured by Si nanograss was more than 98% and reflectance less than 1% in all range of wavelength (300–1100 nm), whereas for a non-textured Si wafer and a pyramid-textured Si, these values were found to be typical. The low reflection value (about 4%) was also observed for Si nanograss. Such nanostructured surfaces in the form of Si micropyramids and SiNWs might exhibit better omnidirectional light-trapping ability by multiple reflections resulting in enhanced optical characteristics and will find a wide variety of significant applications in solar cells, photodetectors, and optoelectronic devices.

References

1. Chapin DM, Fuller CS, Pearson GL (1964) A new silicon p-n junction photocell for converting solar radiation into electrical power. J Appl Phys 25:676
2. Green MA (2001) Crystalline silicon photovoltaic cells. Adv Mater 13:1019–1022
3. Goetzberger A, Hebling C, Schock HW (2003) Photovoltaic materials, history, status and outlook. Mater Sci Eng R 40:1–46
4. Swanson RM (2006) A vision for crystalline silicon photovoltaics. Prog Photovolt 14:443–453
5. Zhao J, Wang A, Green MA, Ferrazza F (1998) 19.8% efficient "honeycomb" textured multicrystalline and 24.4% monocrystalline silicon solar cells. Appl Phys Lett 73:1991–1993
6. Knechtli RC, Loo RY, Kamath GS (1984) High-efficiency GaAs solar cells. IEEE Trans Electron Devices 31:577–588
7. Tanabe K (2009) A review of ultrahigh efficiency III-V semiconductor compound solar cells: multijunction tandem, lower dimensional, photonic up/down conversion and plasmonic nanometallic structures. Energies 2:504–530
8. Hoppe H, Sariciftci NS (2004) Organic solar cells: an overview. J Mater Res 19:1924–1945
9. Yerokhov V, Ierokhova O (2016) Improved porous silicon-based multifunctional materials for the solar cells antireflection coating. 2016 International Conference on Electronics and Information Technology, EIT 2016 – Conference Proceedings, pp 49–52
10. Yerokhov V, Ierokhova O (2016) Coatings of the black-silicon type for silicon solar cells. Modern problems of radio engineering, telecommunications and computer science, Proceedings of the 13th International Conference on TCSET 2016, pp 388–391
11. Kim JY, Lee K, Coates NE, Moses D, Nguyen TQ, Dante M, Heeger AJ (2007) Efficient tandem polymer solar cells fabricated by all-solution processing. Science 317:222–226
12. Peng K, Xu Y, Wu Y, Yan Y, Lee ST, Zhu J (2005) Aligned single-crystalline Si nanowire arrays for photovoltaic applications: review. Small 1(11):1062–1067
13. Sun CH, Min WL, Linn NC, Jiang P, Jiang B (2007) Templated fabrication of large area subwavelength antireflection gratings on silicon. Appl Phys Lett 91:231105
14. Chu AK, Wang JS, Tsai ZY, Lee CK (2009) A simple and cost-effective approach for fabricating pyramids on crystalline silicon wafers. Sol Energy Mater Sol Cells 93:1276–1280
15. Wang HP, Lai KY, Lin YR, Lin CA, He JH (2010) Periodic Si nanopillar arrays fabricated by colloidal lithography and catalytic etching for broadband and omnidirectional elimination of Fresnel reflection. Langmuir 26:12855–12858
16. Han SE, Chen G (2010) Toward the Lambertian limit of light trapping in thin nanostructured silicon solar cells. Nano Lett 10(11):4692–4696
17. Wang FY, Yang QD, Xu G et al (2011) Highly active and enhanced photocatalytic silicon nanowire arrays. Nanoscale 3:3269–3276
18. Lin H, Xiu F, Fang M et al (2014) Rational design of inverted nanopencil arrays for cost-effective, broadband and omnidirectional light harvesting. ACS Nano 8:3752–3760
19. Wang HP, Lin TY, Tsai ML et al (2014) Toward efficient and omnidirectional n-type Si solar cells: concurrent improvement in optical and electrical characteristics by employing microscale hierarchical structures. ACS Nano 8:2959–2969

20. Peng KQ, Lee ST (2011) Silicon nanowires for photovoltaic solar energy conversion. Adv Mater 23:198–215
21. Parida B, Choi J, Lim G, Park S, Kim K (2014) Formation of nanotextured surfaces on microtextured Si solar cells by metal-assisted chemical etching process. J Nanosci Nanotechnol 14(12):9224–9231
22. Cui Y, Wei QQ, Park HK, Lieber CM (2001) Nanowire nanosensors for highly sensitive and selective detection of biological and chemical species. Science 293:1289–1292
23. Maryamova I, Druzhinin A, Lavitska E, Gortynska I, Yatzuk Y (2000) Low temperature semiconductor mechanical sensors. Sens Actuators A Phys 85:153–157
24. Druzhinin A, Lavitska E, Maryamova I (1999) Medical pressure sensors on the basis of silicon microcrystals and SOI layers. Sens Actuators B Chem 58:415–519
25. Cui Y, Zhong ZH, Wang DL, Wang WU, Lieber CM (2003) High performance silicon nanowire field effect transistors. Nano Lett 3:149–152
26. Peng KQ, Jie JS, Zhang WJ, Lee ST (2008) Silicon nanowires for rechargeable lithium-ion battery anodes. Appl Phys Lett 93:033105
27. Goldberger J, Hochbaum AI, Fan R, Yang PD (2006) Silicon vertically integrated nanowire field effect transistors. Nano Lett 6:973–977
28. Chen LJ (2007) Silicon nanowires: the key building block for future electronic devices. J Mater Chem 17:4639
29. Druzhinin A, Ostrovskii I, Kogut I (2006) Thermoelectric properties of Si-Ge whiskers. Mater Sci Semicond Process 9:853–857
30. Zheng G, Patolsky F, Cui Y, Wang WU, Lieber CM (2005) Multiplexed electrical detection of cancer markers with nanowire sensor arrays. Nat Biotechnol 23:1294–1301
31. Chan CK, Peng HL, Liu G, McIlwrath K, Zhang XF, Huggins RA, Cui Y (2008) High-performance lithium battery anodes using silicon nanowires. Nat Nanotechnol 3:31–35
32. Hochbaum A, Chen R, Delgado RD, Liang W, Garnett EC, Najarian M, Majumdar A, Yang P (2008) Enhanced thermoelectric performance of rough silicon nanowires. Nature 451:163–167
33. Druzhinin A, Ostrovskii I, Kogut I, Nichkalo S, Shkumbatyuk T (2011) Si and Si-Ge wires for thermoelectrics. Phys Status Solidi C 8(3):867–870
34. Fang H, Li X, Song S, Xu Y, Zhu J (2008) Fabrication of slantingly-aligned silicon nanowire arrays for solar cell applications. Nanotechnology 19:255703
35. Garnett EC, Yang P (2008) Silicon nanowire radial p−n junction solar cells. J Am Chem Soc 130:9224–9225
36. Kelzenberg MD, Turner-Evans DB, Kayes BM, Filier MA, Putnam MC, Lewis NS, Atwater HA (2008) Photovoltaic measurements in single-nanowire silicon solar cells. Nano Lett 8:710–714
37. Stelzner T, Pietsch M, Andra G, Falk F, Ose E, Christiansen S (2008) Silicon nanowire-based solar cells. Nanotechnology 19:295203
38. Druzhinin AA, Yerokhov VY, Nichkalo SI, Berezhanskyi YI, Chekaylo MV (2015) Texturing of the silicon substrate with nanopores and Si nanowires for anti-reflecting surfaces of solar cells. J Nano Electron Phys 7(2):02030-1–02030-6
39. Zhu J, Yu Z, Burkhard GF, Hsu CM, Connor ST, Xu Y, Wang Q, McGehee M, Fan S, Cui Y (2008) Optical absorption enhancement in amorphous silicon nanowire and nanocone arrays. Nano Lett 9:279–282
40. Li J, Yu H, Wong SM, Zhang G, Sun X, Lo PGQ, Kwong DL (2009) Si nanopillar array optimization on Si thin films for solar energy harvesting. Appl Phys Lett 95:033102
41. Ramanujam J, Shiri D, Verma A (2011) Silicon nanowire growth and properties: a review. Mater Express 1(2):105–126
42. Ma DDD, Lee CS, Au FCK, Tong SY, Lee ST (2003) Small-diameter silicon nanowire surfaces. Science 299:1874–1877
43. Tsakalakos L, Balch J, Fronheiser J, Korevaar BA, Sulima O, Rand J (2007) Silicon nanowire solar cells. Appl Phys Lett 91:233117

44. Tsakalakos L, Balch J, Fronheiser J, Shih MY, LeBoeuf SF et al (2007) Strong broadband optical absorption in silicon nanowire films. J Nanophoton 1:013552
45. Druzhinin A, Yerokhov V, Nichkalo S, Berezhanskyi Y (2016) Micro- and nanotextured silicon for antireflective coatings of solar cells. J Nano Res 39:89–95
46. Nichkalo S, Druzhinin A, Evtukh A, Bratus' O, Steblova O (2017) Silicon nanostructures produced by modified MacEtch method for antireflective Si surface. Nanoscale Res Lett 12:106
47. Li X, Li J, Chen T, Tay BK, Wang J, Yu H (2010) Periodically aligned Si nanopillar arrays as efficient antireflection layers for solar cell applications. Nanoscale Res Lett 5:1721–1726
48. Kelzenberg MD, Boettcher SW, Petykiewicz JA et al (2010) Enhanced absorption and carrier collection in Si wire arrays for photovoltaic applications. Nat Mater 9:239–244
49. Pei Z, Hu H, Li S, Ye C (2017) Fabrication of orientation-tunable Si nanowires on silicon pyramids with omnidirectional light absorption. Langmuir 33(15):3569–3575
50. Huang Z, Geyer N, Werner P, de Boor J, Gösele U (2011) Metal-assisted chemical etching of silicon: a review. Adv Mater 23:285–308
51. Han H, Huang Z, Lee W (2014) Metal-assisted chemical etching of silicon and nanotechnology applications. Nano Today 9:271–304
52. Schmidt V, Senz S, Gösele U (2005) Diameter-dependent growth direction of epitaxial silicon nanowires. Nano Lett 5:931–935
53. Druzhinin AA, Ostrovskii IP (2004) Investigation of Si-Ge whiskers growth by CVD. Phys Status Solidi C 1(2):333–336
54. Druzhinin A, Evtukh A, Ostrovskii I, Khoverko Y, Nichkalo S, Dvornytskyi S (2015) Technological approaches for growth of silicon nanowire arrays. Springer Proc Phys 156:301–307
55. Yuan G, Aruda K, Zhou S, Levine A, Xie J, Wang D (2011) Understanding the origin of the low performance of chemically grown silicon nanowires for solar energy conversion. Angew Chem Int Ed 50:2334–2338
56. Lin C, Povinelli ML (2011) Optimal design of aperiodic, vertical silicon nanowire structures for photovoltaics. Opt Express 19:A1148–A1154
57. Bao H, Ruan X (2010) Optical absorption enhancement in disordered vertical silicon nanowire arrays for photovoltaic applications. Opt Lett 35:3378–3380

Chapter 20
Effects of Amino Acids Protonation on Double-Layer Parameters of the Electrode/Chlorates(VII) Interface, as well as Kinetics and Mechanism of Bi(III) Ion Electroreduction in the Aspect of the "Cap–Pair" Effect

Agnieszka Nosal-Wiercińska, Mariusz Grochowski, and Małgorzata Wiśniewska

20.1 Introduction

Irreversibility of metal ions electroreduction in the solutions of poor complex creative properties makes their determination by simple electrochemical methods difficult or even impossible. Knowledge of the organic substances effect on the rate of electrode reactions is of significant importance in analytics and elaboration of technological characteristics. Organic substances can inhibit, accelerate the electrode process or not affect its course. The inhibition of electroreduction processes by organic substances has been described in many papers [1–4]. It was shown that the adsorption of organic substances on the electrode blocks its surface, changing the surface charge and the potential of the internal Helmholtz plane. This affects the mechanism and kinetics of the electrode reaction, especially in the area of large covering of the electrode surface with adsorbate [5]. The first papers on the catalytic activity of organic substances discussed:

- Lowering the detection limit of Cd(III), Cu(II) and Ni(III) ions by hexamine [6]
- Increase of the peak currents Pb(II), Sn(II), Tl(I), Bi(III) and In(III) in the presence of methylene blue and thymol blue [7]

A. Nosal-Wiercińska (✉) · M. Grochowski
Department of Analytical Chemistry and Instrumental Analysis, Faculty of Chemistry, Maria Curie-Sklodowska University, Lublin, Poland
e-mail: anosal@poczta.umcs.lublin.pl

M. Wiśniewska
Department of Radiochemistry and Colloids Chemistry, Faculty of Chemistry, Maria Curie-Sklodowska University, Lublin, Poland

O. Fesenko, L. Yatsenko (eds.), *Nanooptics, Nanophotonics, Nanostructures, and Their Applications*, Springer Proceedings in Physics 210,
https://doi.org/10.1007/978-3-319-91083-3_20

– Catalytic effect of toluene ions – p-sulfonate on electroreduction of Cr(III) ions and 1,10-phenanthroline and chinaldine acid on electroreduction of Eu(III) ions [8, 9]

Systematic research of the catalytic activity of organic substances on the selected electrode processes carried out by Sykut et al. [10] led to the formulation in 1978 of the "cap–pair" rule, determining the conditions which must be satisfied by the depolarizer–supporting electrolyte – organic substance system for the electrode processes to be accelerated by organic substances. It was stated that an important role in accelerating the electrode process is played by the ability of the organic substance – undergoing adsorption on the electrode – to create in the by-electrode layer complexes with a depolariser and locate the potential of depolarizer's reduction in the area of the labile equilibrium of the adsorption organic substance [10].

Research on this issue allowed to explain catalytic mechanism of the activity of organic substances on the Zn(II) electroreduction [5, 11–14]. This mechanism is complex and includes both electron transition and chemical steps. According to Dalmata [5] acceleration of the first electron transition is associated with the formation of an active complex involving the catalysing substance. Accelerating activity of different organic substances, also surfactants, in relation to Zn(II) ions was pointed out to the papers [15, 16]. They also defined a mechanism describing depolarisation of Zn(II) ions. Studies of electroreduction of Eu(III) [17], In(III) [18, 19] and Bi(III) [20] ions showed that accelerating substances facilitate the exchange of electrons by creating an active complex with a depolarizer on the electrode surface. This complex can formed in the adsorption layer for In(III) and Bi(III) or outside the adsorption layer for Eu(III).

Lovrič and co-workers [21] studied the electroreduction of Bi(III) ions in a chlorate(VII) medium applying the methods of square wave voltammetric and cyclic voltammetric as well as DC polarography, demonstrating considerable influence of water activity on the slope of the experimental curves. They assumed that in the first stage of Bi(III) electroreduction, cations of the supporting electrolyte participate and adsorb as well as release water molecules in their hydration spheres which can be explained by the inversely proportional dependence of the reaction rate on water activity [21].

The studies on the influence of water activity and selected amino acids on the mechanism and kinetics of Bi(III) ion electroreduction [22–26] allowed to develop interpretation of the mechanism of "cap–pair" effect including the effect of basic electrolyte concentration and kind polarographically active accelerating organic substances. It was proved that a decrease in water activity gives the same effect as an increase in the concentration of catalytic organic substances, e.g. methionine, cysteine or cystine [23–26]. This fact is undoubtedly connected with the formation of active complexes on the electrode surface playing an intermediary role in electron transfer which results from the presence of both amino acids and water in the molecules of this complex. The fact that the complex forms only on the electrode surface is connected with amino acid adsorption which enhances electrode activity without limiting the electrode surface [23–26].

It was also proved that the reaction environment has significant influence on the form in which particular amino acids occur and on the electron process. The Bi(III) ions hydrolysis limits the pH range that allows to investigate their electroreduction [22].

It was found that change in the chloric(VII) acid/sodium chlorate(VII) ratio in the basic electrolyte affects the structure of both the bismuth aqua ion and the molecule of the selected amino acid (homocysteine (HCE), homocystine (HCY) and ethionine (ET)) and, consequently, the course of the electrode process [27–30].

The studied chemical compounds play an important role in maintaining homeostasis in living organisms. For example, an elevated level of homocysteine in blood plasma (hyperhomocysteinemia) is considered to be a risk factor for cardiovascular and neurodegenerative diseases, some types of cancer, as well as miscarriage [31].

Homocystine forms readily as a result of homocysteine oxidation and is a source of sulphur in the human body. It is excreted in urine as a result of homocystinuria [31].

Ethionine, which is an ethyl analogue of methionine, has very destructive influence on the livers of living organisms and disrupts the activity of many enzymes. It is a strongly carcinogenic metabolite contributing to the development of most types of human cancer [32].

The electrochemical reactivity of cysteine and homocysteine was extensively studied [33, 34]. As follows the reaction of cysteine or cystine with mercury ions is the basis of their electrolytic activity at mercury electrodes. From an electrochemical point of view, the behaviour of homocysteine at mercury electrodes was found to be close to that of cysteine. The investigations of homocystine indicate that while the kinetics of its reduction at mercury differs from that of cystine, the overall reaction in which two thiol molecules are formed from each disulphide is the same [34]. These reactions involve two steps of mercury electrooxidation:

$$\text{to mercury (I) cysteine thiolate} \quad Hg_2^{+2} + 2RSH \Leftrightarrow Hg_2(SR)_2 + 2H^+$$
$$\text{andmercury(II)cysteinethiolate} \quad Hg_2(SR)_2 + 2RSH \Leftrightarrow 2Hg(SR)_2 + 2e^- + 2H^+$$

Both these products are strongly adsorbed on the surface of the mercury electrode [34].

Homocysteine as well as homocystine and ethionine, according to the “cap–pair” rule, catalyses the process of Bi(III) ion electroreduction. Also changes in the double-layer parameters at the electrode/chlorates(VII) interface and correlation between the rate of Bi(III) ion electroreduction in the presence of amino acids and water activity under the conditions of varying protonation of the catalysing substance were observed.

The “cap–pair rule”, which specifies the conditions under which electrode processes are accelerated by organic substances, indicated the directions of the present research.

There were investigated:

– Adsorption of the accelerating substance at the interface of mercury/chlorates (VII)
– Mechanism and kinetics of the Bi(III) ions electroreduction process in the presence of homocysteine, homocystine and ethionine

The proposed experimental research will contribute substantial new and valuable information which will form the basis for addressing the existing and future problems related to, e.g. the use and determination of Bi(III) and the studied amino acids. The preparations containing bismuth, one of the less toxic heavy metals, have found application in various industrial branches. The present research can indicate new opportunities for the application of organic Bi(III) compounds. The use of bismuth in pharmaceutical and cosmetic industries is of particular importance.

20.2 Materials and Apparatus

All reagents: $NaClO_4$, $HClO_4$, $Bi(NO_3)_3{\cdot}5H_2O$ and amino acids (Fluka), were of analytical grade. Water applied to prepare all solutions was purified in the Millipore system.

The 2, 4 and 6 mol·dm^{-3} chlorate(VII) solutions of $HClO_4$:$NaClO_4$ concentration ratios: (1:1) solution A, (1:4) solution B, (1:9) solution C, (4:1) solution D and (9:1) solution E were examined. The enumerated solutions were designated according to the scheme, e.g.

$$\begin{aligned}
&\text{4A designates 4 mol}\cdot\text{dm}^{-3}\text{chlorates(VII), where } HClO_4 : NaClO_4 = 1:1\\
&\text{4B designates 4 mol}\cdot\text{dm}^{-3}\text{chlorates(VII), where } HClO_4 : NaClO_4 = 1:4\\
&\text{4C designates 4 mol}\cdot\text{dm}^{-3}\text{chlorates(VII), where } HClO_4 : NaClO_4 = 1:9\\
&\text{4D designates 4 mol}\cdot\text{dm}^{-3}\text{chlorates(VII), where } HClO_4 : NaClO_4 = 4:1\\
&\text{4E designates 4 mol}\cdot\text{dm}^{-3}\text{chlorates(VII), where } HClO_4 : NaClO_4 = 9:1.
\end{aligned}$$

The amino acids solutions ($1{\cdot}10^{-4}$ mol·dm^{-3}, $5{\cdot}10^{-4}$ mol·dm^{-3}, $1{\cdot}10^{-3}$ mol·dm^{-3}, $3{\cdot}10^{-3}$ mol·dm^{-3}) were prepared immediately before the measurements. The concentration of Bi(III) ions in the studied solutions was always $1{\cdot}10^{-3}$ mol·dm^{-3}. The solutions were sonicated due to the weak solubility of $Bi(NO_3)_3$ in chlorates(VII). The measurements were performed in a three-electrode cell containing a dropping or hanging mercury – electrode with a controlled increase rate and a constant drop surface (0.014740 cm^2), as a working electrode (MTM Poland); Ag/AgCl as a reference electrode; and a platinum spiral, as an auxiliary electrode. All electrochemical measurements were performed in the thermostated cells at 298 K using an Autolab Fra 2/GPES (version 4.9) frequency response analyser (Eco Chemie, Utrecht, Netherlands).

20.3 Experimental Methods

20.3.1 Adsorption Measurements

The double-layer capacity (C_d) was measured using the electrochemical impedance spectroscopy (EIS). For the whole polarisation range, the capacity dispersion was tested at different frequencies between 200 and 1000 Hz. To obtain proper equilibrium values of differential capacity, linear dependence of capacity on the square element from frequency was extrapolated to zero frequency. This procedure assumes that the impedance of the double layer is equivalent to a series capacity-resistance combination and the rate of adsorption is diffusion controlled [35–39].

The potential of zero charge (E_z) was measured for each solution by the method of streaming mercury electrode [35, 36]. The interfacial tension (γ_z) between the mercury and the electrolyte solutions at E_z was measured with a conventional maximum bubble-pressure capillary electrometer described earlier [35, 36]. The adsorption parameters were derived by the back integration of capacity-potential dependencies. No corrections for the effects of the medium on the activity of the supporting electrolyte and the activity coefficient of the adsorbate were made [35, 36].

20.3.2 Kinetic Studies

In the DC polarography, square-wave voltammetry (SWV) and cyclic voltammetry (CV), the optimal experiment operating conditions were as follows: step potential 2 mV for DC, pulse amplitude 20 mV, frequency 120 Hz and step potential 2 mV for the SWV and scan rate 5–1000 mV s^{-1} and step potential 5 mV for the CV. The electrochemical impedance spectroscopy data were collected at 26 frequencies in the range from 200 to 50,000 Hz within the faradaic potential region with 10 mV intervals.

The values of the kinetic parameters αn_α and k_s were calculated from the cyclic voltammetry measurements.

The transfer coefficients αn_α of the electroreduction reaction of Bi(III) ions in the studied systems were determined using the equation:

$$\alpha n_\alpha = \frac{0.048}{E_{\mathrm{pc}/4} - E_{3\mathrm{pc}/4}} \tag{20.1}$$

where $E_{\mathrm{pc}/4}$ and $E_{3\mathrm{pc}/4}$ – the potentials of one-fourth and three-fourths of the cathodic peak height, respectively.

The values of standard rate constants k_s were calculated using the Nicholson method based on the equation:

$$\Psi = \left(\frac{D_{\text{ox}}}{D_{\text{red}}}\right)^{\alpha/2} \frac{k_s(RT)^{1/2}}{(\pi nFvD_{\text{ox}})^{1/2}} \tag{20.2}$$

The function Ψ was determined from the product of electron number exchanged in the electrode process (n) and the difference between the potentials of anodic and cathodic peaks ($E_{\text{pa}}-E_{\text{pc}}$), and its dependence on $n(E_{\text{pa}}-E_{\text{pc}})$ was tabled [40].

For the given values of product $n(E_{pa}-E_{pc})$, the values of Ψ function were found in Table 5.7 [40] which allowed to determine k_s using Eq. (20.2).

For the irreversible processes, the values k_s, which are dependent on the kinetic parameters, are described by Eq. (20.3):

$$E_{\text{pc}} = E_f^0 - \frac{RT}{\alpha n_\alpha \text{F}}\left[0.78 - \ln k_s + \ln\sqrt{D_{\text{ox}}b}\right] \tag{20.3}$$

where

$$b = \frac{\alpha n_\alpha Fv}{RT} \tag{20.4}$$

The values of the apparent rate constants k_f of Bi(III) ion electroreduction in the chlorates(VII) solutions as a function of the potential were calculated from the impedance measurements.

k_f values were computed from the R_{ct} values as a function of DC potential [22].

$$R_{\text{ct}} = \frac{RT}{n^2F^2c_0k_fS}\cdot\frac{a_0/k_f + 1 + r_s\exp(b)}{\alpha a_0/k_f + r_s\exp(b)} \tag{20.5}$$

where

$$a_0 = D_0^{1/2}(3\pi t/7)^{-1/2} + D_0r_0^{-1} \tag{20.6}$$

$$r_s = \frac{r_0 + (3\pi tD_0/7)^{1/2}}{r_0 - (3\pi tD_R/7)^{1/2}} \tag{20.7}$$

$$b = (nF/RT)\left(E - E_{1/2}^r\right) \tag{20.8}$$

$$\alpha = -(RT/nF)(d\ln k_f/dE) \tag{20.9}$$

S – the electrode surface area ($S = 0.014740\ \text{cm}^2$).

In these equations t is the drop lifetime, r_0 is the drop radius at that time, and c_0 is the concentration of depolarizer, while $c_R = 0$ in the experiments.

20.4 Results and Discussion

The changes of determined adsorption parameters as a function of increasing concentration of organic substances and a function of water activity under the conditions of varying protonation of the amino acids indicate a growth in adsorption of the studied amino acids to mercury.

The differential capacity curves (Fig. 20.1) obtained in the studied systems for different amino acids point out to the changes in capacity values compared with the supporting electrolyte. In the region of "hump" potentials, appearing in 6 mol·dm^{-3} chlorate(VII) at the $HClO_4/NaClO_4$ concentration ratio of 1:4 (B) without the amino acids (≈ -600 mV), after the introduction of homocysteine, homocystine and

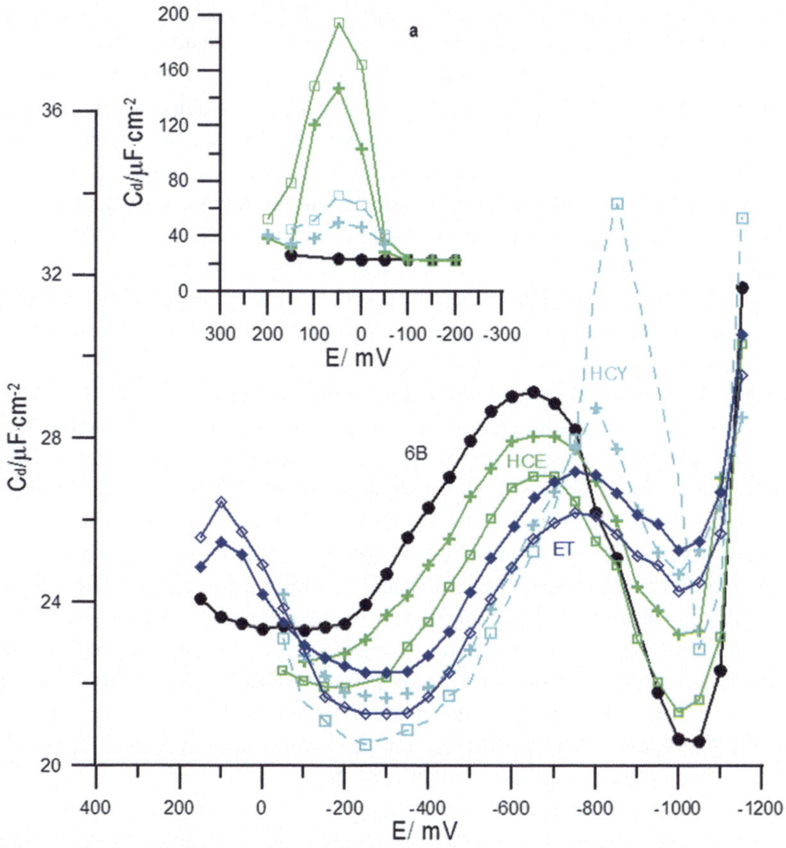

Fig. 20.1 Differential capacity–potential curves of double-layer interface Hg/6 mol·dm^{-3} chlorate (VII) with various concentrations of homocysteine (—), homocystine (- -) and ethionine (—) (in mol·dm^{-3}): (●) 0, (+) $5{\cdot}10^{-4}$, (□) $1{\cdot}10^{-3}$, where $HClO_4/NaClO_4 = 1:4$ (B)
(**a**) Insert shows data for homocysteine (—) and homocystine (- -)

ethionine to the solution, the height of the "hump" decreases. The increase in homocysteine and ethionine concentration causes a further decrease in differential capacitance, whereas the increase in the homocystine concentration causes an increase of differential capacitance in the region of "hump" potentials. At the same time, in the presence of HCY and ET, the "hump" moves significantly towards the negative potentials. The presence of HCE causes only a slight shift of the "hump" towards the negative potentials.

In the region of higher potentials (≈ 0 mV), the peaks occur in the presence of HCE and HCY (Fig. 20.1a) which results from their above-mentioned electrochemical reactivity on mercury. The replacement of mercury (I) cysteine thiolate by mercury (II) cysteine thiolate (products of mercury oxidation by cysteine on the anode) at the electrode surface is accompanied by a pronounced change of the electrode capacity [41]. The pronounced peaks (Ep ≈ 0 mV), which increase with the increasing HCE and HCY concentrations in all studied chlorate(VII) solutions, are associated with adsorption of mercury (II) cysteine thiolate $Hg(SR)_2$.

The area of more negative potentials (from -300 to -1000 mV) reflects the adsorption of the oriented surface layer of mercury (I) cysteine thiolate $Hg_2(SR)_2$ [41]. The differences in the capacity curves (there is observed increased differential capacity for homocystine, whereas that for homocysteine is decreased) result probably from formation of mercury (I) cysteine thiolate at the mercury surface.

It should be also noted that ethionine is polarographically inactive. Therefore the peak appearing at the positive potential (≈ 0 mV) at $C_d = f(E)$ (Fig. 20.1) curves shows the adsorption peak which increases with the increasing of ethionine concentration in the supporting electrolyte. No desorption peaks were observed on the capacity curves for all studied amino acids.

The course of the differential capacity curves on mercury in the chlorate(VII) solutions depends also on the changes in the $HClO_4/NaClO_4$ concentration ratio in the basic electrolyte and water activity [28, 29]. Such changes indicate a considerable influence of water on the surface properties of the interfaces [28, 29, 41]. The increase in the basic electrolyte concentration also shifts favourably the adsorption equilibrium due to high water structure destruction capacity of ClO_4^- ions.

For all studied concentrations of chlorates(VII) after the addition of homocysteine and homocystine as well as ethionine into the basic electrolyte, the values of the zero charge potential E_z shift towards the more negative potentials (Table 20.1a–c) which is the characteristic of anions adsorption [36]. It should be also noticed that the increase in sodium salt concentration in the basic electrolyte solution results in the shift of the E_z towards more positive potentials, whereas the increase in the chloric (VII) acid concentration results in the shift of the E_z in the opposite direction. Such changes accompanying the change in the amino acids protonation in the basic electrolyte solution confirm the altered arrangement of the cysteine mercury thiolates or ethionine on the electrode surface [35, 36, 42, 43]. The surface tension values (Table 20.1a–c) at the potential of zero charge γ_z decrease which also confirms the phenomenon of amino acids adsorption [41] on the mercury electrode.

Table 20.1 **(a–c)** Potential of zero-charge E_z vs. Ag/AgCl electrode and surface tension γ_z for E_z of chlorates(VII) solutions of concentration ratio $HClO_4$: $NaClO_4$ (1:1) A, (1:4) B, (1:9) C, (4:1) D, (9:1) E + $1 \cdot 10^{-3}$ mol·dm^{-3} homocysteine [41], 1·10–3 mol·dm^{-3} homocystine [41] and $1 \cdot 10^{-3}$ mol·dm^{-3} ethionine systems

a

	0		$1 \cdot 10^3$ c_{HCE}/mol·dm^{-3}		$1 \cdot 10^3$ c_{HCY}/mol·dm^{-3}		$1 \cdot 10^3$ c_{ET}/mol·dm^{-3}	
2 mol·dm^{-3} chlorate (VII)	$-E_z$/V	γ_z/mN·m^{-1}	$-E_z$/V	γ_z/mN·m^{-1}	$-E_z$/V	γ_z/mN·m^{-1}	$-E_z$/V	γ_z/mN·m^{-1}
A	0.483	469.5	0.487	466.3	0.489	467.4	0.492	470.42
B	0.481	470.3	0.487	458.9	0.485	461.4	0.487	469.50
C	0.480	474.4	0.483	467.9	0.484	466.2	0.492	466.75
D	0.494	467.1	0.497	465.4	0.504	451.6	0.509	462.17
E	0.496	466.3	0.499	462.9	0.509	461.3	0.513	465.84

b

	0		$1 \cdot 10^3$ c_{HCE}/mol·dm^{-3}		$1 \cdot 10^3$ c_{HCY}/mol·dm^{-3}		$1 \cdot 10^3$ c_{ET}/mol·dm^{-3}	
4 mol·dm^{-3} chlorate (VII)	$-E_z$/V	γ_z/mN·m^{-1}	$-E_z$/V	γ_z/mN·m^{-1}	$-E_z$/V	γ_z/mN·m^{-1}	$-E_z$/V	γ_z/mN·m^{-1}
A	0.497	0.483	0.518	466.3	0.520	463.6	0.512	465.84
B	0.496	0.481	0.503	464.6	0.500	461.3	0.507	465.84
C	0.487	0.480	0.499	438.8	0.500	462.7	0.494	466.75
D	0.536	0.494	0.540	4.646	0.545	458.1	0.539	463.08
E	0.547	0.496	0.553	455.6	0.550	458.9	0.553	463.08

c

	0		$1 \cdot 10^3$ c_{HCE}/mol·dm^{-3}		$1 \cdot 10^3$ c_{HCY}/mol·dm^{-3}		$1 \cdot 10^3$ c_{ET}/mol·dm^{-3}	
6 mol·dm^{-3} chlorate (VII)	$-E_z$/V	γ_z/mN·m^{-1}	$-E_z$/V	γ_z/mN·m^{-1}	$-E_z$/V	γ_z/mN·m^{-1}	$-E_z$/V	γ_z/mN·m^{-1}
A	0.528	471.2	0.535	466.3	0.554	463.3	0.546	462.17
B	0.522	468.7	0.528	460.5	0.533	457.2	0.537	462.17
C	0.519	464.6	0.526	458.1	0.535	358.4	0.535	459.42
D	0.581	466.3	0.588	462.9	0.605	454.8	0.586	458.50
E	0.607	465.4	0.619	460.5	0.621	458.7	0.603	457.58

The presence of homocysteine and homocystine as well as ethionine in the basic electrolyte solution containing $1 \cdot 10^{-3}$ mol·dm^{-3} Bi(III) causes an increase in the SWV peak current for Bi(III) ion electroreduction and a shift towards positive potentials (Fig. 20.2) as well as a simultaneous reduction in the width of SWV peaks at half their height. This indicates an increase in the reversibility of Bi(III) electroreduction in the presence of the studied amino acids. As can be seen from Fig. 20.2, the greatest catalytic effect was obtained in the presence of homocysteine. The magnitude of this effect depends on the amino acids concentrations as well as changes in the $HClO_4$/$NaClO_4$ ratio in the chlorate(VII) solutions with varying water activity [28–30].

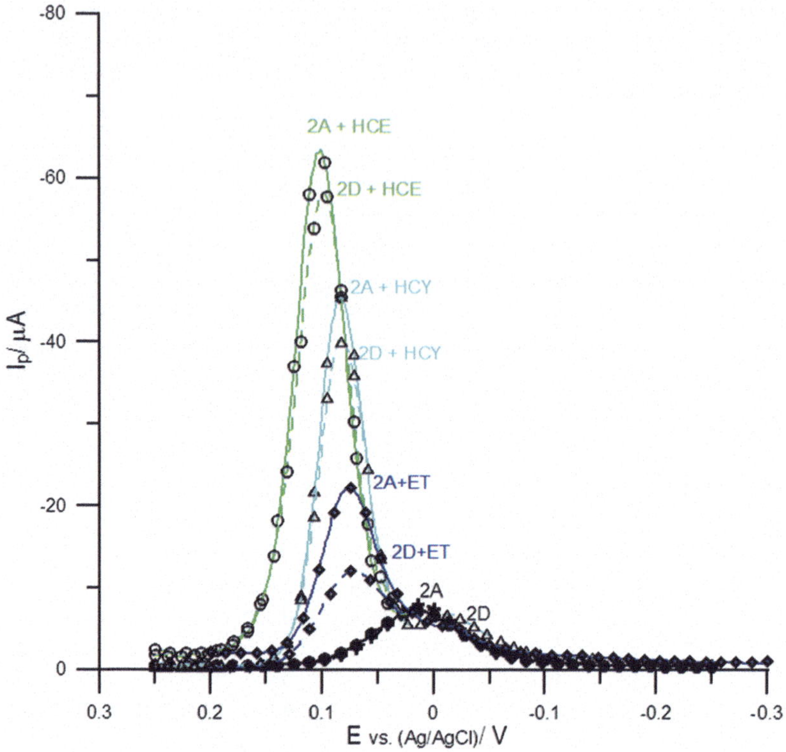

Fig. 20.2 The SWV peaks of the electroreduction of $1 \cdot 10^{-3}$ mol·dm^{-3} Bi(III) in 2 mol·dm^{-3} chlorates(VII) (—) and in the presence of $1 \cdot 10^{-3}$ mol·dm^{-3} homocysteine (—), homocystine (- -) and ethionine (—), where $HClO_4$:$NaClO_4$ = 1:1 (2A) and $HClO_4$:$NaClO_4$ = 4:1 (2D)

The influence of amino acids and changes in the $HClO_4$/$NaClO_4$ ratio, as well as water activity on the reversibility of Bi(III) ion electroreduction, also follow from the course of cyclic voltammetry curves (Fig.20.3a, b). The increase in both amino acids [28–30] and $NaClO_4$ concentrations in the supporting electrolyte results in the decrease of the distance between the anodic and cathodic peaks ($\Delta E_{a\text{-}c}$) (Fig. 20.3) in all chlorate(VII) solutions which confirms the increase of reversibility of the Bi (III) ions electroreduction processes. However, the increase of $HClO_4$ concentration in the chlorates(VII) solution causes reverse dependencies $\Delta E_{a\text{-}c}$. With an increase in the concentration of chlorates(VII) from 2 to 6 mol·dm^{-3} in solutions C of basic electrolytes, an evident decrease in the $\Delta E_{a\text{-}c}$ is observed (Fig. 20.3a). The investigations showed a multistage character of the electrode process also in the presence of HCE, HCY and ET as a chemical reaction which represents a stage that controls the rate of Bi (III) ion electroreduction [28–30]. The dependences $\ln k_f = f(E)$ are non-linear, and the slope of the curves changes with a change in potential and the degree of amino acids protonation [28–30]. The above-mentioned reaction is probably the formation of active complexes on the electrode surface which mediate in electrons transfer. Adsorption of amino acids on the electrode shifts favourably the

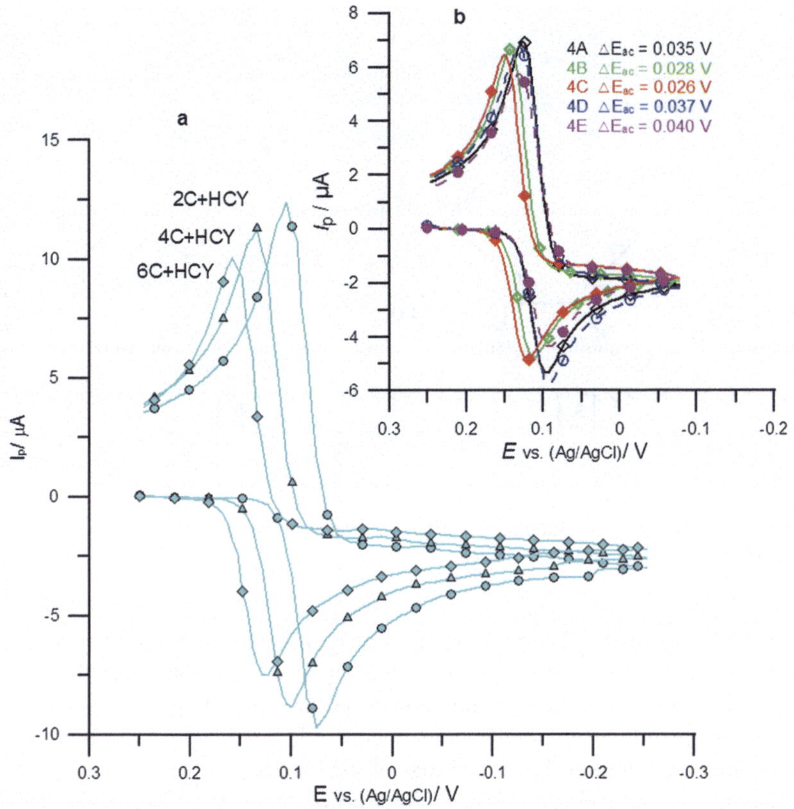

Fig. 20.3 (**a**) The cyclic voltammograms of $1 \cdot 10^{-3}$ mol·dm^{-3} Bi(III) in 2 mol·dm^{-3} chlorate(VII) (○), 4 mol·dm^{-3} chlorate(VII) (Δ) and 6 mol·dm^{-3} chlorate(VII) (◇) with $1 \cdot 10^{-3}$ mol·dm^{-3} homocystine, where $HClO_4/NaClO_4 = 1{:}9$ (C). (**b**) The cyclic voltammograms of $1 \cdot 10^{-3}$ mol·dm^{-3} Bi(III) in 4 mol·dm^{-3} chlorates(VII) in the presence $3 \cdot 10^{-3}$ mol·dm^{-3} ethionine, where $HClO_4$: $NaClO_4 = 1{:}1$ (—)(◇) 4A; $HClO_4/NaClO_4 = 1{:}4$ (—)(◇) 4B; $HClO_4/NaClO_4 = 1{:}9$ (—)(♦) 4C; $HClO_4/NaClO_4 = 4{:}1$ (---)(○) 4D and $HClO_4/NaClO_4 = 9{:}1$ (---)(●) 4E

equilibrium of Bi(III) ions complexation. It does not confine the electrode surface but additionally increases its activity. However, it should be noticed that to the already mentioned electrochemical reactivity of both homocysteine and homocystine, one can point out to the complexes of Bi–$Hg_2(SR)_2$ or Bi–$Hg(SR)_2$. Moreover the studies show [28–30] that bismuth is more likely to react with mercury (II) cysteine thiolate $Hg(SR)_2$. However, in the case of polarographically inactive ethionine, the complexes of the Bi–ethionine type are formed on the electrode surface [30]. The active Bi–ET and Bi–$Hg(SR)_2$ complexes are sure to be located inside the adsorption layer [28–30] as presented in Fig. 20.4. Arrangement of adsorbed $Hg(SR)_2$ (in the case of HCE and HCY) as well as adsorbed ET on the mercury electrode surface due to the change of protonation degree on the amino

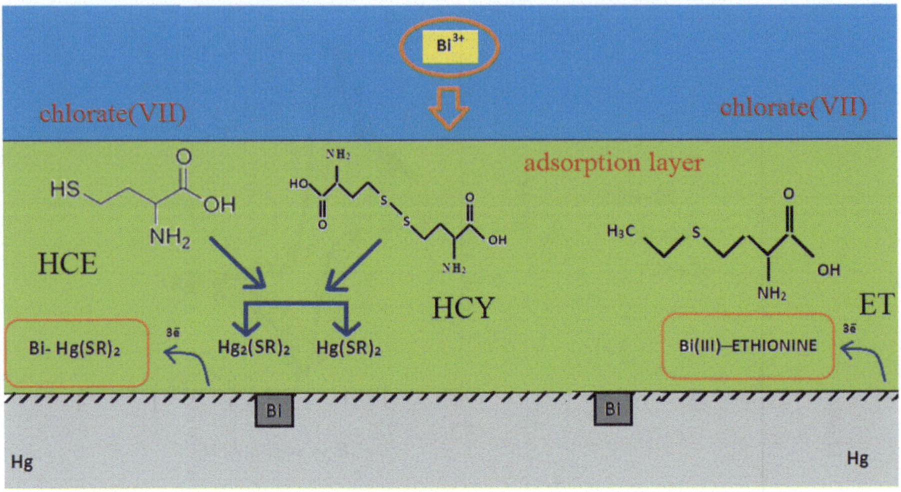

Fig. 20.4 The reaction path obtained for Bi(III) ion electroreduction showing active complexes acting as intermediates in electron transfer

acids in the basic electrolyte solutions plays a significant role in the formation of the above-mentioned active complexes. However, it should be noted that the composition of the active complexes after the transition of subsequent electrons undergoes a change. This results from the Marcus theory [5] predicting the change in the solvation shell of ion after the partial loss of the charge.

The kinetic parameters αn_α and k_s determined using the electrochemical techniques (Table 20.2) indicate the magnitude of the catalytic effect of amino acids. As follows from the correlation of these parameters, both water activity and the presence of differently protonated amino acids affect the Bi(III) ions electroreduction rate. The values k_s indicate that catalytic activity of amino acids increases in the order ET < HCY < HCE of large water activity (2 mol·dm^{-3} chlorate(VII)). For higher concentrations of chlorates(VII) (4 and 6 mol·dm^{-3} chlorate(VII)), there is observed a comparable effect of studied amino acids on the Bi(III) ions electroreduction rate particularly in the case of HCE and HCY. However, the catalytic activity of ethionine is a bit larger compared to that of cysteine and cystine derivatives. This may be associated with a different number of water molecules in the coordination sphere of the aqua ion Bi(III) as confirmed by the increase in the adsorption of studied amino acids with the increasing chlorates(VII) concentration. Therefore the state of the electrode layer plays a less significant role in the mechanism of electrode reaction acceleration by the studied amino acids compared to the composition of active complexes. The standard constants of Bi(III) ions electroreduction rate in the presence of both homocysteine, homocystine and ethionine increase with the increasing amount of $NaClO_4$ (solutions A, B, C) at all studied concentrations of chlorates(VII). However in the solutions of predominant concentration of chloric (VII) acid (solutions D and E), the value k_s decreases compared to the solutions

Table 20.2 The values of cathodic transition coefficients (α) and standard rate constants (k_s) of electroreduction of $1 \cdot 10^{-3}$ mol·dm^{-3} Bi(III) in chlorates(VII) solutions of concentration ratio $HClO_4/NaClO_4$: (1:1) A, (1:4) B, (1:9) C, (4:1) D, (9:1) E [27] and in the presence $3 \cdot 10^{-3}$ mol·dm^{-3} homocysteine [28], $3 \cdot 10^{-3}$ mol·dm^{-3} homocystine [29] and $3 \cdot 10^{-3}$ mol·dm^{-3} ethionine [30]

	α				$10^3 k_s$/cm·s^{-1}			
Chlorate(VII)	Bi(III)	Bi(III) + HCE	Bi(III) + HCY	Bi(III) + ET	Bi (III)	Bi(III) + HCE	Bi(III) + HCY	Bi(III) + ET
	2mol·mol^{-3} chlorate(VII)							
A	0.28	0.61	0.64	0.53	0.143	7.39	6.60	5.35
B	0.33	0.64	0.68	0.60	0.150	9.28	8.34	7.39
C	0.35	0.66	0.70	0.66	0.152	12.7	12.0	9.62
D	0.31	0.64	0.63	0.46	0.152	7.60	5.45	3.07
E	0.27	0.66	0.61	0.42	0.128	8.83	5.32	2.32
	4mol·mol^{-3} chlorate(VII)							
A	0.40	0.65	0.62	0.60	2.44	7.74	6.35	11.70
B	0.41	0.66	0.67	0.61	1.35	9.04	8.05	12.90
C	0.40	0.68	0.68	0.66	1.24	11.4	11.5	15.20
D	0.39	0.60	0.62	0.60	1.04	4.03	5.47	8.53
E	0.39	0.62	0.60	0.88	0.713	4.63	4.61	6.13
	6mol· mol^{-3} chlorate(VII)							
A	0.70	0.68	0.60	0.75	6.17	8.49	6.60	8.63
B	0.60	0.71	0.65	0.77	4.66	9.67	8.00	17.10
C	0.54	0.73	0.67	0.78	2.71	9.99	13.6	18.10
D	0.40	0.61	0.57	0.64	1.09	4.53	4.33	6.66
E	0.39	0.65	0.55	0.56	0.691	4.66	3.71	6.13

where its amount is much smaller (solutions A, B, C). This confirms the assumption of various active complexes mediating in electrons transfer resulting in the differentiated catalytic activity of the studied amino acids. H_2O molecules as well as the amino acid molecule protonated to a different degree will compete in filling the coordination sites of the complex [42–44].

References

1. Kalvoda R (2007) Is polarography still attractive? Chem Anal 52:869–873
2. Barek J (2013) Possibilities and limitations of mercury and mercury-based electrodes in practical electroanalysis of biologically active organic compounds. Port Electrochim Acta 31:291–295
3. Nieszporek J (2013) The mechanism and kinetics of Zn^{2+} electroreduction in the presence of octyltrimethylammonium bromide. J Electroanal Chem 706:108–116
4. Perek-Dlugosz A, Socha A, Rynkowski J (2017) Electrochemical reactions of sodium 2-ethylhexyl sulfate salt. Electrocatalysis 8:270–278
5. Dalmata D (2005) Kinetics and mechanism of Zn(II) ions electroreduction catalyzed by organic compounds. Electroanalysis 17:789–793
6. Weiss D (1962) Chem Zvesti 16:302
7. Kim NS, Li JB (1972) Punsok Hwahak 10:5
8. Popowa LN, Chałturina TJ, Maklakowa ZP (1972) Ż Obszcz Chim 43:1223–1228
9. Alias K, Fawcett WR, Parsons R (1974) Effects of adsorbed ions on simple electrode reactions. Part 1.—Reduction of periodate ion in the presence of toluene-p-sulphonate anions. J Chem Soc Faraday Trans 1(70):1046–1056
10. Sykut K, Dalmata G, Nowicka B, Saba J (1978) Acceleration of electrode processes by organic compounds – "cap-pair" effect. J Electroanal Chem 90:291–302
11. Dalmata G (1996) Catalytic effect of 2,4- or 2,6-Diaminotoluene on the electrochemical reduction of zinc(II) ions. Croat Chem Acta 69:85–94
12. Dalmata G (1997) The catalysis of the reduction of Zn(II) ions by 3,4-diaminotoluene. Electrochim Acta 42:1307–1314
13. Dalmata G (1997) Studies of the effect of anthranilic and thiosalicylic acid on the two-step electroreduction of Zn(II) ions. J Electroanal Chem 431:67–75
14. Dalmata G, Nosal – Wiercińska A, Zięcina T (2004) The influence of diaminotoluene isomers on the two-step electroreduction of Zn(II) ions in acetate buffers at different acidities. Collect Czechoslov Chem Commun 69:267–278
15. Nieszporek J, Gugała-Fekner D, Sieńko D, Saba J, Nieszporek K (2008) Kinetics and mechanism of Zn(II) ion electroreduction in the presence of vetranal. Collect Czechoslov Chem Commun 73:616–626
16. Nieszporek J (2011) Influence of sodium 1-decanesulfonate on the enthalpy of activation of two-step Zn2+ ions electroreduction. J Electroanal Chem 662:407–414
17. Ikeda O, Watanabe K, Taniguchi Y, Tamura H (1984) Adsorption effect of highly polarizable organic compounds on electrode kinetics. Bull Chem Soc Jpn 57:3363–3367
18. Nosal – Wiercińska A, Dalmata G (2002) Studies of the effect of thiourea on the electroreduction of In(III) ions in perchloric acid. Electroanalysis 14:1275–1280
19. Nosal – Wiercińska A (2010) Catalytic activity of thiourea and its selected derivatives on electroreduction of In(III) in chlorates(VII). Cent Europ J Chem 8:1–11
20. Sykut K, Dalmata G, Nieszporek J (1998) The catalysis of the reduction of Bi(III) ions by methionine. Electroanalysis 10:458–461

21. Komorsky – Lovrič S, Lovrič M, Branica M (1993) Effect of ionic strength on Bi(III) reduction from perchlorate medium. J Electrochem Soc 140:1850–1853
22. Nosal-Wiercińska A (2010) The kinetics and mechanism of the electroreduction of Bi(III) ions from chlorates (VII) with varied water activity. Electrochim Acta 55:5917–5921
23. Nosal – Wiercińska A (2011) The catalytic influence of methionine on the electroreduction of Bi(III) ions in chlorates (VII) solutions with varied water activity. J Electroanal Chem 654:66–71
24. Nosal – Wiercińska A (2011) The catalytic activity of cysteine and cystine on the electroreduction of Bi(III) ions. J Electroanal Chem 662:298–205
25. Nosal – Wiercińska A (2012) Electrochemical and thermodynamic study of the electroreduction of Bi(III) ions in the presence of cystine in solutions of different water activity. J Electroanal Chem 681:103–108
26. Nosal – Wiercińska A (2013) The role active complexes in the multistep process of Bi(III) ion electroreduction in chlorate (VII) solutions with varied water activity in the presence of cysteine. Electrochim Acta 92:397–403
27. Nosal – Wiercińska A, Grochowski M, Wiśniewska M, Tyszczuk – Rotko K, Skrzypek S, Brycht M, Guziejewski D (2015) The influence of protonation on the electroreduction of Bi(III) ions in chlorates(VII) solutions of different water activity. Electrocatalysis 6:315–321
28. Grochowski M, Nosal-Wiercińska A, Wiśniewska M, Szabelska A, Gołębiowska B (2016) The effects of homocysteine protonation on double layer parameters at the electrode/chlorates (VII) interface, as well as the kinetics and the mechanism of Bi (III) ion electroreduction. Electrochim Acta 207:48–57
29. Grochowski M, Nosal – Wiercińska A (2017) The influence of homocystine protonation on her catalytic activity in the process of electroreduction of Bi(III) ions in chlorates(VII). J Electroanal Chem 788:198–202
30. Nosal – Wiercińska A, Grochowski M (2017) The catalytic impact of ethionine on the multi-step electroreduction of Bi(III) ions in chlorates(VII) solutions. Electrocatalysis 8:492–497
31. Galík M, Bănică FG, Bănică A, Švancara I, Vytřas K (2010) Homocysteine voltammetry at a mercury electrode in the presence of nickel ions. Electroanalysis 22:1733–1736
32. Read JF, MacCormick KJ, McBain AM (2004) The kinetics and mechanism of the oxidation of DL – ethionine and thiourea by potassium ferrate. Transit Met Chem 29:149–158
33. Heyrovský M, Mader P, Veselá V, Fedurco M (1994) The reactions of cystine at mercury electrodes. J Electroanal Chem 369:53–70
34. Heyrovský M, Vavřička S (1999) Electrochemical reactivity of homocysteine at mercury electrodes as compared with cysteine. Bioelectrochem Bioenerg 48:43–51
35. Nosal – Wiercińska A, Dalmata G (2010) Adsorption of methionine at mercury/aqueous solution of chlorate(VII) interface; dependence on the supporting electrolyte concentration. Electroanalysis 207:198–202
36. Nosal – Wiercińska A, Grochowski M (2011) Adsorption of thiourea and its methyl derivatives from chlorate(VII) with varied water activity, collect. Czech Chem Commun 76:265–275
37. Nosal – Wiercińska A (2012) Adsorption of cystine at mercury/aqueous solution of chlorate (VII) interface in solutions of different water activity. Cent Europ J Chem 10:1290–1300
38. Ostolska I, Wiśniewska M (2015) The impact of polymer structure on the adsorption of ionic polyamino acid homopolymers and their diblock copolymers on colloidal chromium(III) oxide. RSC Adv 5:28505–28514
39. Ostolska I, Wiśniewska M (2014) Comparison of the influence of polyaspartic acid and polylysine functional groups on the adsorption at the Cr_2O_3—aqueous polymer solution interface. Appl Surf Sci 311:734–739
40. Galus Z (1979) Electroanalytical methods of determination of physicochemical constants. PWN, Warsaw, (in Polish)

41. Nosal – Wiercińska A, Wiśniewska M, Grochowski M, Kaliszczak W, Skrzypek S, Brycht M, Guziejewski D, Franus W (2017) The effect of homocysteine and homocystine protonation on double-layer parameters at the electrode/chlorates(VII) interface. Adsorpt Sci Technol 35:396–402
42. Klepka T (2008) Construction of axial-symmetric polymeric extrudates of complex forms. Polymers 53:390–395
43. Klepka T, Dębski H, Rydarowski H (2009) Characteristics of high-density polyethylene and its properties simulation with use of finite element method. Polymers 54:668–667
44. Nosal – Wiercińska A (2014) Intermolecular interactions in systems containing Bi(III) – ClO4- – H2O – selected amino acids in the aspect of catalysis of Bi(III) electroreduction. Electroanalysis 26:1013–1023

Chapter 21
Phase Composition and Nanoporous Structure of Core and Surface in the Modified Granules of NH_4NO_3

A. Artyukhov and J. Gabrusenoks

21.1 Introduction

The ammonium nitrate used as a component of the industrial explosives is based on its capability to disengage oxygen in the process of exothermic reaction. The ammonium nitrate is more widely used as the main oxidation agent in the simplistic explosive mixture [1, 2]. However, the necessary criterion for the successful implementation of ammonium nitrate as a component of the industrial explosives is developed nanostructural porous surface on the surface and in the core. In this case, the question is about porous ammonium nitrate (PAN) [3].

Analysis of studies series [4–7] has shown that the porous surface layer on ammonium nitrate granules producing process with high efficiency can be carried out in granulators of vortex type with variable in height cross-sectional area. Granulators of vortex type due to their versatility, high specific productivity, and environmental safety allow to obtain granules of porous structure with normative levels of quality with relatively low energy costs (as compared with other devices with fluidized bed) and the low concentrations of harmful substances in the exhaust heat transfer agent [8]. It should be noted that the main characteristics of vortex granulators application in obtaining PAN are adequately reflected in studies [4, 9–12], where it studied hydrodynamic and thermodynamic conditions of porous surface layer obtaining energy and environmental characteristics of this process.

A. Artyukhov (✉)
Processes and Equipment of Chemical and Petroleum Refinery Department, Sumy State University, Sumy, Ukraine
e-mail: artyukhov@pohnp.sumdu.edu.ua

J. Gabrusenoks
Institute of Solid State Physics, University of Latvia, Riga, Latvia
e-mail: jg14078@lu.lv

O. Fesenko, L. Yatsenko (eds.), *Nanooptics, Nanophotonics, Nanostructures, and Their Applications*, Springer Proceedings in Physics 210,
https://doi.org/10.1007/978-3-319-91083-3_21

The results of computer simulation of the hydrodynamic conditions of single-phase and two-phase flows motion in the vortex granulator workspace [13, 14] have also confirmed the effectiveness of this type of devices.

Additional interest deals with the detailed study of granules structure (geometrical parameters, pores structure, and their depth), also the question of determination of the phase composition, and crystal structure after humidification PAN granules, and heat treatment remains open.

21.2 Description of Object and Methods of Research

As samples, there were used as

- originate granules of the ammonium nitrate;
- granules of the porous ammonium nitrate, wetted with ammonium nitrate solution and dried at the temperature 230–240 °F to 0.002 kg of humidity/kg of material (humidifier composition – 40% of ammonium nitrate and 60% of water).

21.2.1 Investigation of Phase Composition of Samples

The phase composition of samples was studied using translucent electron microscope PEM – 125 K in the work of light-field and microdiffraction regime without introduced selective diaphragm. Accelerating capacity was 90 kW.

21.2.1.1 Diffraction from the Selected Field (Microdiffraction)

While working in the microdiffraction regime, there is an ability to get diffraction from selected, small area of the sample, size of which is smaller than under usual diffraction. This method allows receive results from the small area of sample, which is important while studying the crystal grid and multiphase samples.

21.2.1.2 Method of Bright-Field Image

This method is a usual regime of PEM work when the object is observed on the microscope screen.

While working in such regime, the image is formed thanks to electrons beam, got through the sample.

21.2.1.3 Preparation of Objects for Microscopic Studies

To locate objects in the object plane of objective lens, PEM specific grids in forms of discs with diameter 3 mm are used. In these researches, support grids of copper (30 × 30 μm) and nickel (50 × 50 μm) are implemented. Choice of material for film base depends on the sample nature, grid material, investigations tasks, preparation method, presence of base thin structure, and background. The carbon films get widespread use. Its gauge was 20 nm in the research.

These small-sized samples were put on the support grid with a carbon film, preliminary fixed in the object holder, thanks to dispersing of suspension ultrasonically. Suspension was received by dissolution of gel (powder) with distilled water. Height from the ultrasonic sprayer to object holder was h ≅ 200 mm and could vary.

Elaboration of electron diffraction research was carried out with the help of the proper software (Fig. 21.1), which let to increase accuracy of measurements.

The algorithm of the program work is based on the generally accepted method of diffraction pattern analysis [15], which can be divided into two stages. At first, one finds an apparatus constant using a sample (aluminum film of 30–40 nm thick). To receive it, an image of the sample diffraction pattern is downloaded into the program, the electron diffraction pattern center is automatically found, necessary lines for calculation are chosen, and computation is performed by the following formula:

$$C = D^0{}_{h\kappa\ell} \cdot d^0{}_{h\kappa\ell}, \tag{21.1}$$

where $d^0{}_{hkl}$ – interplane distance of the sample,
$D^0{}_{hkl}$ – diameters of the sample diffraction fringes.

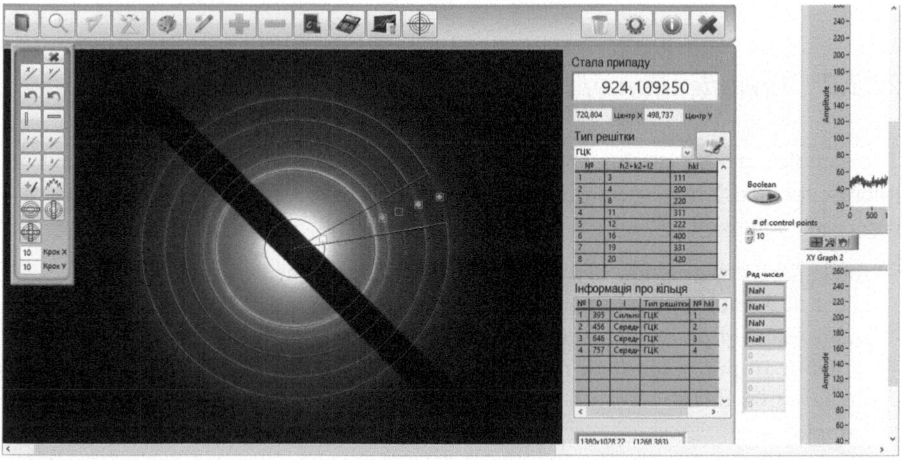

Fig. 21.1 Working surface of the software to decode electron diffraction patterns [15]

Aluminum films of 20–40 nm thick were used, and interplane distances of which are given.

On the second stage, electron diffraction patterns from experimental sample were used. Analogical actions (as with standard sample) are carried out to decode the electron diffraction pattern, an image is downloaded to the program, electron diffraction pattern center is found, and lines for computing are chosen. Difference of two stages consists in the calculation of interplane distances and crystal grid parameters for experimental sample with implementation of its diffraction fringes diameter and apparatus constant. The grid parameters are calculated with the help of formulas

$$d_{hkl} = \frac{C}{D_{hkl}}, \tag{21.2}$$

$$a_{hkl} = d_{hkl}\sqrt{h^2 + k^2 + l^2}, \tag{21.3}$$

where D_{hkl} – the diffraction fringes diameters of unknown sample;
d_{hkl} – interplane distance of the unknown sample;
h, k, l – Miller indices.

While interpreting the electron diffraction patterns, decoding results known table data about interplane distances and relative intensity of lines on the diffraction patterns.

Raman analysis – Renishaw InVia90V727 micro-Raman spectrometer.

Raman spectra were measured at room temperature using an Ar green ($\lambda = 514$ nm) laser excitation. Quite well-quality spectra were obtained with an exposure time of 40 s and laser excitation power density of 338.9 W/cm^2.

21.2.2 Investigation of the Samples Crystal Structure

The samples crystal structure was investigated by SEM -100 U scanning electron microscope and X-ray spectrometer with energy dispersion. As a result of microscopy, one defines:

- total space construction of surface and PAN granule near-surface layers;
- shear of granule and its internal crystal structure;
- main zones of pores localization;
- structure of pores and their nature;
- quantitative distribution of various pores per square unit of the investigated sample.

21.3 Visualization of Results and Discussion

21.3.1 Phase Composition of Samples

Diffractometry results of studied samples (Figs. 21.2 and 21.3) have shown that the phase composition of samples is not affected by humidification with any type of solution and heat treatment. The crystal structure of granules after the humidification and heat treatment has some changes due to the increase of the number of pores. The change of crystal structure, in turn, allows open access to nanopores that are located in the volume of granules. This allows increase the holding capacity indicator of granules. An important result of conducted researches is the establishment of the fact of oxygen molecules "integration" to the crystal lattice of ammonium nitrate that have positive impact on the industrial explosives detonation velocity.

Comparison of samples shows that after heat treatment, PAN has larger crystals than in the initial sample. This fact is supposed to be explained through sintering of some crystals influenced by hot heat transfer agent (drying agent) and "adhesion" of the humidified crystals. PAN diffractogram also allows to make conclusions that oxides are not formed as a result of heat treatment. This fact can prove that all air molecules (and thus, oxygen for future oxidation reaction while the industrial explosive detonation) are inside the granule after humidification and heat treatment.

Results of Raman analysis of PAN structure (Fig. 21.4) demonstrated that it is corresponded with modification II [15].

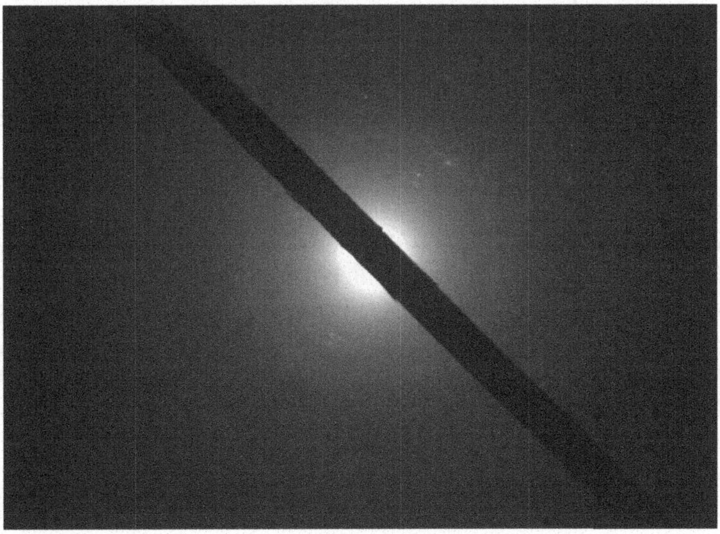

Fig. 21.2 Diffractogram of the ammonium nitrate initial sample before humidification and heat treatment

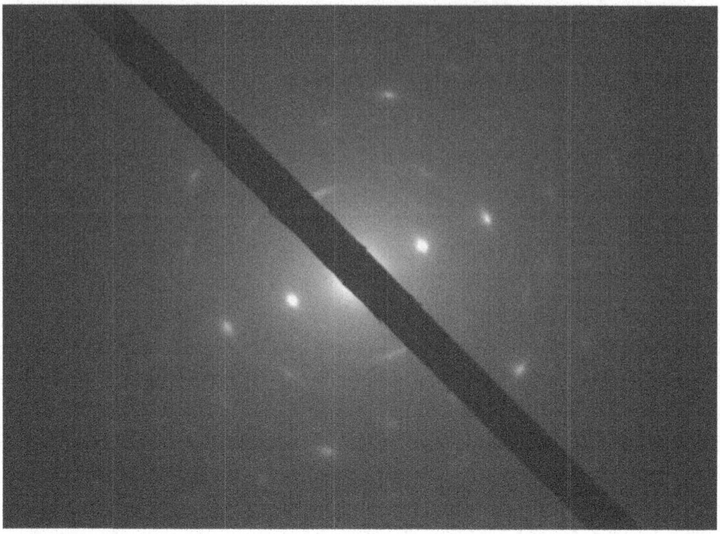

Fig. 21.3 PAN diffractogram after humidification and heat treatment (humidifier–water solution of ammonium nitrate)

21.3.2 Crystal Structure of the PAN Sample

1. General space construction of surface and near-surface layers of PAN granule: three-dimensional pores with level distinguishing on the surface and inside the granule (Fig. 21.5a).
2. Granule shear and its internal crystal structure: monolithic core without mechanical damages, porous near-surface, and surface layers (Fig. 21.5b).
3. Main zones of pores localization: it depends on linear humidification and drying intensity. Under optimal conditions of the process conduct – the whole granule surface (Fig. 21.5c).
4. Pores structure and their nature: straight-line pores on the surface, curvelinear – in the near-surface layers and inside the granule. Coexistence of pores of "mechanic" (destruction of sample after thermal strain) and "modificational" (pores, created thanks to water vapor and gas ammonia outflow from the granule) nature (Fig. 21.5d).
5. Quantitative distribution of various pores per the investigated sample area unit: macropores – granule surface and near-surface layers; mesopores and micropores – internal layers of the granule.

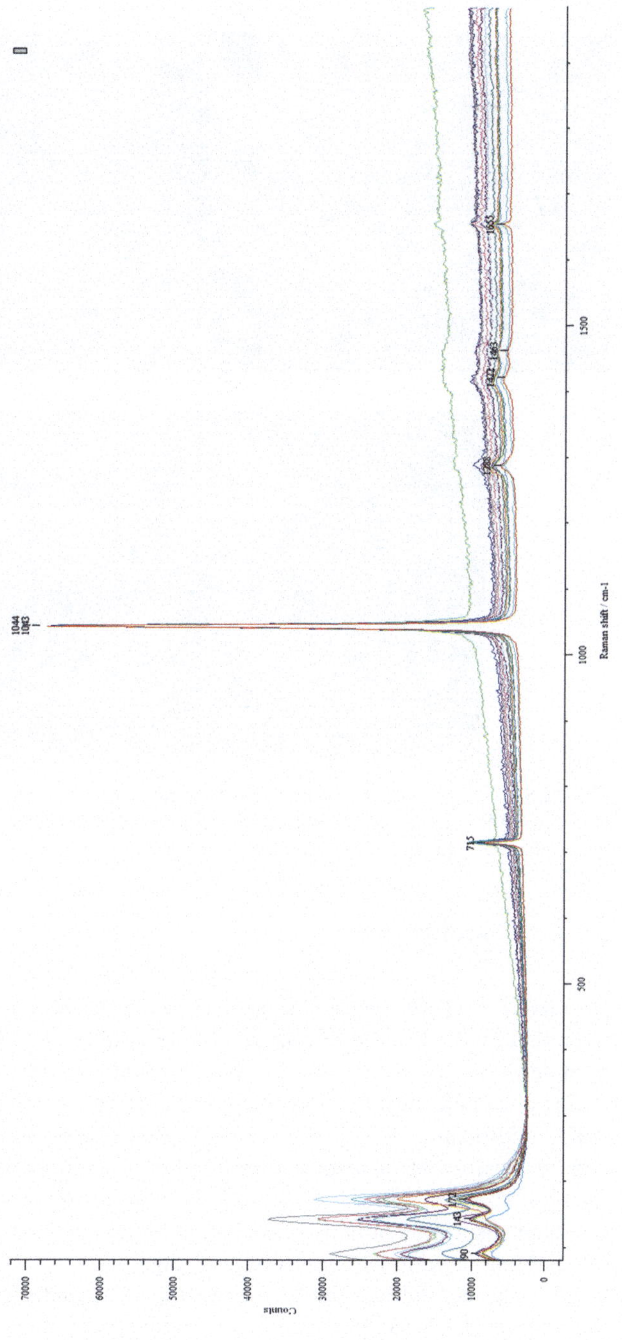

Fig. 21.4 Raman analysis of the PAN granule sample (humidifier–water solution of the ammonium nitrate)

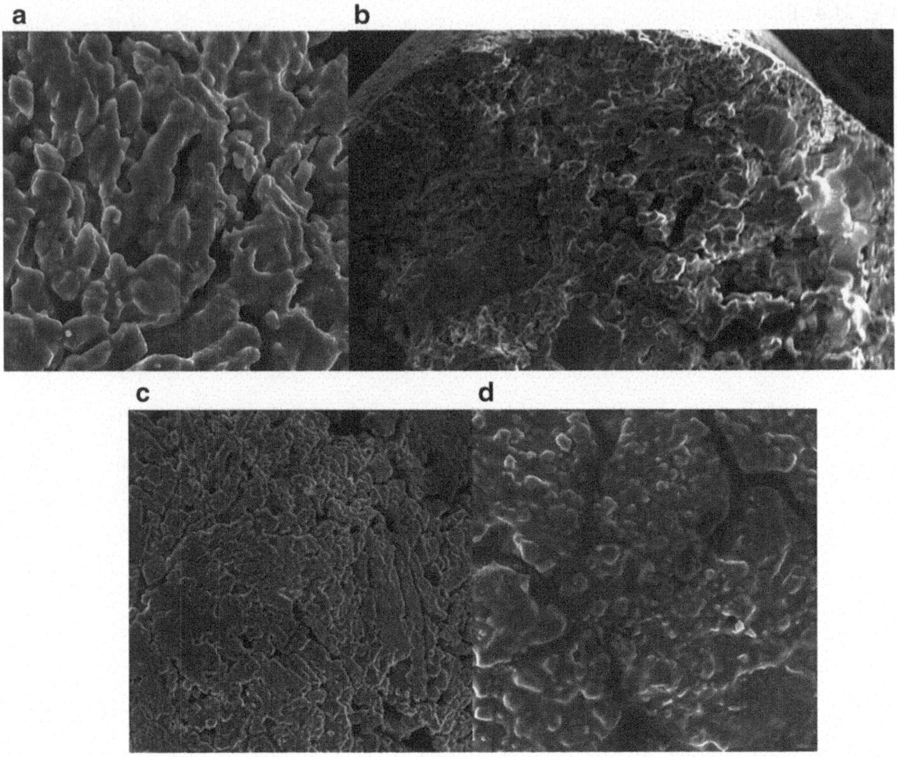

Fig. 21.5 Structure of PAN sample. (**a**) General space construction of surface and near-surface layers of PAN granule, (**b**) granule shear and its internal crystal structure, (**c**) main zones of pores localization, and (**d**) pores structure and their nature

21.4 Conclusions

Results of the PAN phase composition research show that humidification and heat treatment of the ammonium nitrate do not lead to formation of new compounds in granule. It has a positive impact on its explosive features.

Results of electron microscopic samples have shown, that after the heat treatment and humidification, developed porous surface layer is formed on the granules, which consists mainly of nanoscale pores (mezo- and macrolevel) and small amount of micropores. Such surface structure is optimal from the standpoint of the diesel fuel distillate absorption at early stage of industrial explosive preparation and the distillate holding at granules transportation stage.

There are air bubbles that assist starting of the industrial explosive detonation, in the received PAN samples thanks to modern technology and temperature reduction in the process conduct in the core center after drying.

Acknowledgments This work was carried out under the project «Improving the efficiency of granulators and dryers with active hydrodynamic regimes for obtaining, modification and encapsulation of fertilizers», state registration No. 0116 U006812. The authors thank researchers of Processes and Equipment of Chemical and Refining Industries Department, Sumy State University and Institute of Solid State Physics, University of Latvia for their valuable comments during the article preparation.

References

1. Erode GM (2013) Ammonium nitrate explosives for civil applications: slurries, emulsions and ammonium nitrate fuel oils. Wiley-VCH Verlag & Co, Weinheim. 230 p
2. Buczkowski D, Zygmunt B (2011) Detonation properties of mixtures of ammonium nitrate based fertilizers and fuels. Cent Eur J Energetic Mater 8(2):99–106
3. Jackson SI (2017) The dependence of ammonium-nitrate fuel-oil (ANFO) detonation on confinement. Proc Combust Inst 36(2):2791–2798
4. Artyukhov AE, Sklabinskyi VI (2013) Experimental and industrial implementation of porous ammonium nitrate producing process in vortex granulators. Nauk Visnyk Nats Hirnychoho Univ 6:42–48
5. Artyukhov AE, Sklabinskyi VI (2016) 3D nanostructured porous layer of ammonium nitrate: influence of the moisturizing method on the layer's structure. J Nano Electron Phys 8(4):04051-1–04051-5
6. Artyukhov AE, Sklabinskyi VI (2016) Thermodynamic conditions for obtaining 3D nanostructured porous surface layer on the granules of ammonium nitrate. J Nano Electron Phys 8 (4):04083-1–04083-5
7. Artyukhov AE, Sklabinskyi VI (2017) Investigation of the temperature field of coolant in the installations for obtaining 3d nanostructured porous surface layer on the granules of ammonium nitrate. J Nano Electron Phys 9(1):01015-1–01015-4
8. Artyukhov AE, Fursa AS, Moskalenko KV (2015) Classification and separation of granules in vortex granulators. Chem Petrol Eng 51(5–6):311–318
9. Artyukhov AE, Voznyi AA (2016) Thermodynamics of the vortex granulator's workspace: the impact on the structure of porous ammonium nitrate. In: 6th international conference Nanomaterials: Application & Properties (NAP-2016). vol 5, no 2, p 02NEA01
10. Artyukhov AE (2016) Kinetics of heating and drying of porous ammonium nitrate granules in the vortex granulator. In: 6th international conference Nanomaterials: Application & Properties (NAP-2016), vol 5, no 2, p 02NEA02
11. Artyukhov AE, Sklabinskyi VI (2015) Theoretical analysis of granules movement hydrodynamics in the vortex granulators of ammonium nitrate and carbamide production. Chem Chem Techn 9(2):175–180
12. Artyukhov AE, Sklabinskyi VI (2015) Hydrodynamics of gas flow in small-sized vortex granulators in the production of nitrogen fertilizers. Chem Chem Techn 9(3):337–342
13. Artyukhov A (2016) Application software products for calculation trajectories of granules movement in vortex granulator. CEUR Work Proc 1761:363–373
14. Artyukhov A, Sklabinskiy V, Ivaniia A, Moskalenko K (2016) Software for calculation of vortex type granulation devices. CEUR Work Proc 1761:374–385
15. Dobrozhan O, Kurbatov D, Opanasyuk A, Cheong H, Cabot A (2015) Influence of substrate temperature on the structural and optical properties of crystalline ZnO films obtained by pulsed spray pyrolysis. Surf Interface Anal 47(5):601–606

Chapter 22
Interaction-Induced Polarizability Anisotropy Correlations in Ultra-Thin Layer of Neon Atoms Confined Between Graphene Walls: Computer Simulation

Z. Gburski

22.1 Introduction

The behavior of the fluids in a confined space has been increasingly studied in the recent years due to their potential applications in nanoelectronic devices, drug carriers, and molecular sensors [1–5]. Noble gases as nonreactive particles were often used to study the dynamics of atoms in a confined space, the subject of increasing interest in the advent of nanotechnology [6–11]. The interacting pair of neon atoms in a fluid phase induces a short-lived dipole moment, and it can be measured in the depolarized Rayleigh (interaction induced) light scattering experiment [12–21]. To take a closer look at the light scattering phenomenon, the molecular dynamics computer simulations are very convenient. There have been made some simulations of depolarized Rayleigh spectra for noble gases in the bulk fluids and clusters [22–31]. In our previous research, some noble gases between graphite walls were simulated. Those simulations were performed for varying density and number of atoms inside the slot and showed the details of mobility of the noble-gas atoms, with the increasing distance between the graphite slabs.

In this research, we focus on the interaction-induced light scattering phenomenon (depolarized Rayleigh) in the thin neon film confined between two parallel graphene slabs, using fully atomistic classical MD simulations.

Z. Gburski (✉)
Institute of Physics, University of Silesia, Katowice, Poland

Silesian Centre of Education & Interdisciplinary Research, Chorzów, Poland
e-mail: zygmunt.gburski@us.edu.pl

O. Fesenko, L. Yatsenko (eds.), *Nanooptics, Nanophotonics, Nanostructures, and Their Applications*, Springer Proceedings in Physics 210,
https://doi.org/10.1007/978-3-319-91083-3_22

Table 22.1 Lennard–Jones potential parameters

Atom	ε [meV]	σ [Å]
Carbon	2.412873	3.4
Neon	3.213	2.782

22.2 Simulation Details

The system studied is the set of neon atoms placed between two parallel graphene walls. We have modeled the graphene wall, a single layer of carbon atoms, the elementary cell of graphene that was constructed from six hexagonal carbon rings with the distance between near carbon atoms equal to 1.42 Å, and the angle between two vectors drawn from one atom to nearby atoms equal to 120°. To simulate this heavy graphite walls, we have assumed that carbon atoms are not moving during the simulations. The Lennard–Jones (LJ) potential has been used to model the interaction between Ne–Ne and Ne–C pairs.

$$V(r_{ij}) = 4\varepsilon_{ij}\left[\left(\frac{\sigma_{ij}}{r_{ij}}\right)^{12} - \left(\frac{\sigma_{ij}}{r_{ij}}\right)^{6}\right],$$

where r_{ij} is the distance between ith and jth atoms, and ε_{ij} and σ_{ij} are the LJ potential parameters listed in Table 22.1.

The parameters between unlike atoms have been calculated using the Lorentz–Berthelot mixing rule [32]. The molecular dynamics simulations have been performed with three-dimensional orthogonal periodic boundary conditions (PBC), using minimum image convention algorithm [32]. The simulation master cell was set to the edge size equal to 28.3 Å in the x-direction, 25.7 Å in the y-direction and 60 Å in the z-direction. The cutoff radius of potential interaction was set to 20 Å. The initial positions of neon atoms were randomly distributed inside the slot. The classical equations of motion are integrated up to 2 ns by the Velocity Verlet algorithm [32]. The integration time step used in simulation is 2.5 fs, which ensures total energy conservation within 0.01%. The average temperature was adjusted as desired by a process of velocity scaling using the Berendsen algorithm [32]. The system was equilibrated for $6*10^6$ MD steps. The total time of a single simulation is 1 ns.

22.3 Interaction-induced Light Scattering DID Model

The colliding pairs of atoms create a temporary polarizability anisotropy that leads to depolarized Rayleigh light scattering. The polarizability anisotropy is usually described by the dipole-induced-dipole mechanism (DID) [18, 33]. The DID interaction comes from the fact that the incident light beam induces an oscillating dipole on the ith particle, and this dipole generates an oscillating local field at the jth particle. The DID mechanism at its basis is a two-body interaction model, but it can be easy extended to three-, and four-body interaction as the contribution to the total intensity of scattered light. The pair anisotropy β_{ij} in the DID limit is $\beta_{ij}(t) = \sigma^3\left[3x_{ij}(t)z_{ij}(t)/r_{ij}^5(t)\right]$, where x_{ij} and z_{ij} are components of the separation vector r_{ij} between the *ith* and *jth* atoms. The depolarized Rayleigh light scattering spectrum is the Fourier transform of the polarizability anisotropy autocorrelation function *G(t)*,for which a monatomic sample of *N* atoms is

$$G(t) \propto \left\langle \sum_{i,j,k,l=1,\,i\neq j,\,k\neq l}^{N} \beta_{ij}(t)\beta_{kl}(0) \right\rangle,$$

where *i, j, k,* and *l* identify different atoms in simulation cell. The total correlation function *G(t)* can be decomposed into pair, triplet, and quadruplet contributions:

$$G(t) = G_2(t) + G_3(t) + G_4(t), \text{ where}$$

$$G_2(t) \propto \left\langle \sum_{\substack{i,j=1,\\ i\neq j}}^{N} \beta_{ij}(t)\beta_{ij}(0) \right\rangle$$

$$G_3(t) \propto \left\langle \sum_{\substack{i,j,k=1,\\ i\neq j, i\neq k, j\neq k}}^{N} \beta_{ij}(t)\beta_{ik}(0) \right\rangle$$

$$G_4(t) \propto \left\langle \sum_{\substack{i,j,k,l=1, i\neq j, i\neq k,\\ k\neq l, i\neq l, j\neq l, j\neq k}}^{N} \beta_{ij}(t)\beta_{kl}(0) \right\rangle$$

The simulated correlation functions were averaged over $2.5*10^4$ time origins. The time gap between origins was equal to 0.12 ps. The depolarized Rayleigh spectrum is the Fourier transform of the polarizability anisotropy total autocorrelation function *G(t)*.

22.4 Results

We have simulated 256 neon atoms embedded between graphene walls. At the beginning, the neon atoms were uniformly distributed in the confined space between graphene walls. The temperature of the system was set to T = 35 K; therefore, we operate in a very dense fluid where many body interactions become important. The simulations have been performed for two different distances d between graphite walls, equal to $d = 14$ Å ($\rho_{14} = 1.506$ g/cm^3) and $d = 16$ Å ($\rho_{18} = 0.996$ g/cm^3). The simulation started with the distance between graphene walls equal to $d = 14$ Å. The calculated normalized total correlation function $G_n(t) = G(t)/\, G(0)$ at the beginning decays fast losing about 85% of its initial value within first 10 ps and then decays

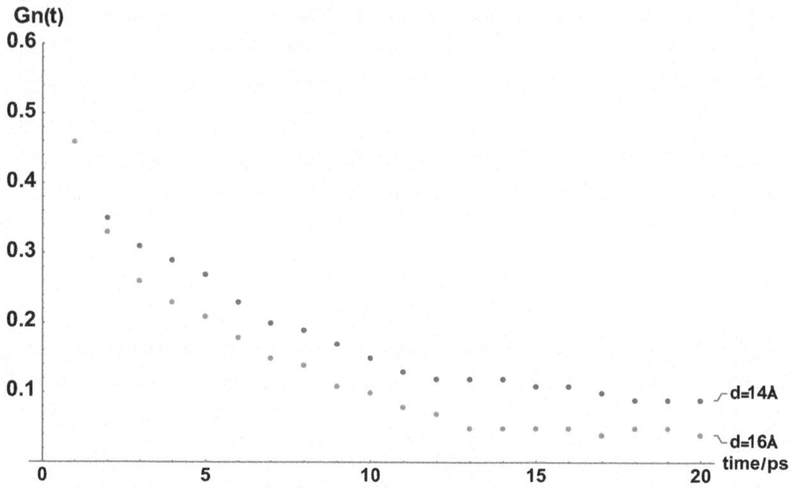

Fig. 22.1 The normalized polarizability anisotropy correlation function $G_n(t)$ of neon atoms confined between two parallel graphene walls, separated by the distance $d = 14$ Å and $d = 16$ Å

very slowly to zero within 220 ps. The correlation function $G_n(t)$ is shown in Fig. 22.1, where the initial value of $G_n(0) = 1$ and the very long time tail were not shown, to increase the readability of the time behavior of $G_n(t)$ correlation function.

The observed behavior of $G_n(t)$ function for $d = 14$ Å slot can be connected with a limited mobility of neon atoms in dense fluid. Neon atoms have not much empty spaces to move in dense fluid and low temperature. The increase of the slot width, while keeping the same number of neon atoms, means that the density of neon atoms confined between graphene slabs decreases. With the increasing of the slot width to $d = 16$ Å, more space is available for neon atoms in the innermost part of the confined space. The total correlation function $G_n(t)$ decays to zero within 50 ps. This situation suggests that the increase of the slot width increases the mobility of neon atoms. By visual inspection of the system (VMD software [34]), we have observed that neon atoms in the innermost (central) part of the slot are more mobile than those located close to the graphene slabs. This observation is also consisted with our previous studies [35], where the values of translational diffusion coefficient D for neon atoms embedded between graphite walls were calculated. It was shown, that the translational diffusion coefficient D is three times larger for the graphite walls distance $d = 16$ Å, in comparison with the corresponding value for $d = 14$ Å. The presented study shows that the situation is somehow similar for the case of neon atoms placed between just one atom carbon thick layers (graphene). The calculated correlation functions $G_n(t)$ might be related to the future experimental work, because the measured depolarized Rayleigh spectrum is the Fourier transform $I(\nu) = \int_{-\infty}^{\infty} \cos(2\pi\nu t)G(t)dt$of the polarizability anisotropy total autocorrelation function $G(t)$.

In this study, we do not attempt to analyze the molecular dynamics simulation results in terms of theoretical model, because the theoretical treatment of the interaction-induced light scattering of noble atoms in confined space is not yet developed. The preliminary MD simulations reported here may serve as the reconnaissance for the future research on the interaction-induced light scattering in confined space and ultrathin layers of noble gases.

References

1. Ma SH, Zhou C (2010) Gas storage in porous metal–organic frameworks for clean energy applications. Chem Commun 46:44–53. https://doi.org/10.1039/B916295J
2. Lee J, Oh H, Jeong S (2014) Investigation of neon–nitrogen mixed refrigerant joule–Thomson cryocooler operating below 70 K with precooling at 100 K. Cryogenics 61:55–62. https://doi.org/10.1016/j.cryogenics.2014.02.006
3. Piatek A, Dawid A, Gburski Z (2006) The existence of a plastic phase and a solid-liquid dynamical bistability region in small fullerene cluster (C-60)(7): molecular dynamics simulation. J Phys Condens Matter 18:8471–8480. https://doi.org/10.1088/0953-8984/18/37/006
4. Nestor JR (1982) Optogalvanic spectra of neon and argon in glow discharge lamps. Appl Opt 21:4154–4157. https://doi.org/10.1364/AO.21.004154
5. Kosmider M, Dendzik Z, Palucha S, Gburski Z (2004) Computer simulation of argon cluster inside a single-walled carbon nanotube. J Mol Struct 704:197–291. https://doi.org/10.1016/j.molstruc.2004.02.050
6. Huang B, Li Z, Liu Z, Zhou G, Hao S, Wu J, Gu BL, Duan W (2008) Adsorption of gas molecules on graphene nanoribbons and its implication for nanoscale molecule sensor. J Phys Chem C 112:13442–13446. https://doi.org/10.1021/jp8021024
7. Dawid A, Gburski Z (1997) Interaction-induced light scattering in Lennard-Jones argon clusters: computer simulations. Phys Rev A 56:3294–3296. https://doi.org/10.1103/PhysRevA.56.3294
8. Dawid A, Gburski Z (1999) Interaction-induced light scattering in xenon clusters: molecular dynamics study. J Mol Struct 482–483:271–276. https://doi.org/10.1016/S0022-2860(98)00668-1
9. Chapeau-Blondeau F, Teboul V, Berrué J, le Duff Y (1993) An experimental and numerical study of high-frequency Raman scattering in argon gas. Phys Lett A 173:153–159. https://doi.org/10.1016/0375-9601(93)90179-4
10. Gburski Z, Zerda T (1980) Vibrational dephasing and intermolecular interactions in liquids. Acta Phys Pol A 57:447–454
11. Gburski Z, Gorny K, Raczynski P (2010) The impact of a carbon nanotube on the cholesterol domain localized on a protein surface. Solid State Commun 150:415–418. https://doi.org/10.1016/j.ssc.2009.12.005
12. Kachel A, Gburski Z (1997) Chain formation in a model dipolar liquid: computer simulation study. J Phys Condens Matter 9:10095–10100. https://doi.org/10.1088/0953-8984/9/46/007 122
13. Bhattacharya A, Chen B, Mahanti SD (1996) Structural dynamics of clusters near melting. Phys Rev E 53:R33–R36. https://doi.org/10.1103/PhysRevE.53.R33
14. Raczynski P, Gorny K, Pabiszczak M, Gburski Z (2013) Nanoindentation of biomembrane by carbon nanotubes – MD simulation. Comp Mat Science 70:13–18. https://doi.org/10.1016/j.commatsci.2012.12.031
15. Mountain RD, Birnbaum G (1987) Molecular dynamics study of intercollisional interference in collision-induced absorption in compressed fluids. J Chem Soc Faraday Trans 2(83):1791. https://doi.org/10.1039/f29878301791

16. Gburski Z (1985) Convergence of memory functions for the vibrational dephasing process in liquids. Chem Phys Lett 115:236–240. https://doi.org/10.1016/0009-2614(85)80687-4 124
17. Dawid A, Gburski Z (2003) Interaction-induced light scattering in a fullerene surrounded by an ultrathin argon “atmosphere”: molecular dynamics simulation. Phys Rev A 68:065202. https://doi.org/10.1103/PhysRevA.68.065202
18. Dacre PD, Frommhold L (1982) Rare gas diatom polarizabilities. J Chem Phys 76:3447–3460. https://doi.org/10.1063/1.443443
19. Dawid A, Gburski Z (2003) Rayleigh light scattering in fullerene covered by a spherical argon film – a molecular dynamics study. J Phys Condens Matter 15:2399–2405. https://doi.org/10.1088/0953-8984/15/14/315
20. Kader MSAE (2002) The depolarized interaction-induced light scattering spectrum and ground state potential curve of gaseous argon. J Phys B Atomic Mol Phys 35:4021–4032. https://doi.org/10.1088/0953-4075/35/19/306
21. Stassen H, Gburski Z (1994) Instantaneous normal-mode analysis of binary-liquid Ar-Kr mixtures. Chem Phys Lett 217:325–332. https://doi.org/10.1016/0009-2614(93)E1390-3
22. Gwizdala W, Gorny K, Gburski Z (2008) Molecular dynamics and dielectric loss in 4-cyano-4-n-pentylbiphenyl (5CB) mesogene film surrounding carbon nanotube – computer simulation. J Mol Struct 887:148–151. https://doi.org/10.1016/j.molstruc.2007.12.045
23. Fan C, Do DD, Li Z, Nicholson D (2010) Computer simulation of argon adsorption on graphite surface from subcritical to supercritical conditions: the behavior of differential and integral molar enthalpies of adsorption. Langmuir 26:15852–15864. https://doi.org/10.1021/la1024857
24. Dawid A, Dendzik Z, Gburski Z (2004) Molecular dynamics study of ultrathin argon layer covering fullerene molecule. J Mol Struct 704:173–176. https://doi.org/10.1016/j.molstruc.2004.01.065
25. Dawid A, Gburski Z (1999) Interaction-induced light-scattering in xenon cluster: molecular dynamics study. J Mol Struct 482-483:271–276. https://doi.org/10.1016/S0022-2860(98)00668-1
26. Gburski Z (1984) Line shape in collision-induced absorption – Mori theory. Chem Phys Lett 106:55–59. https://doi.org/10.1016/0009-2614(84)87010-4
27. Piatek A, Dawid A, Gburski Z (2011) The properties of small fullerenol cluster (C60(OH)24)7: computer simulation. Spectrochim Acta A 79:819–823. https://doi.org/10.1016/j.saa.2010.08.059
28. Gorny K, Dendzik Z, Raczynski P, Gburski Z (2011) Dynamic properties of propylene glycol confined in ZSM-5 zeolite matrix-a computer simulation study. Solid State Commun 152:8–12. https://doi.org/10.1016/j.ssc.2011.10.020
29. Gburski Z, Gray CD, Sullivan DE (1983) Information-theory of line-shape in collision-induced absorption. Chem Phys Lett 100:383–386. https://doi.org/10.1016/0009-2614(83)80292-9
30. Dendzik Z, Gorny K, Gburski Z (2009) Cooperative dipolar relaxation of a glycerol molecular cluster in nanoscale confinement-a computer simulation study. J Phys Condens Matter 21:425101. https://doi.org/10.1088/0953-8984/21/42/425101
31. Dawid A, Gorny K, Gburski Z (2011) The structural studies of fullerenol C-60(OH)(24) and nitric oxide mixture in water solvent – MD simulation. Nitric Oxide Biol Chem 25:373–380. https://doi.org/10.1016/j.niox.2011.08.004
32. Allen MP, Tildesley DJ (1989) Computer simulation of liquids. Oxford University Press, New York
33. Barocchi F, Zoppi M, Proffitt MH, Frommhold L (1981) Determination of the collision-induced depolarized Raman light scattering cross section of the argon diatom. Can J Phys 59:418–1420. https://doi.org/10.1139/p81-187
34. Humphrey W, Dalke A, Schulten K (1996) VMD: visual molecular dynamics. J Mol Graph Model 14:33–38. https://doi.org/10.1016/0263-7855(96)00018-5
35. Dawid A, Gburski Z (2017) Interaction-induced light scattering in thin neon film confined between graphite slabs: MD study. J Mol Liq 245:71–75. https://doi.org/10.1016/jmolliq.2017.06.040

Chapter 23
Sintering Methods of Inkjet-Printed Silver Nanoparticle Layers

O. Kravchuk, R. Lesyuk, Ya. Bobitski, and M. Reichenberger

Resource efficiency of fabrication processes and low price of the devices are becoming increasingly important nowadays. There is a growing interest in the development and production of flexible electronic devices, including displays [1], photovoltaic modules [2], batteries [3] and sensors [4]. Inkjet printing is a low-cost, non-impact and rapid technique with a large potential. Costs are reduced owing to digital imaging, eliminating the multiple process steps (photolithographic processes in the first place). As inkjet printing is a relatively fast technique, it potentially enables fast roll-to-roll patterning of conductive material, e.g., nanomaterials, including metal nanoparticles (generally Ag, Au, Cu, Ni or Co) [5] or metal-organic complexes [6]. Inkjet technology allows the metallization with given topology and high resolution at minimal material consumption.

Nanoparticles are defined as isolated solid-phase objects with a size of up to 100 nm. Nanoparticles have a very large surface-to-volume ratio. Metal nanoparticles are characterized by a number of unique properties, namely the resonant optical behaviour due to the plasmonic oscillations, Coulomb blockade in electrical transport and melting temperature reduction effect [7]. High surface-to-

O. Kravchuk (✉)
Department of Photonics, Lviv Polytechnic National University, Lviv, Ukraine

R. Lesyuk
Pidstryhach Institute for Applied Problems of Mechanics and Mathematics, National Academy of Sciences of Ukraine, Lviv, Ukraine

Institute for Physical Chemistry, University of Hamburg, Hamburg, Germany

Ya. Bobitski
Department of Photonics, Lviv Polytechnic National University, Lviv, Ukraine

Institute of Technology, University of Rzeszow, Rzeszow, Poland

M. Reichenberger
Technische Hochschule Nuernberg Georg Simon Ohm, Nuremberg, Germany

O. Fesenko, L. Yatsenko (eds.), *Nanooptics, Nanophotonics, Nanostructures, and Their Applications*, Springer Proceedings in Physics 210,
https://doi.org/10.1007/978-3-319-91083-3_23

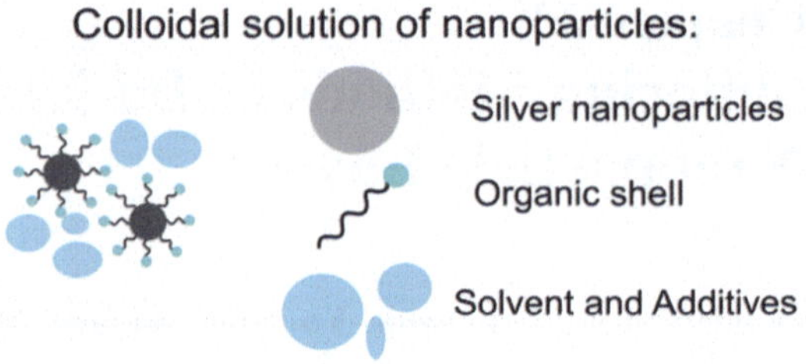

Fig. 23.1 Main components of the colloidal solutions of metallic nanoparticles

volume ratio of nanoparticles explains the significant decrease in the melting temperature compared with large particles, which is also accompanied by a decrease in the sintering temperature [8]. However, there are also problems that can complicate the sintering of nanoparticles. For example, there is a significant risk of particle contamination due to high surface activity. In addition, oxide layers that quickly form on the surfaces of nanoparticles can interfere with the sintering process and worsen electrical properties of resulted structures.

Nanoparticles for functional inkjet printing as a rule have a size <50 nm and are dispersed in solvents. These dispersions (inks), to achieve desired electrical conductivity of printed structures, are usually produced of metal particles, especially silver or copper. It should be noted that silver dispersions for the past decade have been the basic material for printing conductive coatings and contacts, mainly due to the advanced synthesis technology, the good conductive properties of silver and its long-term stability. However, we also consider works on the sintering of copper and gold nanoparticles, which due to their unique properties are also of interest in the context of printed electronics. Nanoparticles of metals in colloidal solutions often have a special organic shell. This shell stabilizes the particles in the solvent and prevents agglomeration and oxidation. Figure 23.1 schematically illustrates the components of colloidal nanoparticulate solutions. However, immediately after printing, the printed structures do not yet have the desired electrical properties. The processing of such solutions can be divided into three stages Fig. 23.2:

1. printing of the desired structures;
2. dispersion drying: evaporation of solvent, removal of stabilizers, agglomeration of nanoparticles (contact stage); and
3. sintering: destruction of the coupling of stabilizers with the surface of the particle and their decomposition, coalescence of nanoparticles.

Applied energy causes the destruction of the organic shell by thermal dislocation, defined evaporation and partial carbonization. As a result, metal nanoparticles begin to contact. To obtain good electrical characteristics, the solvent and other organic dispersion components should be removed before the actual sintering phase. Then diffusion of atoms between nanoparticles can occur. Further aggravation induces the formation of "necks" with a subsequent growth of grains, due to which an increase in the conductivity of the structure can be observed [9, 10].

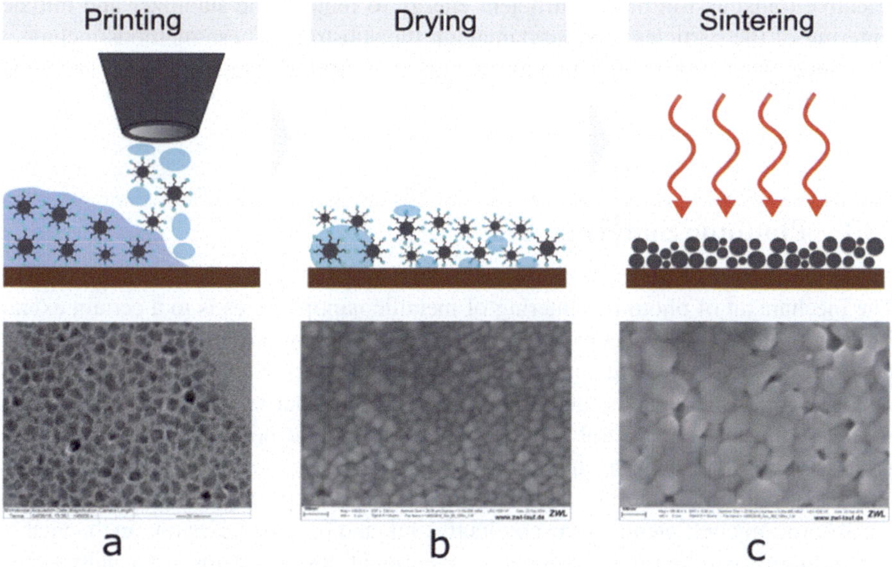

Fig. 23.2 Processing steps for electrical functionalization using nanoparticle colloidal solutions (TEM image of nanoparticles on the carbon/copper grid (**a**), SEM images: dried for 15 min. at 80 °C (**b**); treated at 180 °C for 1 h (**c**))

The above-mentioned organic compounds and potentially present solvent residues can impede the sintering of particles. On the other hand, the residual organic components in the layer are needed to increase mechanical adhesion to the substrate.

The annealing time and temperature of conductive inks used in printed electronics are much lower compared with the conductive pastes of traditional thick-film hybrid technology on ceramics. This fact allows using conductive ink printing not only on ceramics that can withstand temperatures of 800 °C, but also on paper, lavsan, plastics, polyamides and other materials, and some types of inks can be printed even on FR4 fiberglass .

The sintering process results depend on a large extent on whether applied energy is enough and organic components are actually removed. This obviously becomes more important at lower processing temperatures in the range of about 200 °C. The developers of sintering processes face a contradictory task. On the one hand, the annealing temperature should be sufficient to form a dense structure of the conductive layer, reliable sintering/bonding between the particles to achieve a minimum resistance of the formed conductor. On the other hand, the temperature should not overcome the oxidation start temperature or damage the substrate.

Thermal sintering is associated with a number of shortcomings. We can name a limited choice of substrates, a large expenditure of time and energy. To achieve sintering at lower temperatures, efforts are directed to reducing the particle size and selecting organic stabilizers, binders, solvents, and their concentrations. An alternative approach is the applying of non-termic post-processing methods, allowing the selective transmission of the sufficient energy to remove the stabilizer and initiate sintering of the particles without damaging the substrate. These methods include a photonic (flash-lamps and IR-lamps), laser, electrical sintering and chemical processing.

23.1 Photonic Sintering

The mechanism of photonic sintering of metallic nanoparticles is to a certain extent similar to thermal sintering. Absorption of light by printed layers of metal nanoparticles leads to local heating (photothermal effect), which causes evaporation of the solvent and ligands, and sintering of the nanoparticles. Different radiation sources with various spectral characteristics and radiant power can be used in the process, in particular lamps, lasers, or photodiodes.

During photonic sintering with the help of flash lamps, significant difference in the absorption coefficients of the conductive ink and polymer substrate in the visible spectral range can be used. Although conventional polymer films are usually transparent, metal nanoparticles strongly absorb visible light. Thereby, choosing a lamp with the corresponding spectrum, the energy can be transmitted selectively to the printed structures without directly affecting the substrate (after filtration of the UV component). The heat transfer from the heated layer to the cold substrate will obviously lead to an increase in its temperature; however, by applying energy by short pulses, it is possible to avoid overheating and damage of the substrate. Using this technology, it is possible to achieve significant electrical conductivity of structures. Obtained conductivity values are close to those after heat treatment. Thereby, the processing time can be significantly reduced. For the sintering of metal nanoparticles, it is promising to use xenon lamps that generate millisecond pulses of high intensity (30 kW/cm^2) [11]. A short sintering time can also solve the problem of oxidation of copper nanoparticles in a normal atmosphere (the formation of conductive structures using copper nanoparticles usually requires an inert atmosphere).

A 1000 W power xenon flash lamp (Xenon XOP50) with a continuous emission spectrum in the range 350–900 nm was used in Ref. [12]. Already after 5 s of irradiation at a flash frequency of 17 Hz, the same resistivity values were obtained, as after 6 h at 130 °C (1/9 of the resistivity of bulk silver). A decrease in the nanoparticle solution spreading compared with the thermal sintering was also observed.

A xenon lamp with a narrow spectrum of 400–500 nm was also used in Ref. [13] for sintering of silver nanoparticles deposited using aerosol printing (which is close to inkjet printing concerning its working principle). The radiation power was varied from 8 to 20 W. Selective sintering made it possible to use substrate materials with low temperature stability, such as polycarbonate and polymethylmethacrylate. For the silver nanoparticles used in the study, a high value of the absorption coefficient at wavelengths of 400–500 nm was measured. The authors managed to achieve resistivity of about $0.1 \cdot 10^{-7}\ \Omega \cdot$ m (1/8 of the resistivity of massive silver).

The process of photonic sintering is actively studied by NovaCentrix company. The physical models of the process of photon sintering as well as PulseForge sintering equipment based on xenon flash lamps were developed [14, 15]. The sintering process can be controlled by varying of two basic parameters: the radiation power and the duration of the light pulse. Because the effect on the substrate can be minimized during photonic sintering, processing of nanoparticle layers on low-temperature substrates such as paper and polymer films is possible.

In addition, the high speed of the process allows the application of nanoparticles solutions, which, in the case of sufficiently long thermal sintering, are oxidized in a non-inert atmosphere, in particular copper. In this case, requirements are imposed on the radiation power [16]. These advantages can lead to the reduction of production costs by using cheaper substrates and nanoparticles. There is also a potential for integration into the "roll-to-roll" production process, due to the high speed. This sintering technology is applicable for metal nanoparticles (silver, gold, copper) [17–21], dielectric nanoparticles (alumina, zirconium dioxide, barium titanate, hafnium oxide) [22], as well as for soft magnetic materials (cobalt, ferrite and iron-nickel permalloy) [23]. In all these works, special NovaCentrix PulseForge equipment was used.

From the viewpoint of further costs reduction, a commercially available Nikon SpeedlightS B-22 flashlight was used for sintering of silver nanoparticles on polyimide (PI), polyethylene (PE) and photographic paper in Ref. [24]. The flash duration was 1 ms. Authors argue that the higher conductivity was obtained on photographic paper (the measured resistivity of structures is 4.6 times higher than the resistance of massive silver), which, according to their assumption, is due to the low thermal conductivity of this substrate.

Two different devices (6.0 and 0.5 kW) with incandescent lamps, emitting in the infrared region of the spectrum, were used in Ref. [25] for sintering of gold and silver nanoparticles on paper. This method can be described as intermediate between thermal and photonic sintering. The authors recorded a significant decrease in the transmission of light at longer wavelengths after thermal sintering, which is mainly due to an increase in the reflection of thin metallic layers. They determined the need for a compromise between the absorption coefficient of nanoparticles and a paper substrate, because the reflection of the kaolin particles, coating the paper surface increases with the wavelength, whereas metallic nanoparticles absorb more light of shorter wavelengths. Because the paper absorption coefficient grows at large wavelengths (> 3 μm), the heating of the paper can be limited by using incandescent lamps with a higher temperature. The lowest resistivity values obtained in [25] are $2.5 \cdot 10–7\ \Omega \cdot$ m for gold and $1 \cdot 10–7\ \Omega \cdot$ m for silver after 20 s of irradiation.

In Ref. [26], a fast R2R IR drying and sintering method ink-jet printed silver layers is reported. An IR module was integrated in R2R printing system. The authors managed to drastically reduce the processing time by achieving low sheer resistance values of 0.30–0.35 Ω/square at the polyethylene naphthalate (PEN) substrate (15% of the bulk Ag conductivity). The peak wavelengths of the IR radiation of the applied lamps are 1200 nm, 1060 nm and 940 nm. By optimization of the processing parameters, drying and sintering were achieved at high web velocities up to 1 m/s with an exposure time of less than 0.5 s. IR irradiation was used in Refs. [27, 28] for sintering of silver or gold [29] nanoparticles on paper substrates. The IR sintering, however, is less suitable for inexpensive plastic substrates, such as PET, which are rapidly deformed due to large coefficients of thermal expansion and low glass transition temperature and melting points [30].

23.2 Laser Sintering

Nanoparticles can be sintered using a power source such as a laser beam [31–46]. The advantages of local laser sintering are the limitation of the thermal influence zone and more efficient selective energy transfer. Therefore, it is promising to use it for creating conductive structures on polymer substrates.

For laser radiation, a Gaussian intensity distribution with a maximum at the centre is usually characteristic. So, only a small area is exposed to radiation. The processing depth when using a laser beam is also limited. Lasers that generate short pulses are commonly used for micromachining and ablation. In both cases, a small zone of thermal influence is decisive, which is also important for the process of selective sintering. High intensity of laser pulses allows achieving greater depth of processing than in the case of continuous wave lasers. Using continuous generation lasers, materials are heated more smoothly, but more heat is transferred to the substrate. Typically, the sample is scanned with a laser beam so that the entire printed structure is processed evenly.

The effectiveness of laser sintering depends on the radiation intensity, material absorption coefficient and the interaction time. The important parameters are the shape of the beam and the diameter of the focal spot. All this determines the optimal processing speed, that is, the scanning speed of the laser beam. It should be also mentioned that the structures consisting of different number of printed layers need to be handled with different parameters (for example, several passes or scanning speed reduction are necessary). The sintering results for combinations of small and large scanning velocities and intensities also differentiate. At the same time, low speed with small beam intensity does not guarantee good electrical properties and adhesion. Part of the beam energy is transferred to the substrate and surrounding materials. Consequently, the speed limits the practical maximum power and sintering temperature.

Two different sources of laser radiation were used in work [31]: a continuous-wave diode laser with a wavelength of 940 nm (near IR), a power of 9–180 W and a scanning speed of 600 mm/s, as well as a pulsed laser with $\lambda = 515$ nm (visible area), with a power of 30–600 mW and a scan rate of 10–300 mm/s. The resistivity values of laser-processed conductive tracks were similar or slightly better than of the sintered using traditional thermal method at 220 °C/1 h. The best results were obtained for a diode laser (according electrical characteristics and processing speed). One of the reasons is the larger diameter of the beam and the intensity of the radiation. The output of a pulsed laser is limited by the intensity of the peak of the laser pulse, the size of the spot and the frequency of the pulses. A diode laser with a wavelength of 940 nm was also used in [32] for sintering of silver nanoparticles on flexible substrates (PI). The diameter of the focal spot was 1 mm, and the radiation power was 20–50 W. The obtained resistivity is close to the value for massive silver.

The efficiency of energy transfer during laser sintering can be increased due to a strong absorption peak when using the surface plasmon resonance of metallic nanoparticles. For example, for gold nanoparticles, laser radiation with a wavelength near 520 nm can be used. In Refs. [33, 34], the authors used a 514.5 nm argon-ion laser beam. In [33], gold nanoparticles were utilized to fabricate electrodes for field-effect transistors made from organic materials.

In Ref. [35], a laser with a wavelength of 488 nm and a power of 50 to 250 mW was used to sinter gold conductors. The beam diameter was 17 μm, and the velocity of its movement was 2 mm/s. The obtained resistivity of structures is 6 times higher than the resistance of massive gold.

The sintering of copper nanoparticles, as already noted, has certain peculiarities. In a non-inert atmosphere, copper nanoparticles should be sintered very quickly to avoid their oxidation. So, the use of two technologies seems to be promising: sintering with the help of flash-lamps or laser radiation. A continuous 808.6-nm diode laser was used in work [36] for sintering copper nanoparticles on a PI. The beam size in the sintering plane was 1.1 × 0.4 mm. The laser radiation power was changed in the range from 2 to 32 W, and the scan rate was 50–400 mm/s. In this case, the value of the surface resistance of the treated layers is about 90 mΩ per square. The authors increased the scanning speed to reduce the negative effect of variation in the thickness of the printed layers. They have also noted good reproducibility of the results of laser sintering. In addition, it was found that secondary scanning with a laser beam (at constant intensity and speed) does not lead to an increase in conductivity.

The ability of selective sintering of printed structures was used in Ref. [37]. With the help of a laser beam that moved along a certain predetermined pattern, selective sintering of the chosen regions was performed. The focus on the printed layer beam energy causes local heating, which leads to rapid sintering (Fig. 23.3). Thus, it is

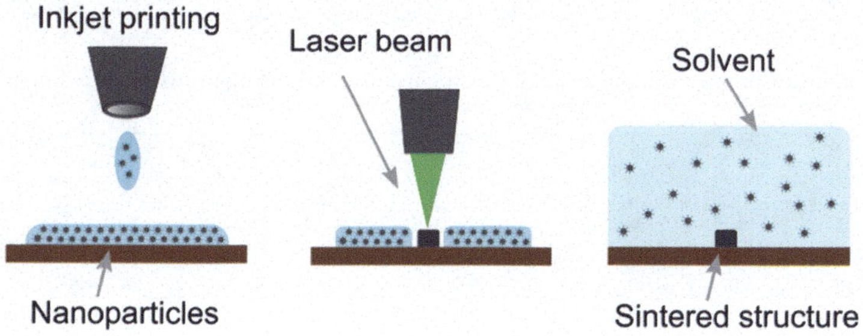

Fig. 23.3 Principal scheme of the selective laser sintering process

possible to obtain fine conductive structures with high resolution, which is limited only by the optics of the scanning system and the thermal conductivity of the target material. High scanning velocity can minimize the effect of heat diffusion and allows the creation of metal tracks just a few micrometers wide without damaging the substrate. To complete the process, the non-sintered material is removed, generally by rinsing and washing in an appropriate solvent. The authors used Nd: YAG (second harmonic 532 nm) laser for sintering copper nanoparticles. The diameter of the focal spot was 30 μm, the scanning speed was within 0.3–5 mm/s and the radiation power was 200–600 mW. The resulting resistivity of the layers was two to three times higher than the value for massive copper (which is close to the best values for thermally sintered in an inert atmosphere structures). Using the "direct write" technology of laser sintering, authors managed to achieve a very high resolution. Conductive tracks with a width of up to 5 μm were created on the PI film. The advantages of this method can be successfully used for the metallization of glass and temperature-sensitive substrates for a wide range of applications. In [33], structures with a line width of up to 10 μm were obtained by irradiation of printed layers of gold nanoparticles with a 3 kW*cm^{-2} argon-ion laser beam with a spot diameter of 3.5 μm.

To improve the resolution, a process of ablation of deposited layers of gold was used in [38] by applying of a short-pulsed (3–5 ns) YAG: Nd laser (second harmonic). Tracks were printed with a large width, and with the help of a laser, the edges of the lines were burned out – so the tracks became thinner (up to 8 μm).

Real-time monitoring of the laser sintering process is described in [32]. The authors measured the intensity of a laser beam reflected by a layer of nanoparticles, providing a mechanism for analyzing the degree of sintering in situ.

Another approach is sintering of colloidal silver printed on a ceramic substrate (Al_2O_3) under the conditions of photoinduced substrate heating with a laser (1064 nm), which excludes the drying stage of the structure (Fig. 23.4).

Thus, laser sintering is a promising alternative method. Lasers of various types are used for sintering and annealing of nanoparticles: argon laser ($\lambda = 488$ nm [40], 514 nm [34, 38, 41–43]) in continuous and pulsed mode, YAG: Nd laser in continuous and pulsed (532 nm, 1064 nm [38, 39, 44]) modes, diode laser in

Fig. 23.4 Optical microscope images of sintered silver conductive trace after laser sintering in the photoinduced substrate heating mode [44]

continuous mode (940 nm [45]) and erbium laser in continuous mode (1550 nm [46]). The process is characterized by high speed, but it is expedient to use it for the processing of small areas. It should be noted that too high intensity of radiation can lead to evaporation of printed layers. Also, if the layer is too thin or there is considerable heterogeneity of its thickness, defects can be traced. Another limiting factor is the low speed of the ray traversing the entire structure.

23.3 Sintering by Microwave Radiation

Treatment with the help of microwave radiation is widely used for the sintering of dielectric and conducting materials and in synthetic chemistry. The advantage of this method is uniform, rapid and volumetric heating. Results of the use of microwave radiation for sintering layers of silver nanoparticles deposited by inkjet printing are reported by Perlaer et al. in Refs [47, 48]. It was found that metallic nanoparticles can be quickly and efficiently sintered using microwave radiation. The dielectric response to the field is given by the complex dielectric permittivity:

$$\varepsilon_n = \varepsilon' + i\varepsilon'' = \varepsilon' + i\frac{\sigma}{\omega\varepsilon_0}, \tag{23.1}$$

where the real component ε' describes caused by the polarization decrease of the electric field in matter, and the imaginary part ε'' describes the currents arising in matter in the alternating electric field; σ is the conductivity; and ω is the angular frequency. The ratio of the imaginary to the real parts of the dielectric constant determines the ability of the material to dissipate energy and is known as the tangent of the dielectric loss angle:

$$\tan\delta = \frac{\varepsilon''}{\varepsilon'}. \tag{23.2}$$

Depending on the characteristics of losses and, consequently, conductivity, the materials can be non- transparent, transparent or absorbing. For massive metals that are good conductors, no internal electric field is generated, and the induced electric charge remains on the surface of the sample. So, metals reflect microwaves. But powders with a particle size of about a micrometer demonstrate significant absorption [49].

It is believed that the interaction of the conductive particles with microwave radiation, that is, the inductive coupling, is mainly based on the Maxwell–Wanger polarization, which is a consequence of charge accumulation on the material separation surfaces, electrical conductivity and Foucault currents. However, the main reason for the successful heating of metal particles by microwave radiation has not yet been fully elucidated [48].

The penetration depth d is defined as the distance in the material at which the falling power decreases to 1/e (36.8%) of the value at the surface:

$$d = \frac{c\varepsilon_0}{2\pi f \varepsilon''} = \frac{1}{\sqrt{\pi f \varepsilon_m \sigma}}, \tag{23.3}$$

where c is the speed of light; f is the frequency of the microwave radiation; ε_m is the permittivity of the material and ε_0 is the permittivity of free space. Usually high-conductivity materials (metals) have a very small penetration depth. For example, the penetration depth of microwaves at a frequency of 2.45 GHz for metallic powders of silver and copper is 1.3 and 1.6 μm, respectively. In contrast to the relatively high absorption of microwaves by conductive particles, the polarization of dipoles in thermoplastic polymers below the glass transition temperature is limited; hence, the polymer film is transparent to microwave radiation. Microwave sintering can be successful only if the size of the object perpendicular to the plane of incidence is commensurable with the depth of penetration. If inkjet-printed conductive structures satisfy this condition, then it is expected that this technology can be adapted for production. It can be expected (for thick lines), that the heating of the track with the help of microwaves will be inhomogeneous. However, in comparison with a polymer substrate, silver conducts have high thermal conductivity. Therefore, the silver path warms evenly.

The absorbed power per unit volume can be calculated as

$$P = \sigma E^2 = 2\pi f \varepsilon'' E^2, \tag{23.4}$$

where σ is the electrical conductivity; ε'' is the dielectric loss factor; and E is the amplitude of electromagnetic field. At normal incidence, the amplitude is an exponentially decaying function of the distance, characterized by a penetration depth d.

In [47], microwave radiation with a power of 300 W was used for sintering structures from silver nanoparticles on PI. The processing time was 240 s. The temperature on the surface of the sample reached 200 °C. The obtained resistivity value was $3.0 \cdot 10^{-7}\ \Omega \cdot \mathrm{m}$, which is 5% of the resistance of massive silver. The authors believe that the described process can be used for manufacturing of antennas, radio frequency (RFID) tags and collector grids of solar cells.

An original approach to sintering of conductive structures on polymer films using the microwave radiation was proposed in [50]. Special antenna structures made of silver nanoparticles were inkjet printed and heat treated at 110 °C (60 min). They were used to measure the resistance of a conductive path and receive electromagnetic waves. A single conductive track was then printed and dried at 110 °C for 1–5 min to remove the liquid solvent. The tracks were irradiated for 1–60 s at a power of 1 W. As a result, a decrease in resistance was observed, the exact value of which depends on the initial resistance. The antenna effect, reflecting the ability of the microwave absorption in the material, was systematically investigated by changing the surface area of the electrodes. An inverse relation between the size of the electrodes and the resistance decrease rate caused by microwave irradiation was found. This is explained by the fact that the increase of the electrode surface area improves the absorption of microwaves. The data obtained in the absence of antennas demonstrate that the energy absorption by the printed lines is negligibly small. The absorption of microwave radiation in the presence of an antenna can be improved due to a smaller impedance mismatch between the air and the sample. The impedance Z of the structure relative to free space

$$Z = Z_0\sqrt{\varepsilon^*} \tag{23.5}$$

where Z_0 is the impedance of the free space ($Z_0 = 377\ \Omega$); ε^* is the complex permeability of the structure relative to the free space m. The complex dielectric constant is related to the dielectric permittivity and the loss factor in accordance with (23.1).

Reflection, that is impedance mismatch Z/Z_0, scales with the square root of the complex permittivity ε^*. Thus, the reflection depends on the resistance of the electrodes and, therefore, their total surface area.

The authors managed to achieve specific conductivity in the range of 10–34% of the value of massive silver. They found that the microwave sintering of tracks printed on polymer substrates depends significantly on the total area of the antenna structures, drying time and geometry. The presence of conductive antennas unambiguously promotes sintering of nanoparticles in pretreated ink tracks, to an extent that depends on the total area of the antenna. When using metal antennas, it takes about 1 s to get noticeable sintering with the help of microwave radiation. The degree of sintering for these short exposure periods, however, depends significantly on the initial resistance of the previously dried track. In addition, increasing the track width accelerates the increase of conductivity due to improved energy absorption due to an increase in surface area. Advantages of the method: high speed and low temperature, reduction of costs, and the possibility of using common polymer films, such as PET.

Another reported method is hybrid sintering of silver nanoparticles using microwave radiation at a frequency of 6.425 GHz (bandwidth – 0.75 GHz) [51]. In this case, the silicon substrate worked as a radiation receiver. The authors have investigated the resistivity and Young's modulus of structures. Hybrid microwave sintering can be used to obtain relative high conductivity in a very short time.

23.4 Plasma Sintering

This type of sintering can be used both for metallic nanoparticles and for organometallic composites on thermally sensitive polymeric substrates. It occurs due to the effect of the low-pressure argon plasma or the electron-cyclotron resonant (ECR) plasma on the printed structures. Plasma can be used for densification of powder materials, as well as etching and sintering of thermally sensitive materials. The intensification of diffusion at the grain boundaries of ceramics provoked by high-energy plasma is widely studied. Therefore, it can be assumed that such a treatment

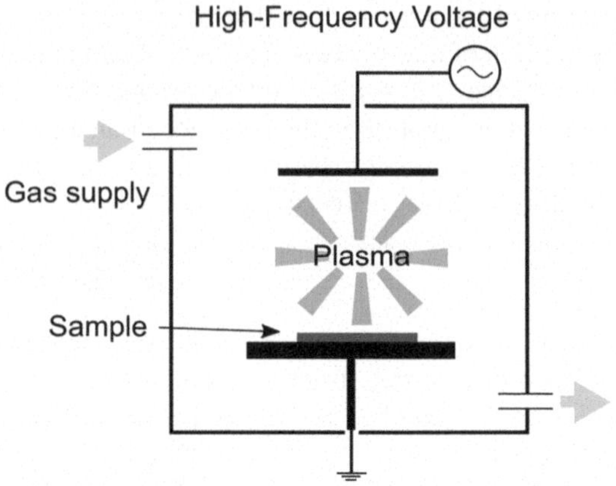

Fig. 23.5 Principal scheme of the low pressure plasma sintering of printed structures

may be suitable for sintering metal nanoparticles. Plasma treatment is widely used in modern production processes (for example, for the manipulation of surface energy).

Rheingold et al. [52] demonstrated for the first time that low-pressure argon plasma processing can cause sintering of silver nanoparticle layers printed on glass, and polymer films of polycarbonate and PET were used as a substrate. The substrate temperature in this case increased to 70–80 °C, depending on the substrate material. The RF power varied from 40 to 80 W. Argon was used to generate the plasma, because inert gases do not lead to deformation of polymer substrates, in contrast to oxidizing gases such as oxygen [53].

The applied system is schematically shown in Fig. 23.5. The plasma discharge is established between two electrodes, one of which is fed with high frequency power, whereas the other is grounded. The sample to be processed is placed on one of the electrodes. High-energy particles generated in the plasma bombard the surface of the sample and cause further coalescence of silver particles into large clusters through the atomic diffusion. Charged particles (ions and electrons) of high energy and UV radiation from the plasma lead to an effective splitting of the surface polymer chains into volatile fragments with low molecular weight that are easily removed in a high vacuum. As a result, the metal nanoparticles begin to directly contact each other. Due to the high surface energy of the nanoparticles and energy, transferred by excited charged particles, diffusion along the grain boundaries and compaction occurs. Thus, a dense network of conductive material is formed.

The results for polymer substrates indicate an exponential decrease in resistance with increasing energy. This is due to an increase in the degree of ionization of the plasma gas. It is also indicated that the plasma penetration depth is limited, so if the thickness of the layer is greater, sintering of the lower layers may not occur. Reducing the thickness of the layer and its cross section, recorded after plasma treatment, indicates the shrinkage and sintering. Densification rate reached about 50% after 60 min. and after that did not change significantly. After about 60 min. at a power of 80 W, the minimum resistivity was reached, that is slightly higher than the value obtained after thermal sintering (220 °C/1 h). The resistivity of structures after treatment with argon plasma reaches a level three times higher than the value of the resistance of massive silver. In [54], films of polypropylene and polyethylene PE were used as a substrate. The decomposition temperature of polypropylene is below 90 °C. Consequently, the processing temperature is an important limiting factor in the case of using this material. The authors planned to compare the results of the treatment using argon and oxygen plasma, but even the first experiments with nanoparticles made showed that oxygen plasma induces silver oxidation on the surface of the layer. Therefore, only argon was used for further experiments. It is reported that the surface resistance of a thin silver layer has decreased by two orders of magnitude, regardless of the used substrate (the resistance obtained is an order of magnitude higher than the value of massive silver). The power of the plasma unit was 75 W at a pressure of 9.5 Pa.

The use of plasma treatment for sintering of silver nanoparticles layers using low-pressure argon plasma in the system of reactive ion etching (RIT) was also investigated [55]. The influence of nanoparticle size at a constant stabilizing surface layer thickness was studied. Plasma sintering was applied on silver nanoparticles layers (nanoparticles of two different diameters: 23 and 77 nm) on PI and PEN films. The effect of plasma parameters on the morphology, composition and electrical resistance of the treated layers was also studied. The plasma sintering process was carried out for various plasma energies and processing times while controlling gas pressure and flow rate. The lowest resistivity values of the silver nanoparticle layers of $8.73 \cdot 10^{-8}$ and $1.87 \cdot 10^{-7}\ \Omega \cdot \mathrm{m}$ were obtained after sintering at 900 W for 60 min (for nanoparticle diameters of 23 and 77 nm, respectively). These values are approximately 5 and 12 times higher than the value of the resistivity of massive silver. The maximum temperature of the substrate during the treatment reached 150 °C. To obtain similar resistivity values using thermal sintering, a temperature of 350 °C was needed. The authors also noted that because the total surface energy is associated with smaller particle sizes, the layers consisting of smaller particles demonstrate better sintering results than of large ones.

Local plasma sintering of printed silver nanoparticles layers at atmospheric pressure is described in [56]. Because plasma treatment leads to the decomposition of a stabilizing organic coating of nanoparticles, it also affects polymer substrates. The disadvantages of the process are the need for complex vacuum equipment and relative long processing time. It can be also accompanied by substrate matting or colour change. In order to overcome these problems, a process of local non thermal argon plasma sintering has been developed, which can be performed at atmospheric pressure and room temperature. Atmospheric plasma equipment is often used for surface modification, thin film deposition, plasma polymerization, purification, as well as biological and medical-biological applications.

The printed structures were selectively treated using plasma flow to produce conductive paths in ambient conditions without damaging the substrate. Nanoparticles were inkjet printed on PEN and PI films. Argon plasma was lit by turning on the high-frequency voltage from 2 to 6 kV at 1.1 MHz. The nozzle was moved over the sample to be processed. The obtained conductivity values of the structures reaching up to 12% of the conductivity of bulk silver. It has been found that the sintering results (considering morphology and electrical conductivity of printed structures) depend significantly on the dispersion properties, such as the polarity of solvent, metal content and particle size. The processing time was reduced from more than 30 min to 4 s. Due to the selective effect on printed structures (diameter of the effect zone – 3 mm), as well as short exposure times, etching, matting and colour changes of polymer substrate materials were avoided. In addition, the process is potentially suitable for the "roll-to-roll" production. The authors reported the importance of the effect of plasma parameters and nanoparticle dispersion on sintering results, such as the morphology and electrical conductivity of printed structures.

23.5 Room-Temperature "Sintering" Using Chemical Agents

Chemically induced sintering at room temperature is demonstrated on metal nanoparticles by charge neutralization [57], dissolving stabilizing ligands by salts [58] or methanol [59] or through local oxidation of protective organic ligands [60], as well as interaction with a special porous substrate coating [61]. The method for achieving conductivity at room temperature as a result of the process of spontaneous three-dimensional coalescence, which occurs on solid substrates, was described in Ref. [57]. The approach is based on provocation of tight packing of nanoparticles after deposition on the substrate, which ultimately leads to their coalescence. The resulting structures have a high electrical conductivity value. First, the authors evaluated the ability of polydialyldimethylammonium chloride (PDAC), a positively charged polymer, to cause aggregation in a dispersion of negatively charged silver nanoparticles stabilized with sodium polyacrylic acid (PAA). The PDAC solution was added to the silver nanoparticle dispersion, and the zeta-potential and the average particle size were measured. It was found out that the zeta-potential of the original nanoparticles is −47 mV, and its negative value decreases with increasing PDAC concentration.

The experiments were then repeated for nanoparticles deposited on the substrate by inkjet printing. After drying, the polycation solution (PDAC, 1% by weight) is also applied to the silver nanoparticle layer by means of inkjet printing. It was revealed that spontaneous coalescence of all silver nanoparticles without heating occurred in the zone of deposited polycation drops, which leads to a significant increase in the conductivity of the printed structure. To quantify the role of the polycation in the coalescence process, droplets of PDAC solutions of different concentrations were applied to the surface of the silver nanoparticle layer. The particle size was examined with an electron microscope. As the authors established, at a mass ratio of PDAC/Ag to 0.01, the polycation does not affect the size and morphology of the particle. In the case of the higher mass ratio of PDAC/Ag, the particle sizes increase with PDAC concentration, and the formation of necks is distinctly observed between the particles. In the case of a mass ratio of PDAC/Ag

0.05, the average particle size significantly increased (more than 120 nm) and the mass ratio of 0.2 a continuous structure of large particles was formed. The electron microscope image of the cross section of the sample (at the 0.2 PDAC/Ag mass ratio) confirms that sintering takes place over the entire thickness of the layer (~0.5 μm) and also on the surface.

The coalescence process can be carried out in a reverse order if the droplets of a solution of silver nanoparticles are applied to a precoated solution of the polycation substrates. Glass and PET films were used as a substrate in [57]; therefore, it was confirmed that "sintering at room temperature" does not depend on the chemical nature of the substrate. The obtained resistivity values are 6.8 μΩ cm for photographic paper (20% conductivity of massive silver). The authors also successfully applied the described method for processing layers of copper nanoparticles [62].

Another chemically based method worth mention is the substrate-activated sintering [63] based on the chemical removal of the ligand, which stabilizes the nanoparticles due to interaction with a special coating of the substrate. After the ligand coating is removed, physical contact occurs between the nuclei of the nanoparticles. "Sintering" of nanoparticles occurs due to diffusion and the release of surface energy.

The coating of the substrates for inkjet printing generally includes a layer absorbing the solvent and a surface layer providing particle fixation. It was found that the coating layer of such a substrate can also cause coalescence of the nanoparticles at room temperature. The authors investigated photo paper for inkjet printing and transparent films based on PET. Comparison of the four commercially available inkjet substrates allowed to reveal significant differences in the resulting conductivity of the printed silver nanoparticles structures even when all the substrates had a silica-based coating. In addition, the effect of humidity on the process was also noted: in the case of a high humidity value, an increase in the conductivity was observed. Substrates with an optimized surface layer provide conductivity of about 27% of the conductivity of bulk silver in the case of storage at high relative humidity (in this case, the minimum conductivity was achieved after storage for several days). The dependence of the conductivity on the relative humidity confirms that water additionally facilitates the removal of the ligand; however, high humidity in itself does not provide "sintering." The process can be further improved by better mutual matching of the chemical composition of the ink and the coating layer of the substrate.

The development of the ligand removal method was proposed by Ford and co-workers [50]. The printed structures were locally treated with a NO_2 gas stream. It has been established that the oxidation of protective thiol ligands and the exothermic decrease in surface area is an important factor in the case of the sintering of layers of thiol-stabilized gold nanoparticles. A significant reduction in resistance was recorded, and after 50 min, the film resistance becomes close to the values obtained after heat treatment. The thiol-stabilized group particles are used very often due to their solubility in organic solvents at a relatively low temperature (~200 °C), which is necessary for rapid sintering. The stabilizing ligands that provide the energy barrier before sintering are generally composed of two parts: an anchoring group that forms an S-Au bond with the surface of the particles and a spacer component that can vary from alkane chains to aromatic structures or even polymers. The particles are sintered if the thermal energy of the system is enough to overcome the activation energy determined by the stabilizing ligand. The sintering of many particles leads to the formation of a relatively continuous layer of gold.

Nanoparticles are sintered through the mechanism of the "neck" formation, in which the two particles are first combined into a peanut-subarctic structure, after which the surface area is minimized and larger particles are formed. The calculations show that as a result of minimizing the surface area, energy is released. Thus, the authors have proved the possibility of "sintering" a gold nanoparticle at room temperature with a simple chemical treatment.

The decrease in resistance at contact with 0.01–0.2 mol/l NaCl solution, after which occurred the heating up to 95 °C, was recorded in [58]. A dependence of the resistance of inkjet-printed conductor tracks on the concentration of sodium chloride in the solution was revealed. As a substrate, several types of paper with a special coating were used. The effect of sodium chloride on the printed solutions of silver nanoparticles was further investigated in Refs. [64–66].

23.6 Electrical Sintering

In this method, the sintering of nanoparticles is achieved by applying a constant or alternating voltage to the printed structure. The current flowing through the printed layer leads to its local heating and sintering. The main advantages of this method are a short sintering time (from microseconds to seconds) and a reduced thermal load of the substrate. This method of sintering can also be applied to organometallic composites inks, if combined with heat treatment.

During the electric sintering process, two electrodes are applied to the printed structure (Fig. 23.6). When a voltage U is applied to the electrodes, a current flows through the printed layer (represented by arrows). This current causes local heating in the layer, initiating the coalescence of the nanoparticles. The resistive heating

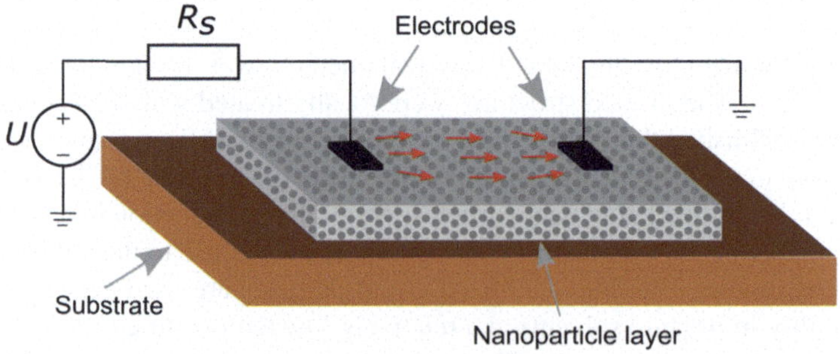

Fig. 23.6 Principal scheme of the electrical sintering process

(in accordance with the Joule-Lenz law) caused by this resistance decrease of the printed structure during sintering leads to a sharp increase in current, causing even more heating. As a result, a sharp increase in conductivity can be observed. Resistor Rs limits the maximum current when the film is sintered. The method of contact electric sintering was used in [67, 68].

In particular, in [68] printed on photographic paper tracks of silver nanoparticles were sintered by transmission a direct current between the contact electrodes. The initial current and power dissipation in the layer of nanoparticles result from tunneling between the particles. As the temperature of the printed structure rises, the resistance R_{tr} decreases due to the decrease in the distance between the particles, leading to an increase in the dissipated power P_{tr}:

$$P_{tr}(t) \sim U^2 \frac{R_{tr}(t)}{|Z_s + R_{tr}(t)|^2}, \tag{23.6}$$

where Z_S represents the impedance of the voltage source, including contact resistance, which causes strong positive feedback, accelerating sintering process. The experiments showed that after 3 ms, application of the voltage, sharp increase in current and a decrease in the resistance of the sample by more than four orders of magnitude was observed (the main transition occurs in less than 2 µs). In addition, a clear dependence of the morphology of the structure of the sintered layer on electric power was revealed.

The applied voltage was 90 V, and the maximum current was limited to 0.45 A. The achieved conductivity values are approximately 50% of the conductivity of massive silver. In [69], direct current was used for sintering of sensor structures printed on glass substrates with solutions of silver nanoparticles. Depending on the initial resistance of the structures, a voltage of 60–95 V was applied. The current value, used for sintering, was 0.147 A. The authors proved that the structure resistance can be measured during the sintering process. Thus, the process can be terminated when the target value is reached. A high sintering rate was registered, though process lasted only a few seconds. In addition, very small deviation of the obtained resistances of the structures from the set value was achieved.

Thus, the following advantages of this method can be highlighted:

- control of the final resistance by adjusting the process parameters, such as the bias resistor value;
- sintering of the specific areas.

The use of an alternating electric field allows electrical sintering without physical contact with the layers of nanoparticles. In work [70], this is achieved by installing electrodes over the sample, and a capacitive coupling with the printed layer occurs. The electrodes are of such a shape and arrangement that they can provide a nearly uniform electric field throughout the whole non-sintered structure. Sufficient capacitive coupling (at a possible working distance between the electrodes and a layer of nanoparticles) is obtained when operating in the microwave range. Non-contact electric sintering (NcES) allows avoiding of many shortcomings of the contact, especially in the case of sintering of large-area structures.

The system for NcES can be conditionally described as an electrical circuit consisting of an equivalent voltage source UTh with a resistance ZTh (according to Tevhenin's theorem) and a load impedance ZL. The main advantage of NcES over contact one is the possibility to avoid the short-circuit effect, as described in [70]. It is assumed that the nanoparticle layer is divided into a plurality of parallel lines between the electrodes, where R_i represents the line resistance, and C_s is the coupling capacitance between the line and the electrodes. If

$$\frac{1}{\omega C_s} > R_i \tag{23.7}$$

for all R_i, and $\mathrm{I}\,\{Z_{Th}\} = -\mathrm{I}\,\{Z_L\}$ to adjust the capacitive coupling, the total power is dissipated in the nanoparticle layer,

$$P_L \approx \frac{R_{av}/n}{(R_{av}/n + R_{Th})^2}|U_{Th}|^2, \tag{23.8}$$

with $R_{Th} = R\,\{Z_{Th}\}$ and

$$R_{av} = \sum_{i=1}^{n} \frac{R_i}{n}, \tag{23.9}$$

where the power dissipated at R_i,

$$P_i \approx \frac{R_i \cdot P_L}{R_{av} \cdot n}. \tag{23.10}$$

As one can see, by the appropriate establishment of a capacitive coupling, it is possible to achieve a positive feedback in the introduction of power (as well as for contact sintering by direct current [68]). In this case, more power is transferred to the least sintered areas [71]. In [70], contactless electric sintering occurs by moving the sintering heads over the printed layer of nanoparticles or vice versa with a fixed head over the moving sample. The authors carried out the sintering experiment by moving the coaxial sintering head (based on the SubMiniature type A connector) over the lines of silver nanoparticles of various widths printed using the inkjet printing at a speed of 1 $\mathrm{mm} \cdot \mathrm{s}^{-1}$ and at a distance of 10 μm. Effective sintering was also achieved in the case of high processing speeds without significant deviations in the resistance values. Experimental results confirm that the increase in processing speed can be compensated by increasing the power.

23.7 Conclusions

In summary, the inkjet printing, possessing some unique advantages for image and coating formation, has now become a new technology in electronics as a completely additive, non-contact method of precise and local application of materials on surfaces of various shapes and nature. The development of a wide range of functional inks made it possible to apply this technology in a wide variety of fields of science and technology – from electronics to biotechnology. One of the key steps in the conductive structures creation process using inkjet printing of colloidal solutions of nanoparticles is sintering of the deposited on the substrate layers to achieve the desired electrical and mechanical properties. Today, there is a considerable interest in the search for alternative thermal sintering processes that cause the coalescence of nanoparticles.

In recent years, the main attention has been paid to the further processing of deposited metal layers aimed at reduce the influence of temperature on flexible polymer substrates as well as the duration of the treatment. As shown in presented review, different methods of sintering are based on the different approaches of transferring energy to the layers of nanoparticles. When thermal sintering in an oven, energy is transferred through convection, thermal conductivity and thermal radiation. The temperature distribution inside the furnace is usually uniform, resulting in which the printed structures and the substrate are heated equally. In the case of electric and microwave sintering, heating occurs due to the flow of electric current. Therefore, heating is local. In the case of laser sintering, optimized spectral and spatial selectivity can be achieved. In this case, the wavelength of the laser can be chosen such that the optimal energy transfer to the printed layers is obtained. Light pulses of very high energy can be used to sinter nanoparticles over a large area. The process can be controlled by adjusting the pulse width (usually from 0.1 to 10 ms), the intensity of the radiation and the number of pulses. Heat transfer from the surface warms the lower layers. One of the most promising advantages of photonic impulse sintering is the ability of sintering copper nanoparticles under ambient conditions without oxidation. As a disadvantage of the method, we point out that the laser and photon-pulse sintering are methods with negative feedback, because the reflectivity of the layers usually increases with sintering.

The laser sintering method allows heating printed areas on a substrate and even creating a conductive pattern by sintering and ablation. However, the use of a laser is more expensive and less acceptable for large areas; thus, the best results are usually obtained while using low scanning speeds.

Chemical processing methods mainly imply the use of chemical reactions aimed at detaching groups of stabilizing ligands from the surfaces of metal nanoparticles. After this, coalescence of nanoparticles at room temperature takes place due to the release of excess energy with a decrease in the total surface area of the nanostructured system.

The described alternative methods ordinarily require prior drying of the solvent from the applied layers. However, laser sintering has also been successfully demonstrated for wet ink directly after printing.

The possibility of sintering nanoparticles on heat-sensitive substrates is a significant advantage for printed electronics. The best case scenario, as in the case of chemical sintering, both printing and «sintering» are carried out at room temperature. Electric, laser and photon-impulse sintering also allow using of heat-sensitive substrates. The temperature of a thin layer of nanoparticles rises only for a very short period of time, so the thermal energy transferred to the substrate remains within the permissible limits. In the case of microwave and plasma sintering, energy can be also selectively transferred to the nanoparticle layer.

Electrical sintering mainly requires contact with printed structures; however, using high frequency voltage, it can be performed without a contact. Electric sintering with a DC current allows monitoring of the resistance in real time making possible the control of the sintering process to obtain the target resistance value. However, in the case of AC sintering, the control of resistance is significantly more complicated.

Concluding we point out that conventional methods, such as heating in an oven or on a hot plate are perfectly suited for drying, but alternative methods are preferred for sintering afterwards due to the higher selectivity. Because the total duration of the processing is still relatively high, of considerable interest is the idea of integrating the sintering equipment directly into the printing system. In fact, this was realized for instance with the help of a laser (you can specify here the laser or something special about the work you wanted to cite) or IR radiation.

Acknowledgments The financial support of Ministry of Education and Science of Ukraine should be acknowledged (grant DB/Fotonika № 0117U007176).

References

1. Kawahara J, Ersman PA, Nilsson D et al (2013) Flexible active matrix addressed displays manufactured by printing and coating techniques. J Polym Sci B Polym Phys 4:265–271
2. Lin Q, Huang H, Jing Y, Fu H et al (2014) Flexible photovoltaic technologies. J Mater Chem C 2:1233–1247
3. Gaikwad AM, Steingart DA, Ng TN, Schwartz DE, Whiting GL (2013) A flexible high potential printed battery for powering printed electronics. Appl Phys Lett 102:233302
4. Kravchuk O, Reichenberger M (2016) Properties and long-term behavior of nanoparticle based inkjet printed strain gauges. J Mater Sci Mater Electron 27(10):10934–10940
5. Perelaer J, Smith PJ, Mager D, Soltman D, Volkman SK, Subramanian V (2010) Printed electronics: the challenges involved in printing devices, interconnects, and contacts based on inorganic materials. J Mater Chem 20(39):8446–8453
6. Zhang Z, Zhang X, Xin Z, Deng M, Wen Y, Song Y (2011) Synthesis of monodisperse silver nanoparticles for ink-jet printed flexible electronics. Nanotechnology 22(42):425601
7. Castro T, Reifenberger R, Choi E (1990) Size-dependent melting temperature of individual nanometer-sized metallic clusters. Phys Rev B 42(13):8548

8. Troitskii VN, Rakhmatullina AZ, Berestenko VI, Gurov SV (1983) Initial sintering temperature of ultrafine powders. Soviet Powder Metallurgy Metal Ceramics 22:12–14
9. Lesyuk R (2008) Ink-jet formation of switching elements of chips using silver nanoparticles / R. Lesyuk, Ya. Bobitsky, V. Yillek. New technologies. – Vol. 2 (20). – P. 30
10. Kravchuk O, Grunewald K, Bahr J et al (2014) Production of miniaturized sensor structures on polymer substrates using inkjet printing. Adv Mater Res 1038:49–55
11. Hwang HJ, Oh KH, Kim HS (2016) All-photonic drying and sintering process via flash white light combined with deep-UV and near-infrared irradiation for highly conductive copper nano-ink. Sci Rep 6:19696. https://doi.org/10.1038/srep19696
12. Galagan Y, Coenen EWC, Abbel R et al (2013) Photonic sintering of inkjet printed current collecting grids for organic solar cell applications. Org Electron 14:38–46
13. Schuetz K, Hoerber J, Franke J (2014) Selective light sintering of aerosol-jet printed silver nanoparticle inks on polymer substrates. AIP Conf Proc 1593:732–735
14. Guillot MJ, McCool SC, Schroder KA (2012) Simulating the thermal response of thin films during photonic curing. ASME 2012 international mechanical engineering congress and exposition, vol 2, pp 9–15
15. Schroder KA (2013) Mechanisms of photonic curing: processing high temperature films on low temperature substrates. NCC Nano, LLC. 200-B(14):78728
16. Schroder KA, McCool SC, Furlan WR (2006) Broadcast photonic curing of metallic nanoparticle films. Technical proceedings of the 2006 NSTI nanotechnology conference and trade show, vol 3, pp 198–201
17. Carter M, Sears J (2007) Photonic curing for sintering of nano-particulate material. Advances in powder metallurgy & particulate materials. In: Proceedings of the 2007 international conference on powder metallurgy & particulate materials, vol 2, pp 210–213
18. Farnsworth S, Schroder K (2012) Photonic curing for millisecond-drying of thin films. Specialist Printing Worldwide 4:34–35
19. Akhavan V, Farnsworth K, Schroder D et al (2013) Processing thick-film screen printed metalon CuO reduction ink with pulseforge tools. Coating 46:14–17
20. Ando B, Baglio S, LaMalfa S et al (2011) All inkjet printed system for strain measurement. Sensors: Proceedings of the IEEE Sensors Conference, pp 215–217
21. Marjanovic N, Hammerschmidt J, Perelaer J et al (2011) Inkjet printing and low temperature sintering of CuO and CdS as functional electronic layers and Schottky diodes. J Mater Chem 21:13634
22. Tetznera K, Schroderb KA, Bock K (2014) Photonic curing of sol–gel derived HfO2 dielectrics fororganic field-effect transistors. Ceram Int 140:15753–15761
23. Schroder KA, McCool SC, Furlan WR (2006) Broadcast photonic curing of metallic nanoparticle films. Nanotechnologies, Inc. 3:198–201
24. Yung KC, Gu X, Lee CP et al (2010) Ink-jet printing and camera flash sintering of silver tracks on different substrates. J Mater Process Technol 210:2268–2272
25. Tobjörk D, Aarnio H, Pulkkinen P et al (2012) IR-sintering of ink-jet printed metal-nanoparticles on paper. Thin Solid Films 520:2949–2955
26. Sowade E, Kang H, Mitra KY, Weiß OJ, Weber J, Baumann RR (2015) Roll-to-roll infrared (IR) drying and sintering of an inkjet-printed silver nanoparticle ink within 1 second. J Mater Chem C 3:11815–11826
27. Denneulin A, Blayo A, Neuman C (2011) Infra-red assisted sintering of inkjet printed silver tracks on paper substrates. Bras, J. In. J Nanopart Res 13(9):3815–3823
28. Tobjörk D, Aarnio H, Pulkkinen P, Bollström R, Määttänen A, Ihalainen P, Mäkelä T, Peltonen J, Toivakka M, Tenhu H, Österbacka R (2012) IR-sintering of ink-jet printed metal-nanoparticles on paper. Thin Solid Films 520:2949–2955
29. Määttänen A, Ihalainen P, Pulkkinen P, Wang S, Tenhu H, Peltonen J (2012) Inkjet-printed gold electrodes on paper: characterization and functionalization. ACS Appl Mater Interfaces 4:955–964

30. Cherrington M, Claypole TC, Deganello D, Mabbett I, Watson T, Worsley D (2011) Ultrafast near-infrared sintering of a slot-die coated nano-silver conducting ink. J Mater Chem 21:7562–7564
31. Kumpulainen T, Pekkanen J, Valkama J et al (2011) Low temperature nanoparticle sintering with continuous wave and pulse lasers. Opt Laser Technol 43:570–576
32. Laakso P, Kemppainen S, Ruotsalainen E et al (2009) Sintering of printed nanoparticle structures using laser treatment. ICALEO 2009 – 28th international congress on applications of lasers and electro-optics, congress proceedings, vol 102, pp 1360–1366
33. Seung HK, Pan H, Grigoropoulos CP et al (2007) All inkjet printed flexible electronics fabrication on a polymer substrate by low-temperature high-resolution selective laser sintering of metal nanoparticles. Nanotechnology 18:345202–345210
34. Chung J, Ko S, Bieri NR et al (2004) Conductor microstructures by laser curing of printed gold nanoparticle ink. Appl Phys Lett 84:801–803
35. Chungb J, Poulikakosa D et al (2004) Manufacturing of nanoscale thickness gold lines by laser curing of a discretely deposited nanoparticle suspension. Superlattice Microst 35:437–444
36. Halonen E, Heinonen E, Mäntysalo M (2013) The effect of laser sintering process parameters on cu nanoparticle ink in room conditions. Optics Photonics J 3:40–44
37. Zenou M, Ermak O, Saar A et al (2014) Laser sintering of copper nanoparticles. J Phys D Appl Phys 47:025501–025512
38. Ko S, Chung J, Choi Y et al (2005) Fabrication of inkjet printed flexible electronics by low temperature subtractive laser processing. In: Proceedings of the international mechanical engineering congress and exposition, p 80535 (1–5)
39. Lesyuk R, Bobitski Y, Kotlyarchuk B, Jillek W (2010) Laser sintering for conductive traces fabrication for electronics needs. Electronics and Communication (in Ukrainian) 3(56):16–19
40. Bieri NR, Chung J, Haferl SE et al (2003) Microstructuring by printing and laser curing of nanoparticle solutions. Appl Phys Lett 82:3529–3531
41. Chung J, Bieri NR, Ko S et al (2004) In-tandem deposition and sintering of printed gold nanoparticle inks induced by continuous Gaussian laser irradiation. Appl Phys A 79:1259–1261
42. Choi TY, Poulikakos D, Grigoropoulos C (2004) Fountain-pen-based laser microstructuring with gold nanoparticle inks. Appl Phys Lett 85:13–15
43. Chung J, Ko S, Grigoropoulos CP et al (2005) Damage-free low temperature pulsed laser printing of gold nanoinks on polymers. J Heat Transf 127:724–732
44. Lesyuk R, Jillek W, Bobitski Y et al (2011) Low-energy pulsed laser treatment of silver nanoparticles for interconnects fabrication by ink-jet method. Microelectron Eng 88 (3):318–321
45. Laakso P, Ruotsalainen S, Halonen E et al (2009) Sintering of printed nanoparticles structures using laser treatment. In: Proceedings of the 5th international WLT conference on lasers in manufacturing, pp 527–532
46. Alemohammad H, Aminfar O, Toyserkani E (2008) Morphology and microstructure analysis of nano-silver thin films deposited by laser-assisted maskless microdeposition. J Micromech Microeng 18:115015 (1–12)
47. Perelaer J, DeGans BJ, Schubert US (2006) Ink-jet printing and microwave sintering of conductive silver tracks. Adv Mater 18:2101–2104
48. Perelaer J, Schubert US (2010) Inkjet printing and alternative sintering of narrow conductive tracks on flexible substrates for plastic electronic applications. Radio frequency identification fundamentals and applications, design methods and solutions, p 324
49. Cheng DK (1989) Field and wave electromagnetics. Addison-Wesley Co. Inc., Reading, p 155
50. Perelaer J, Klokkenburg M, Hendriks CE et al (2009) Microwave flash sintering of inkjet-printed silver tracks on polymer substrates. Adv Mater 21:4830–4834
51. Cauchois R, Saadaoui M, Yakoub A et al (2012) Impact of variable frequency microwave and rapid thermal sintering on microstructure of inkjet-printed silver nanoparticles. J Mater Sci 47:7110–7116

52. Reinhold I, Hendriks CE, Eckardt R et al (2009) Argon plasma sintering of inkjet printed silver tracks on polymer substrates. J Mater Chem 19:3384–3388
53. Hegemann D, Brunner H, Oehr C (2003) Plasma treatment of polymers for surface and adhesion improvement. Nucl Instrum Methods Phys Res B 208:281–286
54. Solodovnyk AN, Li W, Fei F et al (2012) Involving low-pressure plasma for surface pretreatment and post print sintering of silver tracks on polymer substrates. In: Proceedings of the international conference nanomaterials: applications and properties, vol 1, pp 1–4
55. Maa S, Singler V, Bromberg L et al (2014) Low temperature plasma sintering of silver nanoparticles. Appl Surf Sci 293:207–215
56. Wünscher S, Stumpf S, Teichler A et al (2012) Localized atmospheric plasma sintering of inkjet printed silver nanoparticles. J Mater Chem 22:24569
57. Magdassi S, Grouchko M, Berezin O et al (2010) Triggering the sintering of silver nanoparticles at room temperature. ACS Nano 4:1943–1948
58. Zapka W, Voil W, Loderer C et al (2008) Low temperature chemical post-treatment of inkjet printed nano-particle silver inks. In: Proceedings of NIP24 and Digital Fabrication, pp 906–911
59. Wakuda D, Kim CJ, Kim KS et al (2008) Room temperature sintering mechanism of Agnanoparticlepaste. In: Proceedings of the 2nd electronics systemintegration technology conference, pp 909–914
60. Coutts MJ, Cortie MB, Ford MJ et al (2009) Rapid and controllable sintering of gold nanoparticle inks at room temperature usinga chemical agent. J Phys Chem C 113:1325–1328
61. Allen ML (2011) Nanoparticle sintering methods and applications for printed electronics. Aalto University School of Electrical Engineering, pp 41–44
62. Magdassi S, Grouchko M, Kamyshny A (2009) Conductive inkjet inks for plastic electronics: air stable copper nanoparticles and room temperature sintering. NIP25 and digital fabrication. Tech Program Proc 25:611–613
63. Allen ML, Leppäniemi J, Vilkman M et al (2010) Substrate-facilitated nanoparticle sintering and component interconnection procedure. Nanotechnology 21:475204
64. Andersson H, Manuilskiy A, Gao J et al (2014) Investigation of humidity sensor effect in silver nanoparticle ink sensors printed on paper. IEEE Sensors J 14:623–628
65. Olkkonen J, Leppäniemi J, Mattila T et al (2014) Sintering of inkjet printed silver tracks with boiling salt water. J Mater Chem C 2:3577–3582
66. Andersson H, Manuilskiy A, Lidenmark C et al (2013) The influence of paper coating content on room temperature sintering of silvernanoparticle ink. Nanotechnology 24:455203–455212
67. Allen ML, Aronniemi M, Mattila T et al (2008) Electrical sintering of nanoparticle structures. Nanotechnology 19:175–201
68. Alastalo A, Mattila T, Allen ML et al (2008) Rapid electrical sintering of nanoparticle structures. Mater Res Soc Symp Proc 1113:2–7
69. Werner C, Behrens G, Hellbernd KH et al (2011) Electrical sintering of printed metal structures for mechanical sensors. LOPE-C Proc, pp 192–195
70. Allen M, Alastalo A, Suhonen M et al (2011) Contactless electrical sintering of silver nanoparticles on flexible substrates. IEEE Trans Microwave Theory Tech 59:1419–1429
71. Ko S, Pan H, Hwang DJ et al (2007) High resolution selective multilayer laser processing by nanosecond laser ablation of metal nanoparticle films. J Appl Phys 102:93–102

Chapter 24
Influence of Thermal Treatment on Phase Formation Processes in Amorphous Alloys

Volodymir Lysov, Tatiana Tsaregradskaya, Oleg Turkov, and Galina Saenko

24.1 Introduction

Investigating the structures and properties of disordered systems that include solids characterized by amorphous structure and amorphous metallic alloys in particular is an important field of modern condensed state physics. Modern thermodynamically nonequilibrium materials with amorphous structure are characterized by a number of unique physicochemical properties. However, most amorphous alloys have low thermal stability, which is a considerable drawback. Amorphous metallic alloys transform to a more stable crystalline state when heated to certain temperature. This transition is followed by the loss of characteristic properties, which explains the importance of theoretical and experimental studies of ways to improve the thermal stability of amorphous alloys.

An actual direction of research of metallic glasses is the development of methods to achieve nanostructured state by partial crystallization of amorphous alloys due to external influences. Such influences include heat treatment (the isothermal and non-isothermal annealing at temperatures below the temperature of crystallization, thermocycling, cryotreatment), intensive plastic deformation, and irradiation by particles of various nature. Large attention to heat treatment of alloys with the amorphous structure is explained by an acquisition of the special properties in the nanocrystalline state. Consequently, the research of amorphous alloys properties under the action of external influences attracts great interest, because there are open questions regarding the mechanisms of influence of external factors on the properties

V. Lysov · T. Tsaregradskaya (✉) · O. Turkov · G. Saenko
Department of Physics, Taras Shevchenko National University of Kyiv, Kyiv, Ukraine

O. Fesenko, L. Yatsenko (eds.), *Nanooptics, Nanophotonics, Nanostructures, and Their Applications*, Springer Proceedings in Physics 210,
https://doi.org/10.1007/978-3-319-91083-3_24

of these materials [1–3]. In the series of experiments on amorphous alloys, heat treatment aiming to increase the thermal stability of alloys and developing ways to obtain alloys in the nanostructural state as well as investigating mechanical properties of nanostructural materials were conducted.

24.2 Conclusions of the Thermodynamic Theory of High-Temperature Stability of Amorphous Alloys

The high-temperature thermal stability of amorphous alloys is often associated with the total absence of crystalline phase nuclei. However, numerous electron microscopy investigations confirm that even after ultrafast quenching from the liquid phase at quenching rates of around 10^4–10^6 K/s, crystallization centers frozen into the amorphous matrix occur in amorphous alloys. The authors of [4] determined that the thermodynamic conditions for the emergence of frozen-in crystallization centers in an amorphous matrix for the *i*-th component in a heterogeneous system of amorphous phase and frozen-in crystallization centers are described by the expression

$$\Delta\mu_i(T) = -\frac{X_T^\alpha V_\beta + X_T^\beta V_\alpha}{X_\alpha + X_\beta}\frac{2\sigma}{r_{02}} - \frac{(V_\beta - V_\alpha)\left(\overline{V_i^\beta} - \overline{V_i^\alpha}\right)}{(X_\alpha + X_\beta)\bar{V}}\left[1 - \left(\frac{r_{01}}{r_{02}}\right)^3\right] + \Delta\mu_{0i}(T), \quad (24.1)$$

where $\Delta\mu_i(T) = \mu_i^\alpha - \mu_i^\beta$ is the difference between the chemical potentials of the *i*-th component in the α and β phases allowing for the pressure at the crystalline nucleus amorphous phase interface due to surface tension (where σ is the surface tension coefficient) and elastic stresses; V_α, V_β, X_T^α, and X_T^β are the molar volumes and isothermal compressibility coefficients of the α and β phases, respectively; r_{01} and r_{02} are the radii of crystalline nuclei for pressure and no pressure at the crystalline nucleus–amorphous phase interface; and $\Delta\mu_{0i} = \mu_{0i}^\alpha - \mu_{0i}^\beta$ is the difference between the chemical potentials of pure components, represented by the following expression (as the standard state, we selected the chemical potential of the *i*-th components in pure crystalline metal of type *i*):

$$\Delta\mu_0 = \frac{\Delta H_0^{(i)}(T_{pi} - T)}{T_{pi}} + RT\ln a_i^\alpha(C_i^\alpha,\ T) - RT\ln a_i^\beta\left(C_i^\beta,\ T\right), \quad (24.2)$$

where $\Delta H_0^{(i)}$ is the enthalpy of melting for the *i*-th component; T_{pi} is the melting point of the *i*-th component; and a_i^α and a_i^β are the activities of the *i*-th component in the α and β phases.

Figure 24.1 shows the plot of dependences $\Delta\mu_i(T)$, calculated by formula (24.1) with allowance for (24.2) under different conditions of crystallization in a melt [5].

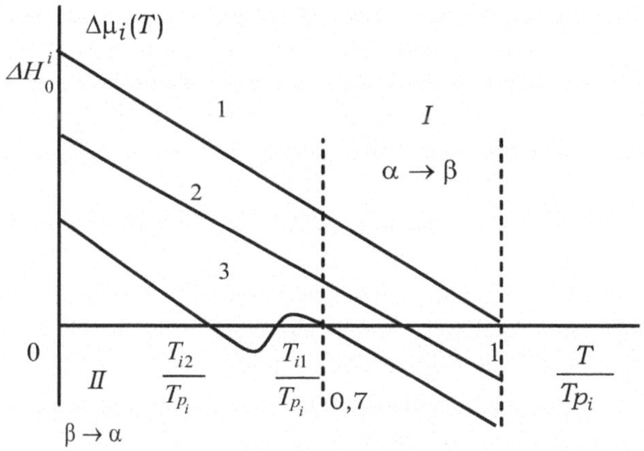

Fig. 24.1 Temperature dependence of $\Delta\mu_i(T)$ for alloys (1) upon slow cooling of the melt $V_\alpha = V_\beta$, $a_i^\beta = a_i^\alpha$, $r_{02} \to \infty$; (2) upon rapid cooling of the melt $V_\alpha \approx V_\beta, a_i^\beta > a_i^\alpha$, $r_{02} = \text{const}$; (3) $V_\alpha \neq V_\beta, a_i^\beta > a_i^\alpha$, $r_{02} = \text{const}$. I is the region of viscous liquid, and II is the region in which the pressure caused by elastic stresses due to $V_\alpha \neq V_\beta$ is considered.

According to the homogeneous nucleation theory, crystalline phase nuclei are formed during the ultrafast quenching of a melt in the range $T \sim 0{,}7\ T_p$ (where T_p is the melting point), with the supercooled liquid transforming into a solid amorphous phase. A heterogeneous system of frozen-in crystallization centers and an amorphous matrix are thus formed in which elastic stresses emerge. Calculations showed that allowing for additional pressure at the crystalline nucleus–amorphous phase interface reduces $\Delta\mu_i(T)$ relative to the region where there are no elastic stresses.

The transition of crystalline nuclei into the amorphous phase is thus possible in the temperature range of T_{i1} to T_{i2} (i.e., in the region where $\Delta\mu_i < 0$). It should be noted that the probability of new crystalline nuclei emerging below temperature T_{i1} is zero, since the frequency of nucleation tends to zero in this region, according to the homogeneous crystallization theory. Only the slow growth of already existing crystallization centers is possible in this region.

24.3 Thermodynamic Analysis of the Possibility of Purifying an Amorphous Phase of Frozen-in Crystallization Centers

Considering the additional pressure on the surface of the interface between a crystallization nucleus and the amorphous phase substantially lowers $\Delta\mu_i$ relative to its value in an area where there is no elastic stress [5]. The size of the frozen-in crystallization centers can then shrink, and they can dissolve in the amorphous matrix. Upward diffusion is one of the physical causes of the dissolution of frozen-in crystallization centers. Below, we perform a thermodynamic analysis of the possibility of achieving ascending diffusion in amorphous alloys [6]. The difference between chemical potentials $\Delta\mu_i$ of the β (crystalline) and α (amorphous) phases, which leads to the emergence of a driving force, is the reason for the displacement of the i-th component in a solution:

$$\vec{F_i} = -\nabla\mu_i. \tag{24.3}$$

Driving force $\vec{F_i}$ behind the diffusional transfer of the i-th component in a multicomponent system is determined by the gradient of the chemical potential of this component rather than the concentration gradient, as is stated in Fick's first law. Atoms demonstrate directed displacement under the influence of force $\vec{F_i}$. This displacement is characterized by the average rate:

$$\langle v_i \rangle_F = u_i F_i, \tag{24.4}$$

where u_i is the mobility of atoms of the i-th component.

As a result, a diffusion flow of atoms of the i-th component arises with a flow density of

$$J_i = c_i \langle v_i \rangle = c_i u_i F_i = -c_i u_i \nabla\mu_i. \tag{24.5}$$

The chemical potential of the i-th component in real solutions is determined as

$$\mu_i = RT \ln a_i + \mu_{0i}, \tag{24.6}$$

where a_i is the activity of the i-th component, and μ_{0i} is the chemical potential of a pure i-th component in its standard state.

We consider that the activity is determined according to the equation

$$a_i = \gamma_i c_i, \tag{24.7}$$

where γ_i is the activity coefficient.

Let us obtain the equation for the chemical potential of the i-th component:

$$\mu_i = RT \ln(\gamma_i c_i) + \mu_{0i}. \tag{24.8}$$

The expression for the density of the diffusion flow of atoms of the i-th component is thus

$$J_i = -u_i RT \frac{\partial \ln a_i}{\partial \ln c_i} \nabla c_i, \tag{24.9}$$

or

$$J_i = -u_i RT \left(1 + \frac{\partial \ln \gamma_i}{\partial \ln c_i}\right) \nabla c_i. \tag{24.10}$$

Comparing (24.10) and Fick's first law, we find that

$$D_i = u_i RT\left(1 + \frac{\partial \ln \gamma_i}{\partial \ln c_i}\right). \qquad (24.11)$$

Equation (24.11) demonstrates the possibility of diffusion against the concentration gradient when

$$\frac{\partial \ln \gamma_i}{\partial \ln c_i} < -1. \qquad (24.12)$$

This condition describes the phenomenon of upward diffusion: the flow of diffusing component is directed toward the increase of concentration gradient; at this crystalline nucleus is going to be dissolved in amorphous phase.

Considering Eq. (24.11), Fick's first law for a real solution can be written as

$$J_i = -D_i \nabla c_i + c_i \langle v_i \rangle F_i. \qquad (24.13)$$

A comparison of (24.10) and (24.13) shows that an additional force emerges in a real solution:

$$F_i = -kT \nabla \ln \gamma_i. \qquad (24.14)$$

This force characterizes the effect of the gradient of the individual potential field of a crystal where a diffusing atom is moving. The driving force of diffusion $(-kT \nabla \ln \gamma_i)$ differs from that of concentration, which corresponds to the entropy of mixing.

Let us examine the possibility of calculating activity coefficient γ_i. The activity coefficient of the i-th component of a regular solution is determined according to the equation

$$\gamma_i = \exp\left(\frac{\Delta H_i}{RT}\right), \qquad (24.15)$$

where ΔH_i is the relative partial enthalpy of a crystal–amorphous phase transition for the i-th component.

The relative partial enthalpies of a crystal–amorphous phase transition for the i-th component can be found from relative integral enthalpy of the alloy:

$$\Delta H_i = \Delta H + (1 - c_i)\frac{d\Delta H}{dc_i}, \qquad (24.16)$$

where ΔH is the relative integral enthalpy of a crystal–amorphous phase transition, which can be determined experimentally.

With a binary solution,

$$\Delta H_1 = \Delta H + (1 - c_1)\frac{d\Delta H}{dc_1},$$
$$\Delta H_2 = \Delta H + (1 - c_2)\frac{d\Delta H}{dc_2}. \quad (24.17)$$

Let us examine the possibility of fulfilling the criterion (24.12) of upward diffusion proceeding in binary amorphous alloys of the Fe–B system. We use the experimental dependences in [7] for the relative integral enthalpy of a crystalline $Fe_{80}B_{20}$ alloy transitioning to a supercooled (metastable) melt at 850 K and the semiempirical formula obtained in [8] for a binary alloy:

$$\Delta H = \frac{\beta c_1(1 - c_1)}{\alpha^2 + (1 - 2\alpha)(1 - c_1)}, \quad (24.18)$$

where α is the position of the extremum of function $\Delta H(c_1)$, and β is the value of function $\Delta H(c_1)$at the point of the extremum.

Calculating the parameter according to formulas (24.12), (24.15), and (24.16) for a crystalline nucleus in $Fe_{80}B_{20}$ alloy at $T = 850$ K and $\Delta H = 8.0$ kJ/mol showed that the condition of upward diffusion was met:

$$\frac{\partial \ln \gamma_{Fe}}{\partial \ln c_{Fe}} = -1,35.$$

The dissolution of frozen-in crystallization centers due to ascending diffusion is thus possible during the isothermal annealing of amorphous $Fe_{80}B_{20}$ alloy in a certain range of temperatures, and it can improve the thermal stability of an amorphous alloy.

Based on our analysis of results from calculations within the theory of high-temperature stability of amorphous alloys, two areas of its practical application were proposed:

Enhancing the thermal stability of amorphous alloys by isothermal annealing in the range of temperatures $T_{i1} \div T_{i2}$, where crystalline nuclei can transform into the amorphous phase

Transforming the amorphous state into a nanocrystalline state via prolonged low-temperature isothermal annealing in the range of temperatures $T_{an} < T_{i1}$, where the growth of existing crystalline nuclei will occur. To reproduce these processes, it is necessary to determine the temperatures T_{i1}, T_{i2}, T_{an} experimentally.

24.4 Method of Experiment

By the parameter, which determines thermal stability of amorphous alloys, there is temperature of the onset of intense crystallization T_k which was determined by means of high-sensitivity dilatometry. The method of the dilatometric researches consists in the following [9]. At heating of amorphous alloy, its volume grows droningly (density, accordingly, goes down); there is diminishing of volume and increasing of density of alloy at the achievement of certain temperature (temperatures of intensive crystallization beginning) that testifies to beginning of process of crystallization. Consequently, it is possible to probe the process of crystallization of amorphous alloy, fixing the change of length of standard and transferring it in volume changes or changes of density. The temperatures of the onset of intense crystallization beginning for the initial amorphous alloys and after the executed thermal or thermomechanical treatment were determined by dilatometry.

For the row of amorphous alloys in the initial amorphous state and after treatment, measuring of microhardness by Vikkers was executed. The method of measuring of microhardness is based on measuring of linear size of diagonal of imprint that appears at pressing of diamond pyramid in the probed material under the certain loading. The device allows to measure microhardness at pressing of diamond pyramid with square basis and corner at a top 136° between opposite verges with the appendix of loadings (2–200 gram). As a result of measurings, the length of diagonal of got imprint is determined by eyepiece micrometer. At measuring of microhardness, volume, which becomes deformed by pressure, must be less than the volume of grain which is measured. Under pressure sample was 10 seconds.

Number of hardness accounts by the formula

$$H = \frac{1854P}{d^2}\left(\frac{\kappa Gf}{mm^2}\right)$$

where P is the pressure $\left(\frac{\kappa Gf}{mm^2}\right)$ and d is the length of diagonal of imprint in microns.

At measurings possibility of values of the microhardness variation was taken into account as a result of influence of nearby structural constituents with other microhardness; measurings were conducted ten times in identical terms; loading was made by 200 gram; and the repeated measurings were executed in a new place of structural constituent.

24.5 Purifying an Amorphous Matrix from Frozen-in Crystallization Centers

A series of experimental investigations was performed to prove theoretical conclusion. The effect of isothermal annealing has on the thermal stability of the multicomponent amorphous alloys $Fe_{80}B_{20}$, $Fe_{80}Si_6B_{14}$, $Fe_{40}Ni_{40}B_{20}$, $Ni_{78}B_{18}Si_4$,

Table 24.1 Temperatures T_k of the onset of intense crystallization for the initial alloys, temperatures $T_{k'}$ of preliminary thermal treatment, temperatures T_k^a of the onset of intense crystallization after thermal treatment at temperature $T_{k'}$, and ranges of change ΔT in the thermal stability of amorphous alloys

Amorphous alloy composition	T_k, K	T_k', K	T_k^a, K	ΔT, K
$Fe_{80}B_{20}$	650	590	685	35
$Fe_{80}Si_6B_{14}$	770	720	810	40
$Fe_{40}Ni_{40}B_{20}$	710	670	730	20
$Ni_{78}B_{18}Si_4$	730	680	750	20
$Fe_{70}Cr_{15}B_{15}$	750	690	780	30

Table 24.2 Values of the temperature of the onset of intense crystallization and the microhardness of initial alloys (I) and alloys after thermal treatment at temperature $T_{k'}$ (II)

Amorphous alloy composition	H, kgF/mm^2		$\frac{(H_a-H_0)}{H_0}$, %
	I	II	
$Fe_{80}B_{20}$	781 ± 41	676 ± 34	13.4
$Fe_{80}Si_6B_{14}$	747 ± 37	654 ± 32	16.4
$Fe_{40}Ni_{40}B_{20}$	526 ± 21	415 ± 26	21.1
$Ni_{78}B_{18}Si_4$	605 ± 30	512 ± 25	15.3
$Fe_{70}Cr_{15}B_{15}$	526 ± 21	439 ± 22	16.5

and $Fe_{70}Cr_{15}B_{15}$ was estimated experimentally using the highly sensitive dilatometric technique. The annealing temperature was determined empirically on the basis of the theoretical theses of high-temperature stability of amorphous alloys. According to this theory, there exists a range of temperatures that corresponds to the negative value of the difference between chemical potentials of an amorphous matrix and frozen-in crystallization centers. The thermodynamic condition of the possible dissolution of frozen-in crystallization centers is thus fulfilled.

The temperatures of the onset of intense crystallization in the initial alloys, their preliminary thermal treatment, the onset of intense crystallization after their preliminary treatment, and the ranges of improvement in the thermal stability of the amorphous alloys are given in Table 24.1. Analysis of the results given in Table 24.1 showed that the introduced regime of thermal treatment expanded the ranges of the thermal stability of our multicomponent alloys by 20–40 K. The improvement in the thermal stability of the amorphous alloys after thermal treatment can be explained by the dissolution of frozen-in crystallization centers in the amorphous phase.

The values of the microhardness of the initial alloys and alloys after corresponding thermal treatment are given in Table 24.2. The value of microhardness fell by 16.4–21.1% after thermal treatment, which indirectly proves the part of crystalline phase in our samples fell (Table 24.2).

Electron microscopic studies of initial alloys and alloys after preliminary thermal treatment were performed to examine this statement. The results from our electron microscope investigations of $Fe_{80}B_{20}$ alloy are presented in Fig. 24.2.

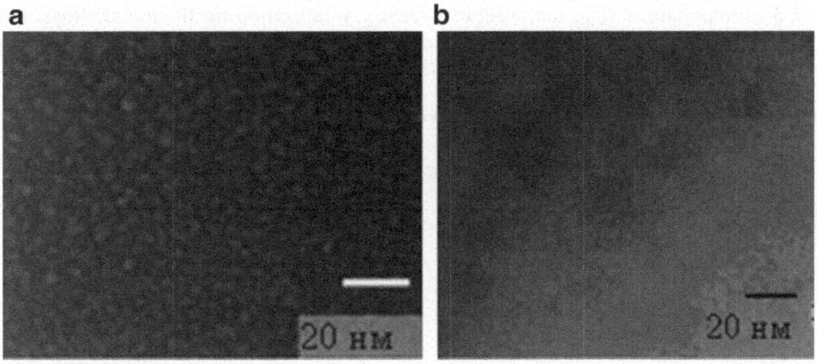

Fig. 24.2 Electron microscopic (dark field) images of $Fe_{80}B_{20}$ alloy in (**a**) its initial state and (**b**) after isothermal annealing for 1 h at 590 K

Analysis of these results showed that the initial amorphous state was characterized by frozen-in crystallization centers with average sizes of 2–5 nm (Fig. 24.2a). There were no crystalline nuclei in the amorphous matrix after annealing at 600 K (Fig. 24.2b). Similar results were obtained for our $Fe_{80}Si_6B_{14}$, $Fe_{40}Ni_{40}B_{20}$, $Ni_{78}B_{18}Si_4$, and $Fe_{70}Cr_{15}B_{15}$ alloys, which is a direct proof of the dissolution of frozen-in crystallization centers in an amorphous matrix. The results from our investigations thus prove the conclusions of the thermodynamic theory of the high-temperature stability of amorphous alloys.

This shows we can shift the phase equilibrium in an amorphous matrix–frozen crystallization centers system using the directed effect of inner factors. It also determines the means by which we can expand the temperature range of the amorphous state's existence.

24.6 Obtaining Amorphous–Nanocrystalline Alloys by Partial Crystallization of Metal Glasses

For amorphous $Fe_{80}B_{20}$, $Fe_{80}B_{14}Si_6$, $Fe_{40}Ni_{40}B_{20}$, $Ni_{78}B_{18}Si_4$, and $Fe_{70}Cr_{15}B_{15}$ alloys, the temperature of processing, in which an amorphous–nanostructural state can be created, was determined empirically. To create an amorphous nanocrystalline state, for each alloy, annealing temperature was determined at which $\Delta\mu_i$ was positive (which is a condition for the growth of frozen-in crystallization centers), but the process of intense crystallization does not yet begin. The empirically determined temperatures of isothermal annealing (at which the subsequent thermal treatment of the initial samples was carried out) are presented in Table 24.3. At the determined annealing temperatures, thermal processing of the initial amorphous alloys was carried out for 1 h, and the volume part of the crystalline phase formed during the annealing was calculated.

Figure 24.3 shows the time dependence of the volume part of the crystalline phase formed during 2 h of annealing at a temperature of 650 K in the $Fe_{80}B_{20}$ alloy.

Table 24.3 Temperatures T_k of the onset of intense crystallization for the initial alloys, temperatures T_k" of preliminary thermal treatment, and volume part of crystalline phase X after thermal treatment at temperature T_k"

Amorphous alloy composition	T_k, K	T_k", K	X
$Fe_{80}B_{20}$	650	615	0.31
$Fe_{80}B_{14}Si_6$	770	750	0.16
$Fe_{40}Ni_{40}B_{20}$	710	685	0.22
$Ni_{78}B_{18}Si_4$	730	705	0.19
$Fe_{70}Cr_{15}B_{15}$	750	730	0.11

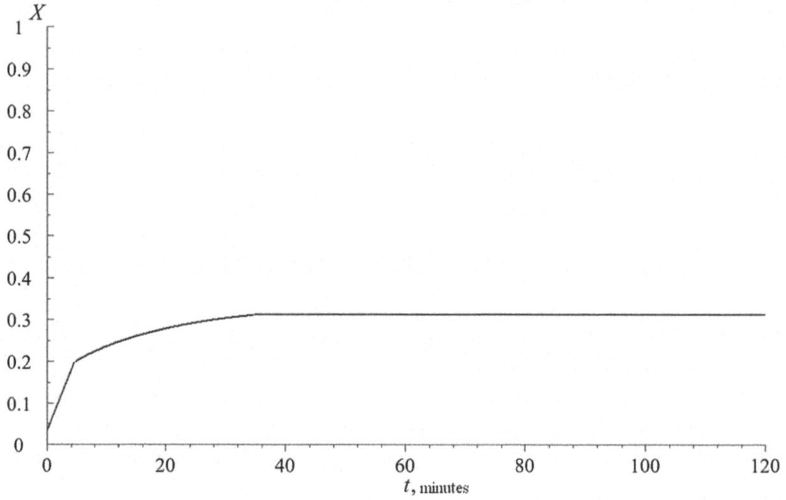

Fig. 24.3 The time dependence of the volume part of the crystalline phase formed during 2 h of annealing at a temperature of 650 K in the $Fe_{80}B_{20}$ alloy

After 30 min, the part of the crystalline phase reached the value $X = 0.31$ and remained constant at further annealing. The heat treatment carried out has led to an increase in the volume part of the crystalline phase X in the all samples; X depending on the composition of the alloy ranges from 11% to 31%.

As a parameter for comparing the mechanical properties of metallic glasses in the initial state and alloys after the heat treatment, the value of the microhardness H was used. The results of the microhardness measurements carried out are presented in Table 24.4.

The microhardness of the received materials increased by 13.8–24.9% compared with the initial amorphous state. An increase in the amount of microhardness can be attributed to the growth of frozen-in crystallization centers and hence to an increase in the fraction of the crystalline phase in samples and the formation of an amorphous nanocrystalline state.

To verify this assertion, electron microscopic investigations of the $Fe_{80}B_{20}$, $Fe_{80}B_{14}Si_6$ $Fe_{40}Ni_{40}B_{20}$, $Ni_{78}B_{18}Si_4$, and $Fe_{70}Cr_{15}B_{15}$ initial alloys were carried out for 1 h (annealing temperatures are given in Table 24.3).

Table 24.4 Values of the temperature of the onset of intense crystallization and the microhardness of initial alloys (I) and alloys after thermal treatment at temperature T_k" (II)

Amorphous alloy composition	H, kgF/mm^2		$\frac{(H_в - H_0)}{H_0}$, %
	I	II	
$Fe_{80}B_{20}$	781 ± 41	976 ± 47	24.9
$Fe_{80}B_{14}Si_6$	747 ± 36	881 ± 40	17.9
$Fe_{40}Ni_{40}B_{20}$	526 ± 21	625 ± 25	18.8
$Ni_{78}B_{18}Si_4$	605 ± 30	731 ± 36	20.8
$Fe_{70}Cr_{15}B_{15}$	526 ± 21	599 ± 26	13.8

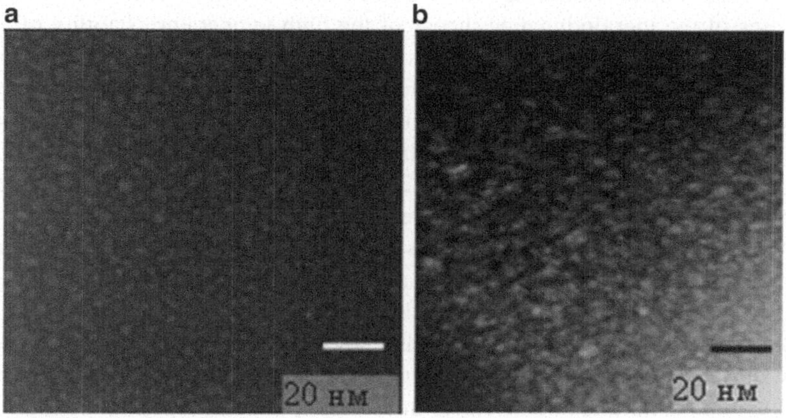

Fig. 24.4 Electron microscopic (dark field) images of $Fe_{80}B_{20}$ alloy in (**a**) its initial state and (**b**) after isothermal annealing for 1 h at 615 K

Figure 24.4 shows the results of electron microscopic investigations for the $Fe_{80}B_{20}$ alloy. The initial state is characterized by the presence in the amorphous matrix of frozen-in crystallization centers with an average size of 2–5 nm (Fig. 24.4a).

The samples after isothermal annealing at 615 K for 1 h are characterized by a structure with an average size of nanocrystals (7–10 nm) (Fig. 24.4b), that is, in this temperature region, the growth of existing crystalline nucleus is observed. Similar results were obtained for alloys $Fe_{40}Ni_{40}B_{20}$, $Ni_{78}B_{18}Si_4$, $Fe_{70}Cr_{15}B_{15}$, and $Fe_{80}B_{14}Si_6$. For all investigated alloys, an increase in the size of frozen in an amorphous matrix of crystallization centers is two to three times higher. Consequently, the results of electron microscopic studies confirm the fact of formation of an amorphous–nanocrystalline state.

24.7 Conclusion

Based on our analysis of results from calculations within the high-temperature stability of amorphous alloys theory, two areas of its practical application were proposed: enhancing the thermal stability of amorphous alloys by isothermal annealing in the range of temperatures where crystalline nuclei can transform into the amorphous phase and transforming the amorphous state into a nanocrystalline state via prolonged low-temperature isothermal annealing. It was confirmed experimentally that external effects which reduce the difference between the chemical potentials of the *i*-th component in an amorphous matrix and in frozen-in crystallization centers (in the temperature region where $\Delta\mu_i < 0$) result in the dissolution of frozen-in crystallization centers and thus to an increase in the thermal stability of amorphous alloy. Prolonged low-temperature isothermal annealing allows us to form nanostructured states. The purification of an amorphous matrix of frozen-in crystallization centers was theoretically verified and experimentally achieved by analyzing the theses of the thermodynamic theory of the high-temperature stability of amorphous alloys, which state that there exists a range of temperatures that corresponds to fulfillment of the condition of the dissolution of frozen-in crystallization centers. It was shown that the proposed regimes of thermal treatment allow us to expand the ranges of the thermal stability of amorphous alloys based on iron by 20–40 K. The value of microhardness then falls by 16–21%, indirectly proving the reduction of the crystalline phase fraction in the samples. The increased thermal stability of the investigated multicomponent amorphous alloys can be explained by the amorphous matrix being purified of frozen-in crystallization centers, as was confirmed by the results from electron microscopic investigations.

Based on the analysis of high-temperature thermodynamic stability of amorphous alloys, the theory proposed a method for producing amorphous–nanocrystalline state from the initial amorphous. The alloys in the amorphous–nanocrystalline state were made, as evidenced by the results of electron microscopic studies. It was found that the microhardness of the obtained materials is increased as compared to the amorphous state to initial due to an increase of the frozen-in crystallization centers' size and the formation of amorphous–nanocrystalline state.

References

1. Glezer AM (2002) Ros Chym J. XLVI(5):57
2. Lysov VI, Tsaregradskaya TL, Turkov OV, Saenko GV (2013) J Phys Stud 7(2):2701.5
3. Ambrosimova HE (2011) Success Phys Sci 181(12)
4. Khar'kov EI, Lysov VI (1987) Rasplavy 1(1):30
5. Lysov VI, Tsaregradskaya TL, Turkov OV, Saenko GV (2014) Rus J Phys Chem A 88 (12):2183–2187
6. Lysov VI, Tsaregradskaya TL, Turkov OV, Saenko GV (2017) Rus J Phys Chem A 91 (12):2322–2226
7. Zaitsev AI, Zaytseva NE (2002) J Phys Chem 76(1)
8. Lysov VI, Tsaregradskaya TL, Turkov OV, Saenko GV (2007) Rus J Phys Chem V 81(10):1765
9. Novikov VN, Kharkov EI (1982) Phys Metals Metall 54(6):1210–1211

Index

A
Absorption, 43–57, 232–234, 237–240, 244, 252, 266, 270, 275, 276, 278, 280, 308, 320–323, 325–327
Absorption and emission of light, 43–57
Alloys, 17–24, 341–352
Amorphous alloys, 341–352
Amorphous composite, 105
Amorphous phase, 342, 343, 345, 346, 348–350, 352
Anikeev, V.V., 175–181
Annealing, 20–23, 268, 319, 324, 341, 346, 347, 349–351
Archie, G.E., 137–140
Arshadi, M., 115
Artyukhov, A.E., 301–308

B
Bandelin, H., 5
BenDaniel, D.J., 219, 221
Bereznyak, Yu., 17–24
Bobitski, Y., 317–336
Bobitski, Ya., 231–240
Bobitski, Y.V., 207–216
Bograchev, D., 127–144
Bonu, V., 266, 267
Bulavinets, T., 231–240
Burak, Ya., 147, 150

C
Catalytic activity, 104, 285, 286, 296, 298
Ceramics, 5, 144, 319, 324, 327
Coagulation, 107
Cobalt, 321
Colloid, 5, 104, 106, 164, 175, 231, 232, 235–239, 318, 319, 324, 335
Composite, 4, 5, 9, 11, 104, 105, 120, 128, 133, 134, 136, 184, 327, 332
Crystallization, 341–352
Crystal structure, 20, 268, 302, 304–306
Current, 22, 26, 59, 91, 129, 147, 177, 183, 222, 246, 275, 285, 325

D
Dalmata, G., 286
Davydov, A.S., 90
Deformation, 21, 23, 27–31, 37, 178–181, 208, 328, 341
Degussa, A.G., 104
Demydenko, Y.V., 243–257
Didukh, L., 73–86
Dielectric permittivity, 246, 325, 327
Diffusion, 9, 17, 26, 109, 117, 129, 137, 138, 185, 186, 193, 194, 196, 202, 289, 314, 318, 324, 327, 328, 331, 343–346
Discrete dipole approximation (DDA), 232, 233, 238–240
Dispersion, 63, 65, 67, 94, 104, 252, 253, 289, 304, 318, 330
Dovhopyaty, Yu., 73–86
Druzhinin, A., 275–281
Duke, C.B, 219, 221
Dyes, 90, 103–106, 112–116, 120
Dzyazko, D., 3–12

O. Fesenko, L. Yatsenko (eds.), *Nanooptics, Nanophotonics, Nanostructures, and Their Applications*, Springer Proceedings in Physics 210,
https://doi.org/10.1007/978-3-319-91083-3

E

Electrical conductivity, 25, 178–181, 268, 318, 320, 325, 326, 330
Electron, 5, 17, 25, 59, 74, 90, 91, 129, 161, 176, 185, 220, 243, 266, 277, 286, 302, 327, 342
Electron beams, 302
Electron spectrum, 74, 79, 219–228
Eren, E., 112
External field, 26, 59, 61–63, 90–101, 186, 247, 257–259, 261

F

Ferrites, 321
Ferromagnetic semiconductor, 43–56
Filevska, L., 265–270
Fitio, V.M., 207–216
Free volume, 175, 176
Frozen-in crystallization centers, 342–352
Fumed and modified silicas, 105
Functionalized MWCNT, 176, 177, 179–181

G

Gabrusenoks, J., 301–308
Gburski, Z., 311–315
Glass transition temperature, 322, 326
Goncharuk, O., 103–120
Grain boundaries, 327, 328
Grains, 5, 318, 327, 328, 347
Green, A.E., 153
Grinevych, V., 265–270
Grochowski, M., 285–298

H

Halutska, I., 3–12
High temperature stability, 342, 343, 349, 352
Hrytsyna, O., 147–159
Hubbard, A.T., 75, 76
Humidifier, 302, 306, 307
Hydrogel, 112, 116

I

Impedance, 289, 290, 327, 333, 334
Impedance spectroscopy, 289
Injected electron, 59, 64, 67, 69, 91–96, 101
Inkjet printing, 317, 318, 320, 321, 325, 326, 329–332, 334, 335
Interparticle interactions, 246, 250, 253, 256
Iron oxide, 103, 108

K

Kempiński, M., 25
Kempiński, W., 25–38
Kinetic equation, 43, 48, 111, 184, 186, 203
Kinetics, 43, 48, 111, 114, 115, 151, 156, 184, 186, 203, 219, 285–298
Koda, V.Yu., 175–181
Kondrat, V., 147–159
Korostil, A.M., 183–203
Kotsiuba, Yu., 207–216
Kramar, O., 73–86
Kravchuk, O., 317–336
Kretschmann, E., 243–245, 247, 250, 256
Krupa, M.M., 183–203

L

Laser radiation, 44, 46, 48–50, 54, 322, 323
Lesyuk, R., 317–336
Lewis, N.S., 104, 105
Linearly-dependent effective mass, 220, 222
Lovrič, S., 286
Luzanov, A.V., 161–172
Lysov, V.I., 341–352

M

Magnon, 43, 44, 46, 49, 50, 52, 54, 55
Maxwell, A., 148, 149, 325
McClure, J.W., 28
Mechanical properties, 17, 25, 178, 335, 342, 350
Meier, C., 265, 267
Metal, 3, 21, 43, 73, 90, 103, 127, 178, 183, 231, 243, 268, 276, 285, 317, 342
Metal ions, 4, 104, 106–112, 120, 285
Mikhalin, A., 127–144
Milyutin, V., 127
Mixed oxides, 104–106, 113, 115–117, 119
Modified silica, 3
Molybdenum, 4
Mykhailova, H.Yu., 175–181

N

Nanocarbon, 4, 112, 177, 178
Nanocomposites, 104–106, 110, 114, 175
Nanoparticles, 4, 105, 144, 161, 231, 243, 265, 317
Nanoporous structure, 301–308
Nanoprism, 236, 238–240
Nanoscale tin dioxide films, 270
Nanosized oxides, 104, 106–120

Nanostructure, 106, 175, 183–203, 221, 227, 238, 240, 266, 267, 275, 279, 280, 335, 352
Nichkalo, S., 275–281
Nicholson, D., 290
Nickel, 303, 321
Nischenko, M.M., 175–181
Nosal-Wiercińska, A., 285–298

O
Odnodvorets, L.V., 17–24
Ostapiv, O., 275–281
Ostwald, W.F., 7, 12
Oxide, 3, 21, 26, 103, 128, 249, 305, 318

P
Palchik, A., 3–12
Pal, U., 113
Park, D., 127–144
Perlaer, 325
Perlova, N., 3–12
Perlova, O., 3–12
Petrovska, H., 207–216
Phase composition, 17, 20, 21, 301–308
Photoluminescence (PL), 265–268, 270
Photonic sintering, 320–322
Photostimulated recovery, 231, 232, 235, 237, 239
Plasmon resonance, 232, 237, 239, 323
Poduremne, D., 17–24
Polarization resistance, 183, 289
Polyacids, 103
Polyalcohols, 103, 116
Polymer/silica interaction, 108, 114
Polypeptide, 59, 60, 62, 67–69
Pores, 8, 104, 110, 114, 128–137, 144, 302, 304–306, 308
Porous ammonium nitrate (PAN), 301, 302, 304–308
Prikhodko, G.P., 175–181
Protein molecule, 59, 66–70
Protsenko, I.Yu., 17–24
Prylutskyy, Yu.I., 25–38

Q
Quantum well, 219–228

R
Rabi splitting, 253, 257
Radchenko, T.M, 25–38
Radio frequency, 326
Rasalingam, S., 114
Reichenberger, M., 317–336
Rheingold, 328
Rivlin, R.S., 153
Rychagov, A., 127–144

S
Saenko, G.V., 341–352
Sagalianov, I.Yu., 25–38
Scanning electron microscope (SEM), 277–279, 304, 319
Semchuk, O., 43–56
Seti, J., 219–228
Shabelnyk, Y., 17–23
Shmeleva, L.V., 59–70
Silica/lignin biocomposites, 3, 103, 105, 106, 108, 114, 116–119, 331
Silver nanoparticles, 231–240, 317–336
Sintering, 305, 317–336
Skorenkyy, Yu., 73–86
Slonczewski, J.C., 28
Smyntyna, V., 265–270
Soft magnetic films, 321
Sosenkin, V., 127–144
Structure, 9, 17, 26, 59, 73, 96–99, 104, 128, 150, 161, 175, 186, 219, 236, 243, 267, 275, 287, 301, 318, 341
Substrate, 18, 28, 113, 243–257, 266, 319–324, 326–333, 335, 336
Suprun, A.D, 59–70
Suprun, A.D., 89–101
Surface plasmon polariton (SPP), 243–246, 251, 253–256
Sykut, K., 286
Synthesis, 4, 5, 8, 38, 60, 69, 104–106, 108, 113, 175, 231–240, 318
Szroeder, P., 25–38

T
Tatarenko, V.A., 25–38
TEM, 5, 8, 9, 176, 178, 179, 319
Temperature, 5, 19, 26, 52, 59, 73, 100, 104, 148, 175, 187, 232, 265, 276, 302, 312, 318, 341
Temperature quenching of luminescence, 268, 270
Thermal properties, 8, 25, 342
Titanium, 103, 104, 106, 108–110
Transmission coefficient, 185, 190, 191, 194, 195, 199, 223, 227
Tsaregradskaya, T.L., 341–352
Turkov, O.V., 341–352

U
Ulug, B., 265
Upward diffusion, 343

V
Vacuum, 129, 130, 157, 246, 268, 328, 329
Varyshchuk, V., 231–240
Vasylyuk, S.V., 90–101
Vikkers, 347
Voitsekhivska, O., 219–228
Volfkovich, Y., 127–144
Vortex granulator, 301

W
Wawrzkiewicz, M., 103–120
Weiss, P.R., 28
Wiśniewska, M., 103–120, 285–298
Wołowicz, A., 103–120

Y
Yaremchuk, I., 231–240
Yashchuk, V.N., 90–101
Yerokhov, V., 275–281

Z
Zirconium, 3–12, 103, 134, 321

T
Tкach, M., 219–228

The manufacturer's authorised representative in the EU is Springer Nature Customer Service Centre GmbH, Europaplatz 3, 69115 Heidelberg, Germany. If you have any concerns regarding our products, please contact ProductSafety@springernature.com

Printed and bound by CPI Group (UK) Ltd, Croydon, CR0 4YY
05/10/2026
02228141-0011